DATE DUE

THE
HUMAN MOTOR

THE
HUMAN MOTOR

Energy, Fatigue, and the Origins of Modernity

ANSON RABINBACH

BasicBooks
A Division of HarperCollins*Publishers*

Library of Congress Cataloging-in-Publication Data
Rabinbach, Anson.
      The human motor : energy, fatigue, and the
origins of modernity / Anson Rabinbach.
          p. cm.
      Includes bibliographical references and index.
      ISBN 0-465-03130-7
      1. Work—History.   2. Work measurement—
History.   3. Fatigue—History.   4. Human
mechanics—History.   I. Title.
HD4904.R33   1990
331.25—dc20   90-80252
                                                    CIP

Contents

CONTENTS

Illustrations

Acknowledgments

THE idea for this book first took shape in a paper I presented at a conference on the work of Norbert Elias at the New York Institute for the Humanities. During 1983–1984 the Institute for Advanced Study in Princeton offered a singularly congenial environment for thinking and writing. Generous support from a John Simon Guggenheim Memorial Foundation Fellowship and from The Cooper Union for the Advancement of Science and Art hastened completion of the manuscript.

I wish to thank Lion Murard and Patrick Zylberman for their valuable suggestions and encouragement at an early stage of the project. Patrick Fridenson, Jacques Rancière, Georges Ribeill, and Yves Cohen offered their welcome advice and stimulating comments. Dietrich Milles of the University of Bremen generously shared his vast knowledge of this subject and provided access to his personal archives. I am also indebted to Gerald Geison, Andreas Huyssen, Herman Lebovics, George L. Mosse, and Wolfgang Schivelbusch for their careful reading of sections of the manuscript.

The Columbia University Seminar on the History of the Working Class, and the Social Knowledge and State Social Policy Colloquium of the Social Science Research Council, under the direction of Dietrich Rueschemeyer and Theda Skocpol, generated useful discussions of the material in this book. The 1983 conference on "Work in France" at Cornell University was important for clarifying many crucial points. My gratitude to Cornell University Press for permission to quote from my contribution to the conference volume: "The European Science of Work: The Economy of the Body at the End of the Nineteenth Century," *Work in France: Representations, Meaning, Organization, and Practice*, Steven Laurence Kaplan and Cynthia J. Koepp, eds. Copyright © 1986 by Cornell University Press.

My colleagues at the Cooper Union—especially Peter Buckley, Anne Griffin, Fred Siegel, Michael Sundell, and Brian Swann—helped in many ways, but particularly by creating a solidarity rare in academic life. Special thanks also to David Abraham, David Bathrick, Fina Bathrick, Atina Grossmann, Helmut Gruber, Miriam Hansen, Michael Krasner, Mary Nolan, Lawrence Pitkethly, Christine Stansell, William R. Taylor, and Sean Wilentz for their intellectual comradeship.

Ulla Volk of the Cooper Union Library was of enormous assistance in securing rare volumes from around the country. Harriet Jackson, Françoise Jouven, and Heidi Lefer all provided generous assistance, and Elena Le Pera improved the writing at the crucial final stage. Steven Fraser, my editor at Basic Books, was unstinting in his support and sage editorial suggestions.

Most of all, my love and appreciation to Jessica Benjamin, whose extraordinary intellect is wonderfully combined with warmth and empathy. And to Jake and Jonah, whose energy and enthusiasm remind me that fatigue can be worthwhile and that the metaphor of the human motor must always fall short.

Anson Rabinbach

Introduction

THIS book is a study of the *human motor,* a metaphor of work and energy that provided nineteenth-century thinkers with a new scientific and cultural framework. Through this metaphor, scientists and social reformers could articulate their passionate materialism, embracing nature, industry, and human activity in a single, overarching concept—labor power. Their vision of a society powered by universal energy offered continental Europe, undergoing its industrial revolution, an exhilarating explanation for its astonishing productivity. In that vision, the working body was but an exemplar of that universal process by which energy was converted into mechanical work, a variant of the great engines and dynamos spawned by the industrial age. The protean force of nature, the productive power of industrial machines, and the body in motion were all instances of the same dynamic laws, subject to measurement. The metaphor of the human motor translated revolutionary scientific discoveries about physical nature into a new vision of social modernity.

In his *Discourse on Method* (1637) Descartes described the animal machine as "made by the hands of God, incomparably better ordered [and] more admirable in its movements than any of those which can be invented by men."[1] He compared the marvelously ingenious mechanical *homunculi,* or automata, constructed by seventeenth-century craftsmen with the living machines produced by nature: but only the human machine was capable of speech and reason—endowments that attested to the presence of the soul. During the nineteenth century, Descartes' animal machine was dramatically transformed by the advent of a modern motor, capable of transforming energy into various forms. For European physicists and physiologists, Descartes' distinction between the animal machine and the human being was no longer meaningful. The human body and the industrial machine were both motors that converted energy into mechanical work. The automata no longer had to be denied a soul—all of nature exhibited the same protean qualities as the machine.

From the metaphor of the motor it followed that society might conserve, deploy, and expand the energies of the laboring body: harmonize the movements of the body with those of the industrial machine. Consequently, European scientists devised sophisticated techniques to measure the expenditure of mental and physical energy during mechanical work—not only of the worker, but also of the student, and even of the philosopher. If the working body was a motor, some scientists reasoned, it might even be possible to eliminate the stubborn resistance to perpetual work that distinguished the human body from a machine. If fatigue, the endemic disorder of industrial society, could be analyzed and overcome, the last obstacle to progress would be eliminated.

This image of the body as the site of energy conservation and conversion also helped propel the ambitious state-sponsored reforms of late nineteenth- and early twentieth-century Europe. The metaphor of the human motor lent credibility to the ideals of socially responsive liberalism, which could be shown to be consistent with the universal laws of energy conservation: expanded productivity and social reform were linked by the same natural laws. The dynamic language of energy was also central to many utopian social and political ideologies of the early twentieth century: Taylorism, bolshevism, and fascism. All of these movements, though in different ways, viewed the worker as a machine capable of infinite productivity and, if possessed with true consciousness, resistant to fatigue. These movements conceived of the body both as a productive force and as a political instrument whose energies could be subjected to scientifically designed systems of organization. Thus, the classical traditions of nineteenth-century social thought, as well as the radical ideologies of the early twentieth century, shared the belief that human society is ultimately predicated on the unlimited capacity to produce and that this "social imperative" mirrored nature's own unlimited capacity for production. The laboring body was thus interpreted as the site of conversion, or exchange, between nature and society—the medium through which the forces of nature are transformed into the forces that propel society. This book is concerned with tracing the origins and implications of this image of "labor power" as the fundamental imperative that links society and nature in nineteenth-century thought.

A central argument of this study is that modern *productivism*—the belief that human society and nature are linked by the primacy and identity of all productive activity, whether of laborers, of machines, or of natural forces—first arose from the conceptual revolution ushered in by nineteenth-century scientific discoveries, especially thermodynamics. Historians and philosophers of science have frequently pointed out that the metaphors and images employed in the great scientific theories of the age, those of Hermann von Helmholtz, Sir William Thomson (Lord Kelvin), and Rudolf Clausius, were shaped by larger theological and social perceptions. However, the impact of their image of nature for a modern conception of work has received little attention. Of particular importance was the contribution of the German physicist and physiologist Hermann von Helmholtz, who elaborated the universal law of the conservation of energy in 1847. Helmholtz, a pioneer of thermodynamics, argued that the forces of nature (mechanical, electrical, chemical, and so forth) are forms of a single, universal energy, or *Kraft*, that cannot be either added to or destroyed. As Helmholtz was aware, the breakthrough of thermodynamics had enormous social implications. In his popular lectures and writings he strikingly portrayed the movements of the planets, the forces of nature, the productive force of machines, and of course, human labor power as examples of the principle of conservation of energy. The cosmos was essentially a system of production whose product was the universal *Kraft* necessary to power the engines of nature and society, a vast and protean reservoir of labor power awaiting its conversion to work.[2]

The remarkable generosity of nature implicit in energy conservation was diminished by the almost simultaneous discovery of the *second law of thermodynamics,* which explains the irreversibility and decline of energy in *entropy.* The second law of thermodynamics, identified by Rudolf Clausius, established that in any isolated system the transfer of energy from a warmer to a colder body is accompanied by a decrease in total available energy. The optimism of energy conservation was thus offset in the 1850s and 1860s by the revelation that, in practical terms, there was also an inevitable dissipation of force, that only a fraction of the total existing energy is available for conversion and that "the entropy of the universe tends to a maximum."[3]

The great discoveries of nineteenth-century physics led, therefore, not only to the assumption of a universal energy, but also to the inevitability of decline, dissolution, and exhaustion. Accompanying the discovery of energy conservation and entropy was the endemic disorder of fatigue—the most evident and persistent reminder of the body's intractable resistance to unlimited progress and productivity. Fatigue became the permanent nemesis of an industrializing Europe.

As a result of these discoveries, the image of labor was radically transformed. It became *labor power,* a concept emphasizing the expenditure and deployment of energy as opposed to human will, moral purpose, or even technical skill. The doctrine of energy thus contributed to a decisive break with the two great traditions that combined to form the Western idea of labor: labor was neither spiritualized as in the Christian worldview, nor deprecated and identified with degradation as in the ancient Greek word *ponos,* which translates as pain or travail.[4] Equally absent here is the ancient craft ideal of labor as an activity not confined to satisfying needs, but as an ennobling, poetic "accomplishment" (a vision that modern socialism translates into labor as the true path to redemption from alienation and the ennobling of human nature). In the energetic image of labor the intellectual, purposeful, or *teleological,* side is incidental. Marx, too, viewed labor power (in contrast to labor) as devoid of purpose and meaning, a purely quantifiable output of force, subject only to abstraction. As mechanical work, as *"Arbeitskraft,"* labor power is entirely indifferent to the nature of its material form.[5]

The discovery of labor power—and its subsequent elaboration in political economy, medicine, physiology, psychology, and politics—was emblematic of a society that idealized the endless productivity of nature. Semantically, this meant that the word *work* was universalized to include the expenditures of energy in all motors, animate as well as inanimate. The Promethean power of industry (cosmic, technical, and human) could be encompassed in a single productivist metaphysic in which the concept of energy, united with matter, was the basis of all reality and the source of all productive power—a materialist idealism, or as I prefer to call it, *transcendental materialism.* The language of labor power was more than a new way of representing work: it was a totalizing framework that subordinated all social activities to production, raising the human project of labor to a universal attribute of nature.

The nineteenth-century distinction between labor and labor power thus expressed a remarkable shift in the magnitude of social explanation. Labor became an ordering principle of both nature and society. The classical political economists of the eighteenth century, like Adam Smith and David Ricardo, clearly foresaw the productive potential of the division of labor and of working machines. But they could not yet grasp that the work of hands, tools, and even complex mechanical devices were insignificant when compared with the technological revolution that produced the very forms of power that propelled industrial progress. Only labor power could adequately express the equivalence of force that was the true *perpetuum mobile* of the nineteenth-century industrial revolution. With the discovery of labor power, work was no longer an anthropological constant or a social and economic activity. Labor power became, in the philosopher Agnes Heller's terms, "the motivating force of human [and we might add, natural] history."[6] Nineteenth-century Europe was transformed by work and energy. It is not surprising, then, that the central doctrine of scientific materialism—the unity of matter and motion in energy—succeeded in erasing the distinction between them.

This book examines a vast, though largely forgotten literature on work that appeared at the end of the nineteenth century, and by the beginning of the twentieth, proliferated into a scientific approach to the working body. Radical in its reduction of work to "economy of force," the language of labor power was not limited to one political or social ideology: it appears in popular science, in Marx's mature theory, in the laboratory fatigue study, in inummerable sociological and psychological treatises on work. Stripping labor of its extraneous social and cultural dimension, and revealing only its objectivity, this language could be found in socialist doctrine, in the arguments of liberal reformers, in parliamentary debates on social issues from the length of the working day to the term of military service.[7]

Nineteenth-century European thought was preoccupied with labor: with its political and economic interests, with its diverse forms of organization, with its intrinsic meaning, and with its productive potential. The emergence of successful working-class movements, the rise of a society propelled by a market economy and new technologies, and the great expansion of the factory system, especially in the last decade of the century, produced a panoply of discourses on labor in which ethics, science, and politics were entwined. Liberal reformers and conservative moralists divided disorderly from orderly and submissive workers, productive and upright workers—usually male—from dissolute and improvident workers, usually female.[8] The same moral and political claims could be enlisted by socialists to condemn the "insalubrity" and risks of certain trades, to assail exploitation and suffering, and to demand an end to the factory system.

By the early 1890s, progressive scientists and reformers were attempting to end this cacophony of moralizing claims and to resolve the "worker question" through science. These experts in fatigue, nutrition, and the physiology of the "human motor" sought to provide a neutral, objective solution to economic and political conflicts arising from labor—one that replaced moral exhortation with experiment and reasoned argument. Science subjected the body's movements and rhythms to detailed laboratory investigation, to new techniques of measurement, and to photographic study, ultimately giving rise to a new discipline: the European science of work.

Nowhere is this attempt to replace moral discourse with science more evident than in the discovery of fatigue by European physiologists, especially after 1870. Though portrayals of fatigue could readily be found in the literary storehouse of *ennui,* lassitude, weakness, and world-weariness—for example, in the poetry of Baudelaire or in the novels of Flaubert, Barbey d'Aurevilly, and Joris Huysmans—they first appeared in medical literature only in the late nineteenth century.[9] Judging from the sheer volume of scientific papers, popular works, and journalistic commentary, we can surmise that the problem of fatigue was epidemic among European workers, students, and even middle-class "brain" workers. An 1892 report on the state of French schoolchildren was hardly unique in its scrutiny of the diverse effects of fatigue: "Muscles without energy painfully support the body; the visage is pale; the carriage is enervated; the posture is weighted down; all of the external aspects of the child give the impression produced by a plant languishing and withering for want of air or sunlight. All of the functions of the organism descend into a characteristic state of decline."[10]

Such vivid illustrations expressed a widespread fear that the energy of mind and body was dissipating under the strain of modernity; that the will, the imagination, and especially the health of the nation was being squandered in wanton disregard of the body's physiological laws. Fatigue thus became the most apparent and distinctive sign of the external limits of body and mind, the most reliable indicator of the need to conserve and restrict the waste and misuse of the body's unique capital—its labor power. Central to this book is the significance of fatigue, which replaced the traditional emphasis on idleness as the paramount cause of resistance to work. Its ubiquity was evidence of the body's stubborn subversion of modernity.

The irreversible decline in force, which scientists and social philosophers had observed in entropy, led to grim predictions of the world's imminent demise—a "heat death," extinguishing all life in an abrupt, chilly end.[11] Less pessimistic spirits, like Helmholtz and the German popularizer of science Ernst Haeckel, resisted such apocalyptic premonitions; but the debate on the heat death of the universe was symptomatic of the scientific and literary anxiety of the age. Even if the cosmic apocalypse might be forestalled, scientific knowledge could not ignore the portentous social effects of fatigue.

Beginning with the "discovery" of fatigue by work-hygienists, physicians, and physiologists in the 1870s, this book traces the emergence of a distinctive European science of work. Etienne-Jules Marey, a remarkable French inventor-scientist, produced the first investigations of motion during the 1870s and 1880s. Marey's techniques, which inscribed the body's movements on smoked paper, gave rise to the laboratory study of fatigue. Angelo Mosso, a Turin physiologist who became the "Galileo" of modern fatigue research, and whose classic *La Fatica* (1891) was enormously influential, attempted to do for the working body what Helmholtz, Clausius, and Thomson had done for the universe: establish the dynamic laws of fatigue by rigorous experiment and new techniques of measurement. The Heidelberg chemist Wilhelm Weichardt's striking announcement in 1904 that he had discovered a vaccine against fatigue ultimately proved a disappointment, but his quest was hardly considered frivolous: the utopian possibility that society might discover a way to eliminate fatigue was much too compelling.

Since the seventeenth century, labor had undergone a major reevaluation in the West. In philosophy and economics, the ennobling of labor as the origin of all wealth and the legitimate basis of property and selfhood was a crucial instrument in the extensive campaign against the "unproductive labor" of the nobility and the idle poor: labor was at once productive, rational, and moral. The transvaluation of labor was invigorated by John Locke and the classical political economists, who drew on Calvinist doctrine to justify its centrality as the source of value, and by the Enlightenment *philosophes,* who publicized its virtues and secrets. "To raise the mechanical arts from the debasement where prejudice has held them for so long" was a chief purpose of Diderot's famous *Encyclopédie.* As William Sewell has shown, the result was a "scientized, individualized, utopian projection of the world of work," made publicly available like all other forms of scientific knowledge.[12] For Hegel, labor became the beginning point of human self-consciousness and autonomy, the source of universal truth. If labor was embraced by reason, its productive power was regarded as wholly rational: the old nemesis of idleness, consistently subjected to reason's disapprobation, was identified with corruption, vice, villainy, and the venality of courts.

Throughout the eighteenth and early nineteenth century, the vast European population of beggars and vagabonds became the subject of countless secular and theological treatises on the sin of idleness, and of a proliferation of laws prohibiting dissolute behavior. These writings and practices were directed not simply against the scourge of idleness, but at raising the moral worth of labor in the eyes of the few remaining skeptics. As Auguste Comte once remarked: "We must invest material labor with a philosophical importance demanded by its social value."[13]

This book describes how, by the last quarter of the nineteenth century, this protracted reevaluation of labor evolved into a far more detailed scientific program for transforming and deploying human labor power. By the 1890s an international avant-garde of fatigue experts, laboratory specialists, and social hygienists created a new field of expertise in which science and politics intersected. For the European science of work, the study of fatigue, of the dynamic movements of the body, and of physical changes during work, became part of a broader strategy of social modernity—one that attempted to solve social problems through empirical research and rational principles.

In this new constellation of knowledge and politics, the state was the "visible expression of the invisible bond that unites all living beings in the same society."[14] Social justice, reformers claimed, would inevitably lead to expanded productivity: "Social intervention to preserve the integrity of the social organism," noted the Belgian hygienist Louis Querton in *The Yield of the Human Motor* (1905), permitted "society to exercise its right to take from it a maximum of efficiency."[15] Social justice, conservation of energy, increased output, and greater efficiency were interrelated since "heredity, milieu, lodging, [and] education exercise a great influence on the personal productivity of the worker."[16] These ideas were attractive to many socialists who also considered exploitation a "social drain" on the productive power of the nation.[17] For late nineteenth-century liberals the economic and social benefit of state policy was to transcend class conflict and substitute scientific neutrality.

This book is an attempt to chart the path from moralism and the old religious proscription on idleness to the new social ethic of energy conservation. In France and Germany *fin-de-siècle* liberal reformers employed what I call a productivist calculus to address the question of how to balance economic well-being with social justice. As the condemnation of idleness was appropriate to a society with its moorings in religious conviction, the calculus of energy and fatigue was syncretic with a more scientific age. Max Weber concluded in *The Protestant Ethic and the Spirit of Capitalism* (1904/5) that a victorious capitalism no longer needed the religious asceticism of the work ethic. But Weber (although he commented elsewhere at length on the "psychotechnics of labor") declared an ending where, in fact, there was a metamorphosis: the traditional work ethic became the ethos of labor power. Weber's elegiac image of the declining work ethic in the modern age proved premature: "The rosy blush of its laughing heir, the Enlightenment, seems also to be irretrievably fading, and the idea of duty in one's calling prowls about in our lives like the ghost of dead religious beliefs."[18] What could be a more appropriate incarnation for this errant spirit in the age of the industrial dynamo than the new calculus of energy and fatigue? With the emergence of energy as the universal principle of work, the old ghost acquired a material carriage, and the image of work was given a scientific pedigree. The metaphor of the human motor succinctly expressed this profound change.

The great epistemological break with positivism—the cognitive monopoly and idealization of the scientific method along with a search for general laws of both nature and society—and with the centrality of labor in European social thought, both began in the early part of this century. Nietzsche provided the credo for the antipositivist revolt when he charged that "the faith on which our belief in science rests is still a *metaphysical faith.*"[19] His protest against the moral and political supremacy of epistemological models—especially that of physics—drawn from the study of nature became the central tenet of the philosophical critique of modernity, which subsequently guided many European intellectuals in their decisive rejection of the hegemony of science in the first decades of the twentieth century.[20]

The great social thinkers at the turn of the century—Weber, Freud, Durkheim—argued for the autonomy of the cultural sciences (sociology, psychoanalysis, anthropology) from older scientific models, though each remained convinced, albeit differently, that labor, defined in largely energeticist terms, was central to their enterprise. Freud transposed the energetic model of nature to sexuality; Weber saw routinized, time-bound labor as the characteristic feature of Western rationality; and Durkheim argued that the division of labor irreversibly destroyed the coherence and integrity of traditional culture. At the same time, however, each began to question the ontological status of labor as the *prime mover* of man and nature: Freud rooted labor in instinctual life; Weber, in asceticism and religious conviction; and Durkheim, in community. Each introduced significant aspects of hermeneutic uncertainty into the interpretation of culture. Despite their shared commitment to the Enlightenment ideal of science, their work ultimately helped to loosen the supreme hold of natural science on intellectual life.

Weber especially remained aware of this issue in his many reflections on the paradoxical implications of positivism. He clearly recognized the extension of instrumental rationality as a style of thought—mathematicization, expertise, bureaucratic and legal formalism—and as a practice—the enormous development of industry and productivity, the extension of market forces and administrative methods to ever-increasing domains of social experience. The advance of instrumental rationality undermined and weakened the capacity of reason to resist effectively the imperatives of increasingly rationalized power. Ironically, its *raison d'être*—neutrality and scientific responsibility—rendered reason powerless to formulate binding values, social ideals, or "ends." Reason was itself implicated in, and subordinated to, larger nonscientific purposes. Weber criticized positivism and scientific naturalism, singling out the social energeticists—Ernest Solvay and Wilhelm Ostwald—for their "umstülpung," or spillover, of the "world picture" of scientific disciplines into the "worldviews" of the social sciences, where they ought not have a place.[21] Yet, for all his prescience, Weber did not investigate the expanding role of social knowledge in the modern state. Always the pessimistic liberal, Weber defended the neutrality of the sciences, as he could conceive of no real alternative apart from ideology or prejudice.

The impact of positivism on social knowledge and on the nineteenth-century ideal of reform politics has somehow escaped the scrutiny of historians.[22] In this work I have focused on the way social modernity emerged from its connection with the insights gained from the scientific discoveries about energy, and their social implications for understanding labor power and fatigue: How laboratory studies of the laws of motion governing the working body contributed to the great political struggles around the "labor question" in the late nineteenth and early twentieth century. By pursuing the tropes of energy and fatigue in the efforts of physiologists, psychologists, economists, social scientists, and reformers, I have traced how scientific concepts became part of a growing body of social knowledge.

This book argues that the science of work transformed the perception of work in Europe. Breaking sharply with earlier doctrines of moral and political economy, the new science focused on the body of the worker. Predicated on the metaphor of the human motor and buoyed by a utopian image of the body without fatigue, the search for the precise laws of muscles, nerves, and the efficient expenditure of energy centered on the physiology of labor. The European science of work promised to overcome the negative effects of badly organized, exploitative, and irregular work. Claiming to transcend class interests, after 1900 its advocates became increasingly involved in politics. In controversies over the length of the workday, occupational accidents, and military training, the science of work sought to deliver an objective and nonpartisan answer to the most vexing social issues. Yet the greatest weakness of the science of work lay in its most compelling assumption, that the body was a motor, and that scientific objectivity and expertise were sufficient to provide an objective solution to the worker question.

The arrival of the American Taylor system in Europe shortly before the First World War dashed those hopes. At the outset the European scientists criticized Taylorism for its primitive conception of the worker and for its crude methods. But ultimately both methods shared a similar image of work: the body—not the social relations of the workplace—was the arena of labor power. After the war there was growing recognition that the two methods were not incompatible and that a rapprochement would benefit both. Nonetheless, the image of a human motor persisted well into the post–World War II period. By the mid-1950s, however, the image of work drawn from that metaphor began to wane. Automation promised to liberate work from the materiality and physicality—muscles, nerves, energy—of the body. With the disappearance of the metaphor of the human motor, the centrality of work in European thought began to disappear as well.

Chapters 1 through 5 expand on that burgeoning discourse on fatigue and energy in the natural sciences, in physiology, in medicine, and in psychology. These chapters deal with the discovery of fatigue as both a pathology and a prophylaxis against the demands of modernity; with the formation of the idea of labor power in Helmholtz and Marx with its metamorphosis into the economy of work by Etienne-Jules Marey and his students; and with the emergence of a European science of work.

Chapters 6, 7, and 8 examine some of the practical social implications of the new energy doctrine after 1890: the growing problems of neurasthenia and mental fatigue in psychology; the anthropology of work; and the institutionalization of social energetics as an influential movement. Some basic differences between the science of work in Germany and in France are relevant here (the French were more physiological; the Germans, more psychological in orientation). The general focus, however, remains broadly Continental—a perspective, I believe, justified by the international character of the movement, and by the simultaneous development in both France and Germany of empirical programs for studying fatigue in industry. A crucial aspect of these chapters is also the calculus of energy expenditure, or productivity, and reform that was applied to economic and social issues emerging from the problem of fatigue: the personal productivity of the worker; the deployment of energy in society; and above all, class conflict. Chapter 8 considers the efforts of physiologists, psychologists, and social reformers to apply the principles of energy conservation, with varying degrees of success, to a spectrum of social problems from inadequate nutrition to the lack of rest pauses, to industrial accidents, and even to the fitness of military recruits.

The final chapters explore the broader political consequences of fatigue research and the science of work in the first half of the twentieth century. Chapter 9 focuses on the reaction of European physiologists and work experts to the American system of scientific management, Taylorism—a challenge that divided European critics on the eve of the First World War and anticipated the assimilation of the science of work to a broader set of managerial strategies in the war's aftermath. The war was also decisive as a laboratory for discovering military and nonmilitary uses of the new energetics and for preparing the way for the widespread use of aptitude testing, the industrial fatigue study, and "psychotechnical training." Chapter 10 examines the interwar period, when "rationalization" became the catchword of various technocratic and political movements. Socialist, communist, and national socialist ideologies, which embodied versions of productivism and adopted aspects of the science of work, paradoxically led to the politicization and professionalization of the science of work during the 1920s and 1930s. The conclusion reconsiders recent debates about "the end of the work-centered society" in terms of the eclipse of the energeticist calculus and the centrality of the body in nineteenth-century visions of work.

My chief concern throughout has been with the problem of labor power in thought and politics. I have tried to show the consequences of the conflation of the natural and social sciences around the problem of labor power and to examine how a knowledge of work became institutionalized in various contexts. These efforts constituted a key element in a new form of social modernity, one in which social control and enlightenment were intertwined. The concepts of energy and fatigue reflected the paradox of this social modernity, at once affirming the endless natural power available to human purpose while revealing an anxiety of limits—the fear that the body and psyche were circumscribed by fatigue and thus could not withstand the demands of modernity.

A WORD ON METHOD

So as not to burden excessively readers disinclined to lengthy prolegomena, I have sequestered my remarks on method to this section. My investigation has concentrated on the intellectual and political implications of certain scientific concepts as they emerged in a zone between the specific concerns of the natural sciences and larger questions of social and political significance. To consider concepts like energy and fatigue as matters of broader social and cultural relevance is not in itself problematic. But, to ask precisely how these concepts operated socially and politically, how they were used, and in what contexts, raises some important questions about the traditional assumptions of both social and intellectual history.

There is no doubt that for some time a deep division has existed between historians of ideas concerned with the problem of the connection between language and meaning, and more conventional social historians who, for more than two decades, have investigated how political actors and social groups disenfranchised from political control and situated outside the purview of elite society articulated their social vision and experience. From the perspective of many social historians, the effort to integrate postmodern theory has resulted in an intellectual history, at best performative and literary, that cannot engage with social and political realities.[23] Intellectual historians like Dominick LaCapra and Hayden White, have convincingly argued, however, that social history is epistemologically and politically naïve: the historian assumes the position of omnipotent chronicler, moving his/her subjects through a carefully constructed narrative, without reflecting on how language, ideology, and emplotment undermine the very image of objective history that is being aimed at. By reducing ideas and events to underlying socioeconomic models that remain unexamined, social history cannot extricate itself from the conundrum that writing about the "real" involves confronting its own representations of the real. Moreover, they charge, social historians frequently engage in a nostalgic enterprise to resurrect from past social struggles a sense of social coherence and meaning. All history, in this sense, is contemporary history, but unself-consciously so.[24]

However, several historians have recently tried to overcome this division and study how social visions and experiences were constituted through language and symbolic systems of meaning. They turned to the work of contemporary European philosophers—most often Michel Foucault, Jacques Derrida, and Jürgen Habermas—to reflect on the ways that problems of narrative and language might call into question the very categories of society, self, and experiencing subject that social historians invoke without hesitation. They argue that historians can productively investigate how class, gender, or any other social identity is constituted by language, and conversely, how language can fragment or subvert even the most stable of identities.[25] My own work adopts the view that one of the most interesting consequences of this approach is that it permits us to see competing systems of knowledge as a central aspect of society; that the definition of society as composed of contending classes, or even of dominant and subordinate groups, is limiting. Ideas are not ancillary or supplementary to social practices and movements. On the other hand, the claim of a neutral knowledge to stand above class conflict is itself historical, a claim that is thematized in this book.

One dilemma that this ongoing debate raises is that both social and intellectual history have diminished the status of knowledge, particularly scientific knowledge, in nineteenth-century culture. For the purely linguistic approach all attempts at totality, most emphatically the scientific one, become entrapped in the paradoxical objectivist metaphysics exemplified by Descartes. The return to writing or language, as a vantage point to undermine the totalizing strategies of the social sciences, has meant not merely the radical subversion of meaning, but a homogenizing tendency to treat all works—even those once accorded the status of scientific knowledge—as textual exemplars of pure *poesis* rather than as socially powerful instruments. This position is not distant from Marxist social historians who judge the claims of science as bound to class and interest. Consequently, both approaches understate the role of science in the nineteenth-century public imagination and often neglect the fusion of scientific claims and social knowledge in its politics.

Even if we attend to the language of the sciences and the myriad attempts to apply scientific knowledge to society, we are still confronted with the other side of the dilemma than the one posed by the linguistically oriented historians of ideas—if metaphor and language determine the concepts and categories through which the world is represented, these concepts also derive from scientific traditions that have worldly consequences far beyond their status as literary expressions. Positivism is not merely a "metaphysic" but an institutionally anchored way of investigating and knowing the world. Although I have organized this study around a central metaphor and have been attentive to the role of language in the texts I have chosen, a fundamental problem for the cultural historian remains how science organizes, represents, and structures its knowledge along sometimes unconscious fault lines, often with unintended political and social consequences.

The point of this book is not to synthesize these two largely incompatible approaches, but rather to work beyond the boundaries that each of these perspectives has set; to offer without either excessive reverence

or fear and trembling, a "third way" of encountering social developments, texts, and discourses. I have attempted to negotiate the byways
of the nineteenth-century discourse of labor power in the context of
scientific developments, and in politics, without discounting its claims
to objectivity, neutrality, and universality. In other words, I have not
reduced these claims to mere ideological ruses through which a hege-
monic class subtly achieves its ends. At the same time I have been
conscious that this discourse has had significant social and political
ramifications, not the least of which were linked to its claim to stand
apart from social conflict. Crucial to its *raison d'être,* this aspiration to
political neutrality is, as we have noted, a pervasive problem in the
evolution of social knowledge.

There is no dearth of social histories documenting changes in the
labor process during the "second industrial revolution" of the late nineteenth century. Labor historians have elaborated a rich and textured
picture of the replacement of skilled artisanal labor by machinery and
factory work; the breakup of the power of local workers' organizations;
the rise of a new industrial working class of unskilled, and often female,
workers; the emergence of a new form of collective action—riots, rituals, and strikes—and more sophisticated techniques of surveillance and
discipline by the state and capital. Historians have attempted to un-
earth how community and kinship have produced affective solidarities
and sustained struggle while sometimes exacerbating patterns of prejudice and traditional modes of dependency. Although attention to family, sexuality, culture, and language served to acknowledge that production was not the alpha and omega of history, these domains were still
often defined by their proximity (outside of, reactive to) to production.
Having abandoned Marxism, many social historians became skeptical of
theory and dismissed their own, earlier productivism as a methodological error. The correction of error, however, frequently evades full analysis of the problem: in this instance, disregard of Marxism's productivist
assumptions as a powerful historical force meriting critical investigation. Rarely interrogated were the constricting effects of productivism
on the labor movement's vision and practices—the ways that labor
movements helped workers to adapt to industrial processes, accelerated improved techniques of production, and excluded significant dimensions of culture and politics not central to production. Most important, the ways in which scientific ideas, epistemological frameworks,
and reform strategies redefined labor (and its practical consequences)
eluded most social historians because they did not emerge directly from
class conflict.

By contrast, historians of ideas have long been aware that a major thrust of modern social theory—both in the German tradition of the Frankfurt School and in French poststructuralism—has been concerned with precisely this question: how both nineteenth-century heirs of the Enlightenment, rationalism in general, and, scientific Marxism in particular, have helped to perpetuate and extend technical and cognitive systems derived from the model of production.[26] As Max Horkheimer and Theodor Adorno, the central figures of the Frankfurt School, asserted in their brilliantly opaque *Dialectic of Enlightenment* (1947), positivism read back onto human nature the attributes of inorganic nature while modeling its method and goals on the social project of conquering and dominating nature.[27] Similarly Marx, in Jürgen Habermas's words, introduced "a principle of modernity that is grounded in the practice of a producing subject" whose historical goal is to realize the potential of technology.[28] Though suggestive, this critique did not provide an adequate historical account of the link between positivism and productivism in nineteenth-century thought; nor did it investigate how that vision became institutionalized in concrete forms of social knowledge specific to the imperatives of industrializing Europe.

This book examines some of these lacunae in social and intellectual history. It argues that the scientific language of labor power and the hegemony of productivism were not merely "corruptions" of Marxism or of the labor movement, but integral aspects of the intellectual framework of nineteenth-century materialism. Furthermore, although Marx's theory plays an important part in this story, it is by no means the only, or even the central one. Rather, by focusing on less well known individuals and texts—whose influence did not resound in the realm of grand theory but in the laboratory, in the social sciences, in economics, or in parliamentary chambers—I hope to show that similar sets of assumptions governed the ideas of physiologists, psychologists, socialists, and liberal thinkers and reformers.

For this reason I have paid particularly close attention to how the language of scientific discovery and scientific practice contributed to the evolution of a body of social knowledge concerned with work. Science constructed a model of work and the working body as pure performance, as an economy of energy, and even as a pathology of work. It produced a vast array of new disciplines concerned with society—social statistics, social medicine and hygiene, and a science of work—which framed political arguments and influenced their outcome. In short, science participated in a much larger web of ethical, social, and political entanglements.

For this insight the cultural historian is indebted to Michel Foucault. His preoccupation with how power generates knowledge and how knowledge in turn governs the way social institutions create and refine their mechanisms and procedures, has begun to influence historians.[29] Foucault asked provocative questions about the way knowledge is embedded in institutional systems of rules and in conceptions of order. At once engaged and stoically detached, his critique of the way reason shapes institutional and social norms is part of a broader philosophic and historical discussion indebted to Nietzsche: a radical questioning of whether "reason" can ever be entirely removed from its dependence on political and instrumental ends.[30]

Foucault's diagnosis of the steady encroachment of reason on the individual has, of course, much in common with Max Weber's classical description of rationalization as the paradoxical consequence of the extension of Western reason—with its inevitable reduction of personal and meaningful relations to abstract, calculable rules and procedures.[31] But Foucault's focus differed from Weber's. If Weber was concerned with the "iron cage" of rationalization, Foucault is interested in how social modernity comes about through knowledge and the self-scrutiny with which human beings investigate, and thereby produce, notions of human nature and society.[32]

Foucault's most radical claim is surely that the desire to fix truth in objective and universally rational systems of thought, such as law, medicine, and social science extends the normative power of knowledge over the self and the institutions that encompass it. Every step toward liberation thus empowers the mechanisms of coercion; every advance toward enlightenment compounds the entanglement of individuals in the social frameworks that constrain them. Humanitarian reform appears as the pretext for an anonymous and coercive strategy designed to penetrate "down to the finest grain of the social body."[33]

This grim perspective, however, also collapses the tension that exists between intention and outcome: between the reformist commitment to relieve injustice and its tendency to generate and extend scientific systems of control. The development of social knowledge during the nineteenth century was not merely an ever-expanding *instrumentarium* of power. Science lent its prestige to conserving the energies of the nation and to deploying its talents most effectively; but it also aimed at reducing suffering and exploitation. For most nineteenth-century thinkers the productivist ethos did not preclude a genuine commitment to humanitarian reform and moral responsibility. What might appear today as a contradiction seemed to them to presuppose each other. The French liberal philosopher Alfred Fouillée, for example, insisted that science and social ethics were both indispensable for realizing the "reparative justice" that could compensate for the defects of an obsolete laissez-faire liberalism.[34] Similarly, the energetic calculus, which viewed fatigue as the objective boundary of the human motor, was not restricted to any single political doctrine or ideology: it represented a widely shared belief that conserving the energy of the working body held the key to both productivity and social justice.

The problem of characterizing the peculiar combination of nineteenth-century social knowledge and politics is reflected in the mutability of the term *discourse,* which many historians, myself included, use to describe the way that epistemology, language, and rhetoric are organized around a common set of political and cultural designs. Though *discourse* has by now entered the conventional vocabulary of historians, it is seldom defined in terms of a coherent body of social thought. Even Foucault's work reveals a distracting lack of consistency. His emphasis on discourse began as an analysis of the rules ordering the organization of knowledge and representation—how the world is mediated by ideas and systems of thought—across epochs but was eventually scaled down to refer to subsystems of ideas, laws, techniques, and practices embodied in particular institutions.[35] Nevertheless, I prefer the earlier model of discourse as an epistemological category because it avoids the reductive position that different forms of knowledge are simply an "endlessly repeated play of dominations."[36] As opposed to ideology, the ambiguity of the term *discourse* with its implication of neutrality ought to evoke the tension that characterizes the relations between knowledge and power, relations that are neither entirely fixed nor predictable.

Finally, the idea for this book was inspired by Walter Benjamin's labyrinthian unfinished study of the origins of modernity, *Paris, Capital of the Nineteenth Century.* The notes he left behind indicate that Benjamin had planned to include a chapter on idleness because it gave birth to the distracted, externally stimulated experience he associated with modernity's triumph.[37] For him, charting the obscure pathways of modernity was an act of redemption: to rescue from oblivion a fragment of the past whose power is still manifest. This study seeks to restore our understanding of an aspect of nineteenth-century thought by exploring a metaphor that, perhaps because of its ubiquity, has become invisible to us: the metaphor of the human motor—a body whose experience was equated with that of a machine.

CHAPTER ONE

━━━━〜〜〜〜〜〜━━━━

From Idleness to Fatigue

THE BODY WITHOUT FATIGUE:
A NINETEENTH-CENTURY UTOPIA

WRITING in 1888, Friedrich Nietzsche asked: "Where does our modern world belong—to exhaustion or ascent?" His characterization of the epoch by the metaphor of fatigue was symptomatic of a general fear shared by the European middle classes that humanity was depleting its "accumulated energy" and falling into that sleep, which was "only a symbol of a much deeper and longer *compulsion to rest.*"[1] In the fading light of a century that "accomplished wonders far surpassing Egyptian pyramids, Roman aqueducts, and Gothic cathedrals," whose industry caused the "revolutionizing of production, the uninterrupted disturbance of all social conditions" (Marx), fatigue came to play a major role in the "mobile army of metaphors" that dominated the language of social description in the late nineteenth century.[2] Exhaustion was the constant nemesis of the idea of progress, the great fear of the "Age of Capital." As George Steiner remarked, "For every text of Benthamite confidence, of proud meliorism, we can find a counterstatement of nervous fatigue."[3]

For Nietzsche, as for many nineteenth-century thinkers, fatigue was identified with modernity itself: "Disintegration characterizes this time, and thus uncertainty: nothing stands firmly on its feet or on a hard faith in itself; one lives for tomorrow as the day after tomorrow is dubious. Everything on our way is slippery and dangerous, and the ice that still supports us has become thin: all of us feel the warm, uncanny breath of the thawing wind; where we still walk, soon no one will be able to walk."[4] Balzac, too, "planned to write a 'pathology of social life,' to show how men's stock of strength was diminished by too much expense of effort, indeed of any kind."[5] Despite his own belief in human progress, Darwin's discoveries, which dethroned the notion of teleology, or design in nature, implied that man's future depended on unending struggle and the vicissitudes of chance. The discovery of entropy attested to a pessimistic view of nature in which the available amount of energy or heat was continuously diminishing, conjuring up the specter of an apocalyptic "state of unchanging death."[6] As the historian Saul Friedländer has pointed out, the *fin-de-siècle* loss of faith in progress brought about a "wholly new vision: that of the total end of man."[7] That vision owed much to the newly acquired prestige of fatigue, not only in literature and philosophy, but in the natural sciences as well.

Fatigue encapsulated, as Nietzsche recognized, the paradoxes of modernity: Was not material progress undermined by the unreasonable demands that it made on the body and spirit? Did not scientific and technological advances produce a dark underside in the physical and psychological exhaustion of modern life? The nineteenth-century obsession with fatigue, both metaphoric and real, located in nature, in the body, and in the psyche the negative dimension of the considerable energies required to service the new productive forces unleashed by nature and harnessed by society. Nietzsche saw little way out since "nothing avails, one *must* go forward—step by step further into decadence (that is *my* definition of modern 'progress')."[8]

Before 1860 almost no medical or scientific studies of fatigue are recorded. By the turn of the century, the U.S. Surgeon General's index listed more than one hundred studies of muscle fatigue, as well as numerous studies of "nervous exhaustion," "brain exhaustion," "asthenia," and "spinal exhaustion."[9] Mental and physical fatigue, along with a range of modern disorders, were classified as "diseases of the will" *(les maladies de l'énergie)* and as the ubiquitous "neurasthenia," and were the object of an outpouring of scientific, medical, and popular literature.[10] Exhaustion was not merely the consequence of physical overexertion, but the cause of a variety of physical and mental pathologies born of the languid and torpid state of men, women, and especially school-age children. Fatigue was also a metaphor for the modern form of psychological suffering, for inertia, loss of will, and depletion of energy. Nietzsche's image of civilization succumbing to fatigue cannot be distinguished from that of many contemporary physicians and physiologists when he attributes "all crime, celibacy, and sickness," in short all vice, to "whatever weakens—whatever exhausts."[11] The importance of fatigue in the mental life of nineteenth-century Europe was, as the historian Theodore Zeldin has shown, more than a hallmark of the scientific mania of the age; it expressed a profound anxiety of decline, social disintegration, and even cosmic death.[12]

The ability of fatigue to move fluently between science and literature reveals the tendency of nineteenth-century thinkers to equate the psychological with the physical and to locate the body as the site where social deformations and dislocations can be most easily observed. A host of social ills could be traced to the consequences of fatigue: from alcohol and opium cravings to miseducation and the loss of social standing; from crime, vice, and the disintegration of the family to the degradations and discontents of industrial work.[13] As a tangible and ever-present mental and physical disorder, fatigue could, however, be distinguished from emotional states, for example—melancholy, *ennui,* and listlessness—which were its subjective manifestations. The physical symptoms of fatigue were regarded as mere "representations" of more profound conditions. As one prominent physician noted:

> The modes of representation are various. Some feel it (ideally) in the muscles; others under a cerebral form. Here are some examples: "the muscular twitchings in the calves of the legs, the back, and the shoulders; the eyes feeling swollen, but no heaviness in the head;" "a feeling of relaxation, of a weight, localised in the shoulders . . . slowness of movement, with a feeling of weight in the head."[14]

As we shall see, physicians and physiologists considered these diffuse states to be objective, measurable, and above all, conquerable. Judging from the explosion of research and interest that accompanied this "discovery" of fatigue, we might conclude that medical science was confronted by a considerable epidemic. The concern with fatigue in nineteenth-century science, and the medical and literary obsession with "degeneration," decline, and cultural decadence in *fin-de-siècle* Europe, was both a consequence of new scientific developments, and, as the historian Eugen Weber has noted, an expression of "the righteous concern for moral regeneration, the social concern for equity and welfare, the 'scientific' interest in physical and psychical well-being (or decline)."[15]

In the last quarter of the nineteenth century, what the historian Robert Nye has aptly called "the medical model of cultural crisis" took root among European scientists and intellectuals and could be registered in politics as well. Increased knowledge, especially the collection of statistics on crime, deviance, suicide, morbidity, and mortality, reinforced the shared perception that cultural crisis was reflected in the rise of both physical and social pathologies. Indeed, as the historian of science Georges Canguilhem has argued, to identify "the normal with the pathological" was a characteristic of nineteenth-century science, a dogma readily accepted by philosophers as diverse as Comte and Nietzsche.[16] "Through the lens of the pathological—as if through a magnifying glass—the truth of the healthy condition is decoded."[17]

In France, the military defeat of 1870 exacerbated this sense of cultural decline, mirrored in "a profound anxiety, a consciousness of weakness, a concern about decadence, an obsession with diminution."[18] Throughout the nineteenth century, new sciences emerged, for example, social hygiene, which gave this preoccupation an empirical and theoretical foundation. Concentrating on the pathological effects of the social milieu and drawing connections among declining population, lack of competitive industry, high rates of crime, suicide, and prostitution, these sciences, located in the terrain between politics and medicine, invoked the metaphors of health and sickness to express national anxiety.[19] "The present generation is born fatigued; it is the product of a century of convulsions," wrote Philippe Tissié, the French advocate of a rational gymnastics in 1887.[20] Maurice Barrés, the founder of the intellectual right in prewar France, appropriately called his nationalist trilogy completed in 1902, *Le Roman de l'énergie national.*[21]

However, concern with national decline was only one aspect of the obsession with fatigue, which was not unique to the French *belle époque*. Although the fatigue mania was equally evident in Wilhelminian Germany, it grew out of a society less fearful of imminent disintegration than of its dizzying ascent to industrialization and economic triumph after 1895. In Germany before the First World War, fatigue became a metaphoric and corporal barrier to progress, economic development, and the implacable onrush of *Zivilisation*.

If the discourse of fatigue transcended national and political boundaries, it proved to be ideologically promiscuous as well. For both French and German thinkers, fatigue (perhaps for reasons that can be explained by the success of competitive capitalism in England during the first half of the nineteenth century, fatigue did not play as large a role in English thought) became an integral part of the redirection of European society away from traditional, agrarian patterns of life toward a productivist order built on the confluence of science and industry. In this sense fatigue was not entirely negative, Nietzsche's assessment notwithstanding. If fatigue could be linked to the body's natural resistance to the demands of productivity, that correlation might also establish the need to reduce the burdens of economic expansion and suggest the way to achieve a just order of work and society. Fatigue also represented the legitimate boundary of the individual's physiological and psychological forces beyond which the demands of society become illegitimate or destructive. Fatigue thus defined both the limits of the working body and the point beyond which society could not transgress without jeopardizing its own future capacity for labor. For this reason fatigue also became the concept and the means through which the industrial body could best be understood and employed. The body without fatigue was the ideal, not only of the industrial bourgeoisie, but of the workers' movement which, albeit differently, imagined a point of maximum productive ouput and minimum exhaustion as the *summum bonum* of modern society.

As a result of developments in late nineteenth-century physiology, especially the mechanics of bodily movement and the problem of physiological heat, physiologists designed elaborate tracing instruments (the ergograph, the aesthesiometer) to register minute changes in the objective course of fatigue during any given occupation or task. By the late 1890s scientists in France, Germany, and Italy had thoroughly investigated such diverse aspects of fatigue as work performance, mental fatigue, workers' diet and nutrition. Extensive studies, which focused on the legal, statistical, and medical aspects of fatigue, were used to calculate and conserve the productive capital of the nation.

This new calculus of fatigue and productivity resulted in a constellation of science and politics concerned with the laboring body. Well before the end of the nineteenth century, a European "science of work" was established at the crossroads of science, medicine, and social policy. Scientists employed the conceptual tools and techniques of the physical sciences to search for the physiological laws of motion of the body at work. The profession of *social medicine* also emerged to contend with the effects of social and occupational relations on health. Legal experts and parliamentary commissions began to debate the role of fatigue in industrial accidents or illnesses. Social hygienists attempted to calculate the amount of energy required for optimal output in order "to determine for each occupation, which influences lead to a reduction of the yield of the worker." Social reformers employed a new energeticist vocabulary in their claims that a "limitation of the hours of work, as well as a minimum salary, is a physiological necessity."[22] Confronted with growing statistics on industrial injury and illness caused by fatigue, medical science was called upon "to provide the tools for action on the terrain of social insurance and social welfare."[23]

Concern with fatigue was also part of a widespread tendency of many nineteenth-century liberal reformers and socialists to link higher productivity with social reform. What united these new forms of social knowledge was a consensus that the health and energy of the worker was a crucial element in a national calculus and that the state was the neutral arbiter of social conflict. By 1900 in both France and Germany, the science of work attempted, with varying degrees of success, to intervene in conflicts over industrial policy. A scientific approach to work was first tested on such issues as the length of the working day, and industrial health and safety. These attempts were based on the shared perception that science could mediate a "neutral," objective resolution of social conflict.

Finally, a profound change in the perception of work and the working body became incorporated in a single metaphor—the frequently invoked "human motor," a striking image that illuminates an underlying affinity between physiology and technology. This image originated in an equally new perception of the universe as an industrial dynamo, or motor, the accomplishment of the thermodynamic physics of the nineteenth century. Although scientists were often reluctant to make such broad analogies, popularizers seized on the obvious parallel between the emerging conceptions of modern thermodynamics and the achievements of industrial technology. Nowhere is this parallel more evident than in the paean to the "law of the conservation of energy," which the German philosopher and protagonist of "Monism," Ernst Haeckel, included in his 1899 scientific bestseller, *The Riddle of the Universe:*

> The sum of force, which is at work in infinite space and produces all phenomena, is unchangeable. When the locomotive rushes along the line, the potential energy of the steam is transformed into the kinetic or actual energy of the mechanical movement; . . . The whole marvelous panorama of life that spreads over the surface of our globe is, in the last analysis, transformed sunlight. It is well known how the remarkable progress of technical science has made it possible for us to convert the different physical forces from one form to another. . . . Accurate measurement of the quantity of force which is used in this metamorphosis has shown that it is "constant" or unchanged. No particle of living energy is ever extinguished; no particle is ever created anew.[24]

In the following chapter we will examine in greater detail how these developments in the natural sciences profoundly affected the language of work and the image of the working body. But it is important to recognize that the metaphor of the machine in nineteenth-century materialism was far more than an extended analogy in which "the machine was a copy of the universe, and the universe itself a machine."[25] It fused the diverse forms of labor in nature, technology, and society into a single image of mechanical work, universalizing and extending the model of energy to a nature conceived of as a vast, unbroken system of production.

The measurement of fatigue thus promised to unlock the principles of the body's energies, to determine its economies of motion, and to reveal the most beneficial methods of organizing the expenditure of energy—both muscular and "nervous"—so that the resources of both the individual and society might be properly deployed. To map out the lines of least resistance to the body's economy of force was to "trace the actions which require the least expenditure of energy, and consequently, the minimum of fatigue."[26] Within the "impassable limits" of the law of the conservation of energy, the science of fatigue would determine how the energies of the human motor could be liberated, with all the concomitant social, economic, and political benefits such a liberation entailed.

THE DISAPPEARANCE OF IDLENESS

In early modern Europe the noble figure of work was constantly threatened by the subversive figure of idleness, whose proscription can be traced to the Christian concept of *acedia*, which was condemned by the hierarchy of vices established in the monastic orders. Siegfried Wenzel's classic study, *The Sin of Sloth: Acedia in Medieval Thought and Literature* (1960) thoroughly documents the persistence of *acedia* as the sin of idleness, despite efforts by the early Church fathers—most prominently St. Gregory in the eighth century—to weaken it and substitute more benign concepts such as *tristitia* or melancholy. Wenzel attributes that failure to the power of the orders, to their insistence on the Bendectine *Regula* and the teachings of St. Cassian, and to the central role of the sanctity of labor in securing the coherence of monastic life. Though the classical definition of *acedia* as idleness and somnolence was upheld throughout the Middle Ages, Wenzel points out that over time its meaning changed considerably and that it sometimes emphasized physical weariness or drowsiness, especially during prayer, and sometimes "tepidity in spiritual pursuits"—restlessness, boredom, lack of fervor.[27] The physical aspect of *acedia* is captured in phrases such as,

"The coming of dawn, at which time *acedia* falls upon us more heavily, must find us upright and busy with reciting the Office." Thomas Aquinas, for example, used the spiritual aspect of *acedia* to reflect "the aversion of the appetite from its own good because of bodily hardships that accompany its attainment."[28]

Although *acedia* remained until about 1200 essentially a monastic vice, a disorder prevalent among those who withdrew from society to take up the contemplative life of the cell, the warning that "idleness is an enemy to the soul" was also preached to lay audiences. Wenzel discovered elaborate sermons against the plague of *acedia* dating from the ninth century.[29] The popular literature of *acedia,* however, which only began to proliferate after the thirteenth century, was more narrowly confined to physical weariness, a profound lack of time sense, and plain laziness *(pigritia). Acedia* had become a largely secular concept, most often paired with the necessity of maintaining a regular, time-bound sense of discipline and labor. The sin of idleness shifted from a physical and spiritual *taedium,* or world weariness, to a disturbance of temporal regularity whose best therapy was work.

The medieval historian Jacques Le Goff has skillfully traced the long transition, beginning at the end of the thirteenth century, from an economy based on labor time ruled by religious festivals, governed by "agrarian rhythyms, free of haste, careless of exactitude, [and] unconcerned by productivity," to a new commercial economy based on chronological clock-time. With this development, Le Goff argues, the imperative not to waste time became "the new measure of life."[30] By the fourteenth century sloth emphasized the neglect of one's economic duties and activities. The sin of *acedia* was transformed into the social proscription on idleness; its proper penance was not merely work but the toil appropriate to one's worldly status.[31]

Thus from the thirteenth century until the middle of the nineteenth century Christian writers, ministers, and middle-class moralists all accorded idleness an esteemed place as the nemesis of an orderly life and the discipline of work. Even the etymology of *idleness* attests to its original relationship to labor time. As Roland Barthes once noted, the Latin adjective *piger,* from which the French word *paresse* was initially derived, means slow. According to Barthes this reduced pace of idleness is "the saddest most negative face of laziness, which is to do things, but poorly, against one's will."[32]

ARISTOCRATIC IDLENESS

Modern thinkers from Max Weber to Michel Foucault, who have contemplated the fate of idleness in the early modern era, have all commented on the antithesis between the order of work with its regular and rational procedures, or rules, and the order of idleness with its disdain for the self-discipline and affective repression necessary for labor. Weber, for example, remarked that for the English Puritans sloth and idleness were mortal sins, the "destroyers of grace" and the "antithesis of the *methodical* life."[33] Similarly, Foucault observed that sloth "had become the absolute form of rebellion" and the most serious of all sins since "it waits for nature to be generous in the innocence of Eden."[34] E. P. Thompson, the British historian, wrote that during the eighteenth century there was "a never-ending chorus of complaint from all the Churches and most employers as to the idleness, profligacy, improvidence, and thriftlessness of labor."[35] Yet these historians have overlooked the extent to which there is also a long tradition of exception to this literature of reproach, one in which idleness was not only free from approbation, but venerated and esteemed.

The emergence of *acedia* out of the monastery and into society was challenged by that part of society absolved from engaging in productive labor: the aristocratic class, whose existence was predicated on the labor of others, did not hold idleness in contempt. The aristocratic social code, to which the European *bourgeois gentilhomme* aspired, valued "honorable behavior and generosity" above commerce and claimed "disinterest" to be superior to propertied interests. The cultural sources of the aristocratic sensibility—by no means identical with either wealth or social origin—derived from the classical tradition rather than the Christian coda and cast leisure in a far more favorable light. The Greek word *argos* refers positively to one "who doesn't work," and it is well known that Plato was contemptuous of physical labor because it deformed both the soul and the body. With their renowned antipathy to labor, the ancients upheld a kind of heroic idleness, a gift to poets from the gods. This form of idleness persisted as an aristocratic prerogative since the Middle Ages, coexisting in an uncomfortable relation to the Christian tradition, which condemned it.

This deep ambivalence in the European discourse of idleness became most manifest during the Enlightenment. Indeed, the *philosophes'* deep love of repose revealed a passion for idleness that the Enlightenment could not easily reconcile with its contempt for the aristocracy and its culture. The *philosophes'* distaste for aristocratic privilege accelerated public contempt for the idleness of the upper classes, frequently excoriated by Voltaire, Condorcet, Sieyès, and countless others. Yet, for the *philosophes* the poet's privileged leisure was not entirely dependent on court society and was thus permitted to retain its exalted status. The *Encyclopédie*, for example, distinguished between two distinct kinds of idleness—*(paresse)* and laziness *(fainéantise)*—because "the former is a lesser vice," a matter of spiritual and corporal temperament, while laziness reflects the character of the soul.[36] This solution reappeared in several other contexts. Rousseau's *Émile* (1762) is totally orthodox in its prescription of the pedagogical rewards of arduous manual labor, and Rousseau often maintained that "he who eats in idleness what he himself does not earn steals." By contrast, Rousseau's *Confessions* (1766–70) exemplify the opposite argument in their frank reveling in a highly productive intellectual sloth between bursts of creativity, a "euphoric idleness," which Flaubert once called "marinating."[37] Gotthold Lessing summed up this ironic distance from the general hostility to idleness in his maxim: "Let us be lazy in everything, except in loving and drinking, except in being lazy."[38]

The enlightened glorification of contemplative idleness reserved for the intellectual muse is also at the root of the important distinction that the German language maintains in its word for leisure *(Musse)* and its opposite, idleness *(Müssiggang),* as in the traditional proverb: "He who enjoys leisure *(Musse)* evades Fortune, he who is given over to idleness *(Müssigang),* falls prey to her."[39] Another German proverb teaches that to play is to "dispense time"; to work is to "hold on to it."[40] As one critic noted, the poet has unfortunately made idleness "the cushion of the devil."[41] Yet, following Diderot's distinction, poetic leisure or idleness was not the same as laziness, since it is neither passive nor dissolute. The general disparaging of idleness did not impinge on the moral dignity accorded to creative insouciance.[42]

IDLENESS AND INDUSTRY

Thus, when the old Christian proscription against idleness was reinvigorated at the end of the eighteenth century, it was usually invoked as a criticism of the *persistence* of older preindustrial behaviors often directed against the new industrial order. The great novelty of the industrial revolution, claims E. P. Thompson, was not so much a strengthening of external compulsion, but the elimination of the "pattern of work and leisure which obtained before the outer and inner disciplines of industrialism settled upon the working man."[43] As vestiges of a less industrious age and from a less "measured" mode of life, idleness was the most pernicious sign of habits that anachronistically and stubbornly refused to loosen their hold on behavior. For Thompson the new order of industry was ushered in by the technological sophistication of the mechanical clock and by the moral exhortations of the Methodist preachers who collaborated with the new order to impose the discipline of the factory on a new and reluctant working class. Time sense and time-keeping developed in tandem. By the end of the eighteenth century, "There was a general diffusion of clocks and watches occurring at the exact moment when the industrial revolution demanded a greater synchronization of labor."[44]

The literature of labor in the first half of the nineteenth century was characterized by this traditional attitude but with a new twist: idleness was literally as well as figuratively the primary sin against industry. In 1807, Napoleon rejected a demand from the Church to prohibit all Christians except those with special dispensation from working on Sunday: "It is contrary to divine law to prevent a man who desires to work on Sunday, as he does on the other days of the week, from acquiring his bread. . . . Besides, the shortcoming of the French people is not working too much." Napoleon set the tone when he concluded: "The more my people work, the fewer vices they will have."[45] When the Prussian reformer Karl August Hardenberg introduced freedom of trade in the same year, he remarked that he had finally removed "all the cushions of laziness."[46] Claude-Lucien Bergery, the French industrialist and founder of a "science of management" in the 1830s, asked: "Is it possible to be happy when your forces are not equal to overcome the fatigues of the trade, when the body suffers from lassitude, when the day appears to be without end, when there is not an instant of gaiety, when finally one feels so overwhelmed that there is only revulsion for work?"[47]

The attack on idleness was zealously applied to both Europeans and non-Europeans. Ethnologists were struck by the *horror laboris* of the "uncivilized races," where "idleness and savagery" were synonymous. One ethnologist summarized the natural tendencies of the indigenous populations of the Americas as follows: "The North Americans, like the Chileans, pass their time in a stupid indolence . . . ; the only happiness that they can imagine is to do nothing. They rest for entire days stretched out on their *hamacs* or sitting on the ground, without changing their position, without moving their eyes, without saying a word. . . . It is almost impossible to extricate them from this habitual indolence. . . . They appear incapable of any vigorous effort."[48] "Natural man, as a whole, often performs no less an amount of work than civilized man; but he does not perform it in a regular manner. . . . Tense, regular work is what natural man avoids" observed the famous German geographer Friedrich Ratzel in 1888.[49] George Sand, vacationing in Mallorca in the winter of 1838, recorded her annoyance that the locals were "never in a hurry" and that their work was accompanied by activities indistinguishable from idleness—for example, women repairing the nets or sewing while chattering and singing.[50]

At the beginning of the nineteenth century, ethnographers attempted to demonstrate that civilization not only promoted the habit of industry, but encouraged physical development as well. As early as 1808, the French naturalist and explorer François Auguste Péron was comparing the strength of Australian natives with that of French and English settlers by a "dynamometer," the newly developed pressure-measuring device invented by Edme Regnier in the late eighteenth century. Though he found the aboriginal natives to be physically inferior to Europeans, Péron discovered that those natives accustomed to work were capable of producing more energy than the settlers (to his surprise he found that the French sailors were physically inferior to the English colonists and explained this phenomenon as the "fatigue that overcame these men following a long navigation").[51]

In his hymn to the moral qualities of *Die deutsche Arbeit* (1861), the nineteenth-century German nationalist and *volkisch* writer, Wilhelm Heinrich Riehl, complained that German folk humor praised only "the holiness of beggars, of do-nothings, of propertylessness," and that "the humor of idleness poetically appeals to the folk more than the virtue of industriousness."[52] The linguistic similitude between the German *Faulheit* (laziness) and *Fäulnis* (putrefaction) confirmed, Riehl believed, the "physical revulsion" that the "living corpse" of idleness would produce in any healthy person.[53] The Anglophile historian Hyppolyte Taine attributed the English success in industry to their "phlegmatic temperament," which permitted them to "eliminate *ennui*" and function with the regularity of a machine.[54]

Ironically heralding "the victory of industry over heroic laziness," Karl Marx provided a succinct commentary on the prevailing attitude toward idleness.[55] Until the last third of the nineteenth century, a two-front battle was waged by the advocates of industry against the unproductive idleness of the aristocracy and against the irregular and desultory work habits of the lower orders. This symmetry of condemnation is most evident in the analogies provided by an 1835 encyclopedia for French schoolchildren:

[T]he result of a natural propensity of many beings to dispose of their lives in the sweetness of doing nothing, filling the asylums of opulence as invariably as the poor fill the convents, cloisters and *hopitaux;* in this sort of religious life, as in a philosophical or artistic career, one is best able to undergo the pleasures of a corporeal indolence joined to a vague freedom of the spirit. These unfortunates find in laziness a consolation for their needs, to the point that they sometimes pass up eating, as do the Negro, the Bedouin, and the Spaniard rather than working.[56]

Climate, class, and idleness were also considered to be closely connected: "Idleness is always the result of a hot climate, often because of its opulence. The upper classes of society, like the equatorial peoples, persist in their idleness. By comparison, cold, like poverty, excites the vigor and activity of other men, and sooner or later dethrones those whom idleness has subdued."[57] Moreover, idleness was also viewed as the cause of physical infirmities brought on by sloth:

Nothing is more pernicious for those flabby and slow people, for those delicate women who lounge ceaselessly on their comfortable divans, on their feathered beds, than this languorous state we call idleness, not only does it fade their attractions, but it disposes them to euccorrhea, to amenorrhea, to migraines, sick nerves, bad stomachs, it rends them pale, depressed, flaccid. Laziness accumulates blood, the lymphes then cause the stagnation of bodily fluids, this is all because of the horizontal life in warm beds, and pillows, amassed under the head, cause humors that form the foundation of apoplexy.[58]

The virtues of order, thrift, and industry were set against the ravages of idleness, laziness, vagabondage, and the dissolute life of crime. The association of "laboring classes and dangerous classes" that belongs to this epoch is rooted in this perception that workers with too much free time turn inevitably to drink and crime. The literature on work written by middle-class reformers is edifying and uplifting, directed against the debilitating effects of idleness. Work was above all a prophylaxis against the dissolute qualities of inordinate leisure and, of course, against the concomitant effects of drink. In 1833 Bergery noted that he had seen an English forger handle a hot iron that ordinarily demanded two French workers and attributed his vigor to the fact that he drank no wine and ate more meat.[59]

Work and worktime had more than one dimension. For the industrial worker the new discipline of clock-regulated worktime meant an end to the longer traditional workday punctuated by periods of leisure;

for the employer it meant the calculation of productivity in terms of hours. Yet, clock-regulated worktime also meant the end of the irregular, extended, and sometimes chaotic workday and the imposition of a new division between labor and leisure.[60] For some reformers this moral order of work could not be imposed soon enough: Writing in the 1830s, Bergery prescribed for young children of both sexes "work proportionate to their age and their constitution."[61] In his address to the First International in 1868, Karl Marx advocated, though not without reservations, age nine as the time to begin work.[62] The model factory-colony Kuchen in Geislingen, Wurtemberg, opened by Arnold Staub in 1858, was modeled on a medieval cloister. Staub's young workers were provided with a written and posted regime, strictly regulating punctuality and behavior in remarkable detail—for example, by punishing eating during work and other infringements. If workers left their workplace before the whistle signaling permission to exit the factory gate, they were also frequently fined.[63] Often the strict regulation of work was accompanied by an idyllic vision of the factory system, and surely none is as unalloyed as this passage, which Walter Benjamin found in Edouard Foucaud's 1844 *Paris Inventeur:*

> Quiet enjoyment is almost exhausting for a working man. The house in which he lives may be surrounded by greenery under a cloudless sky, it may be fragrant with flowers and enlivened by the chirping of birds; but if a worker is idle he will remain inaccessible to the charms of solitude. However, if a loud noise or a whistle from a distant factory happens to hit his ear, if he so much as hears the monotonous clattering of the machines in a factory, his face immediately brightens. He no longer feels the choice fragrance of flowers. The smoke from the tall factory chimney, the booming blows on the anvil, make him tremble with joy. He remembers the happy days of his work that was guided by the spirit of the inventor.[64]

It is as easy to exaggerate the contrast between the unregulated, sometimes disordered worktime of the preindustrial worker and the modern time-regulated workplace, as it is to glorify the premodern experience of work. But it can hardly be disputed that the moral proscription against idleness played a major role in the transformation of work. After the revolutions of 1848, the campaign against idleness was accelerated, and the struggle to eliminate the almost universal day of respite—the notorious Saint Monday, *Saint Lundi* and *blaue Montag*—attested to the persistence of the Monday holiday well into the century.[65] In 1851 the statistician Moreau de Jonnès calculated that on any given day "French workers' wives" were robbed of 2,110,000 francs "by the detestable custom of *faire de lundi*, a vestige of unhappy times, when the serfs became intoxicated on Sunday to forget their miserable condition and prolong into the next day their degradation of the night."[66] The German manufacturer Friedrich Harkort addressed his famous "Bienenkorb-Brief" of May 1849 to the *manual laboring class*, which he defined as "those people, raised by decent parents, who come to ruin through the seduction of large cities; depraved and drunk, who hold *blaue Montag* to be holier than Sunday—lost sons without rest—for whom law and order is an abomination."[67]

In the 1840s the famous public surveys of workers (Enquête industrielle) were hardly neutral instruments but part of a general effort to draw up a taxonomy of working-class experience according to the moral framework of reformers. The most famous of these, Louis Villermé's *Tableau de l'etat physique et moral des ouvriers employés dans les manufactures* (1840), stressed the moral corruption, physical deformity, and improvidence that resulted from the mixing of sexes, long hours, and the practice of extending loans to workers.[68] Joan Wallach Scott's analysis of the political underpinnings of the *Statistique de l'industrie à Paris, 1847–48*, underscores how the reformer's emphasis on the risks of employing unmarried women in industry often functioned as a code for establishing a link between women's sexuality and the absence of industrial discipline and a sense of thrift.[69] In 1857, during a discussion in the French Chamber of Deputies on the causes of poverty, the liberal deputy M. de Perceval remarked that he fully agreed with the economists of his day that work was the only remedy for the dissolute life of poverty among women: "If one studies, in the bowels of great cities, the causes which lead unhappy women to fall into the ultimate degree of abjection, we find, in the first place, and for the most part, the misery that is the fruit of idleness. If, however, we open these women to the new vista of work and endear them to it, it shields them against vice and moralizes all of society."[70]

"No work is dishonorable, only idleness dishonors" wrote the Paris professor Pierre Foissac in his 1863 *Hygiène philosophique de l'âme:* "There is no more certain guarantee of the social order, its most vigilant sentinel." Foissac went so far as to proclaim that despite its value as a gift of nature, "Sleep is nevertheless an enormous waste of time, and the principal obstacle to the study of the sciences and to the achievement of works in a life that is already too short."[71]

This chorus of voices in praise of work reached its crescendo in a book published in 1870 by Denis Poulot, a French *patron* horrified at the fecundity of spirit displayed by the French worker in evading the discipline of work. Poulot created a moral taxonomy of the characteristics of the French working class in which the "sublime" worker is a foil for his unabashed exaltation of "l'ouvrier vrai," the virtuous worker who neither drinks, nor refuses to work, nor is insubordinate to his employer. "The number of days which the workers work in a year is an almost certain *criterion* for their classification," a criterion that placed Poulot's *vrai sublime* (true sublime) at the bottom of the moral hierarchy.[72] As late as 1871, Jules Simon, the French liberal and reformer despaired that "moral reform is more desirable and more difficult than the reform of industry."[73] Even at its best, medical accounts of the occupational hazards of the artisanal trades and the *Enquête industrielle* remained enveloped in a discourse concerned with maintaining a morally virtuous labor force rather than a healthy one.[74]

The morality of work was invoked not only by disgruntled entrepreneurs like Poulot or liberal reformers like Simon, but even more frequently by the early socialists, such as Fourier, St. Simon, and Proudhon. "Morality teaches us to love work." wrote Charles Fourier. "Let us know, then, how to render work lovable, and first of all, let it introduce luxury into husbandry and the workshop. If the arrangements are poor, repulsive, how arouse industrial attraction?" For this most imaginative of the early socialists, the solution was "perfected hygiene, coupled with variety of employments," which he predicted, "will accustom them not to get fatigued in their labors."[75] Proudhon, perhaps the most representative of this venerable tradition, did not consider work merely to be moral but built his rationale for sharing the common wealth of industry based on the universal moral imperative to expand productivity.[76] By no means a prerogative of socialist intellectuals, William Sewell has shown that in the 1830s and 1840s militant artisan worker-poets composed songs for their trades exalting the creative mission of labor.[77]

The most striking exception to the litany of idleness—an essay-pamphlet that even today remains a scandal to the work ethic—is Paul Lafargue's irreverent treatise, *Le Droit à la paresse (The Right To Be Lazy)*, which in 1880 announced (to the eternal shame of his father-in-law, Karl Marx) that "work is the cause of all intellectual degeneracy, of all organic deformity."[78] Ever since, Lafargue—a true ancestor of Tristan Tzara and Herbert Marcuse—has held an unchallenged place of honor in the socialist pantheon as the first, and perhaps the only, nineteenth-century socialist to refuse to bow and scrape before the altar of industriousness. It is not surprising that his work found a new audience in the international student revolt of the 1960s.[79] Economists and moralists have "cast a sacred halo over work," Lafargue declared. They created "a disastrous dogma" preaching to the proletariat who, "betraying its instincts, despising its historic mission, has let itself be perverted by the dogma of work."[80] Lafargue not only condemned this proletarian productivism but eloquently proclaimed the overthrow of the doctrine of idleness: "But to arrive at the realization of its strength the proletariat must trample under foot the prejudices of Christian ethics, economic ethics and free-thought ethics. It must return to its natural instincts, it must proclaim the Rights of Laziness, a thousand times more noble and more sacred than the anemic Rights of Man concocted by the metaphysical lawyers of the bourgeois revolution. It must accustom itself to working but three hours a day, reserving the rest of the day and night for leisure and feasting."[81]

Yet Lafargue remains not only an anomaly but an anachronism, constructing his treatise wholly within the antinomy of work and idleness. His defense of laziness, his hymn to the virtues of worklessness, and his idyll of a working class dedicated to consumption and luxury turns sin to virtue, culminating in a gluttonous rapture: "Instead of eating an ounce or two of gristly meat once a day, when it eats any, it will eat juicy beefsteaks of a pound or two; instead of drinking moderately of bad wine, it will become more orthodox than the pope and will drink broad and deep bumpers of Bordeaux and Burgundy without commercial baptism and will leave water to the beasts." Whereas the old system forbade idleness, Lafargue preached that "work ought to be forbidden and not imposed."[82] In the end Lafargue conjured up the heroic figure of idleness that served the ancients in their utopia of intellectual leisure: "O Laziness, mother of the arts and noble virtues, be thou the balm of human anguish!" are the final words of his infamous text.

WORK AND HYGIENE

Toward the end of the nineteenth century, idleness began to wane as the predominant mode of conceptualizing resistance to labor. The reasons for this decline can be enumerated: the old Christian proscription on idleness was losing its appeal for urban workers and industrialists; the technology of the factory system required more than externally imposed discipline and direction, but rather an internally regulated body ancillary to the machine. Consequently, the ideal of a worker guided by either spiritual authority or direct control and surveillance gave way to the image of a body directed by its own internal mechanisms, a human "motor." Almost simultaneously, fatigue and energy emerged as a more modern conceptual framework for expressing the relations between work and the body. Accordingly, the concept of work underwent a crucial transformation: fatigue, not idleness was the primary discontent of industrial labor. By the 1860s and 1870s a new literature stressing the hygienic aspects of work began to appear. Often the product of physicians, occupational hygienists, or even obscure *savants,* this literature was at first difficult to distinguish from the older variety of moral-medical literature. A careful examination of these texts reveals, however, that a more scientific evaluation of work, often materialist in emphasis, gradually displaced the old moral discourse. More important, these texts placed the working body at the center of attention and treated the labor of the body as a physiological process abstracted from the conditions of work or political economy and from the specific qualities of a trade or occupation.

To be sure, moralizing writers continued to condemn idleness and to write about the virtues of work, but the destructive effects of irregularity and overwork were new and powerful themes in the growing number of treatises concerned with industrial labor. Though the change is halting at first, this new literature by work-hygienists considered the physiological and moral qualities of work as complementary— each aspect balancing and reinforcing the other to create an internal equilibrium between the needs of the body and the soul, an economy of physiology and morality.[83] Such hygienic treatises departed from the narrow focus of earlier medical experts on the "insalubrity" of the workplace and were skeptical of work as a therapy against vice and profligate behavior. In 1862, Apollinaire Bourchardat, professor of hygiene at the faculty of medicine at the University of Paris, warned an audience of skilled workers who attended lectures at the Association Polytechnique on the harmful physiological effects of work. Science, Bourchardat claimed, could now offer irrefutable proof of labor as "a condition of health, of morality, and of indefinite progress."[84] Work involved a necessary "expenditure of energy," he said, but it also required a "recuperation" of energy through adequate nutrition, rest, and sleep. Regular work established a norm of health and well-being, which was threatened by the shock of overwork. Bourchardat cautioned his listeners that as idleness produced anorexia, gout, or obesity, overwork accelerated the onset of stomach cancer, senility, and various forms of mental alienation. The excessive deployment of "vital forces" was prejudicial to the maintenance of organic equilibrium. As a crucial part of the human economy, work was at once moral, physical, and political; its absence produced "what the English call *'spleen,'* the French *'ennui de la vie.'* "[85]

The normative dimension of work is no less evident here, but the most striking aspect of Bourchardat's rhetoric is his emphasis on an economy of energy as opposed to the traditional condemnation of idleness. In 1870 Simon noted that despite Rousseau's harsh judgment against idleness as crime, such absolutist attitudes could only be maintained from the viewpoint of "moral law but were excessive, and therefore false, for a political or coded law." A society in which everyone works is good, Simon wrote, but it is equally important that everyone work well: "Between the error of not working at all and those who do not work as well as they might, there is a difference of degree." Moreover, productive work diminished waste and "enriched the sum of the forces over which humanity disposes."[86] Though it remained an ethical imperative, the rational regulation of work was beginning to supercede the older emphasis on the morality of work, irrespective of its costs to the worker.

For this reason, it is important to consider Martin Méliton's relatively unknown treatise, *Le Travail humain,* which appeared in 1878. With originality and verve, Méliton's work extended to the physiological domain the normative aspects of work encountered in the writings of moralists and reformers. *Le Travail humain* is a curious mixture of moralizing philosophy and an attempt to consider labor in the framework of the physiology of human sensation and of the body as a "human motor."[87] Méliton theorized that "discomfort was the principle of movement," the "motor of activity." But he quickly qualified that only the savage works because of "corporal discomfort." Civilized populations, particularly the majority of European workers, are motivated by a desire to produce: "It is not a question of the body but of the spirit." For Méliton the body is capable of moral self-regulation: "The desires are the regulators and the grooves of the human machine." The "education of the will," which is the principal task of any science of work, will produce the "good worker." Méliton's text thus dressed up the old sentiment of idleness in the "rigorous precision of scientific language," but his language also recast the body as a working machine.[88]

These early and obscure treatises on the hygienics of work are ambiguous from several points of view. Politically, their authors are difficult to classify: These texts are more neutral than previous works, which are easily categorized by their authors' liberal, socialist, or religious convictions; theoretically they fall somewhere between science and moral exhortation. The obscurity of their authors is not insignificant. We can imagine that they were typical of a certain kind of nineteenth-century *savant* who casts a plague on all political houses. When seen in the light of later developments, they are still the products of a consciousness that has not yet cast off the dark images of idleness but that already recognizes the significance of changing scientific ideas for the perception of work. In these treatises the strictly moral coloration of idleness as the source of indolence and criminality is clearly waning. The long tradition of industrial edification in which the moral, intellectual, and spiritual benefits of work are opposed to the debilitating effects of sloth and the endemic laziness of barbaric peoples, is beginning to lose its discursive power. The language of labor is evolving into a language of labor power as a quantifiable force of production, localizable in the economies of energy distributed within the body and the psyche.

THE DISCOVERY OF FATIGUE

The first signs of a change in the perception of work appears in the medical literature of the late 1870s, which began to consider overwork, overexertion, and fatigue in the taxonomy of modern disorders. Although this literature often did not explicitly deal with labor, it considered fatigue as the chief sign of the body's refusal to bend to the disciplines of modern industrial society. If fatigue existed in the premodern era, it did not yet appear as a medical term, nor did it receive significant attention. In the 1870s, however, a new medical discourse began to chart the topography of fatigue and to place landmarks in its previously unexplored terrain.

In 1875 George Poore published a brief article in the London medical journal *Lancet* distinguishing between general and local, acute and chronic symptoms of the disorder.[89] In France, M. Carrieu in his pioneering study, *De la fatigue et de son influence pathogénique* (1878), noted that fatigue did not appear in any of the great medical dictionaries and that all prior efforts to define fatigue had been notoriously subjective. Carrieu bemoaned that "we can see what abuse the word fatigue has been subjected to in the language of all epochs, [since] its exterior manifestations, in effect, are nothing but the result of different types of experiences among individuals who are themselves distinct and diverse."[90] To end the confusion and to find "the essential idea of fatigue," Carrieu provided the following auspicious definition: "Un trouble dans l'activite des éléments anatomiques, causé par un fontionnement exagéré au point que la réparation y est momentanément impossible (a disorder in the activity of anatomical elements, caused by excessive functioning until repair is momentarily impossible)."[91] It was not uncommon for physicians to attribute the emergence of illness or deficiencies of character to the effects of fatigue, particularly of the passions or the intellect. But fatigue also bred social discontent: "Worn by fatigue, man is abandoned to sadness, discouragement, misanthropy, guilt-ridden lamentations against Providence, and bitter recriminations against the social order."[92]

Significant in these early perceptions of fatigue is the identification of a difficulty or a disorder and ultimately the breakdown of the mental and physical system. This association of fatigue with pain, and especially with the depletion of bodily or mental forces, contrasts sharply with a much older, more benign perception of fatigue as the necessary accompaniment of work. "Do not continue any day's journey to fatigue," preached the ubiquitous Dr. Johnson, in a rhetoric that emphasized fatigue as the point beyond which unmeasured exertion would be unwise.[93] Diderot describes fatigue simply as "the effect of considerable work" and notes that fatigue and travail were often used interchangeably, "though one was the cause and the other the effect."[94]

Here we can see the faint echo of a highly spiritualized fatigue sometimes encountered in the medieval literature of the monastic orders: "The strength of the soul enters through the *fatigacion* of the body."[95] This fatigue is explicitly *not* identified with *acedia*, but rather with the weariness produced by the work of the community. Fatigue is a sign of limit, of the point of rest, even of spiritual awakening. A similar notion of fatigue as a welcome sign of the body's need for restoration persists into the nineteenth century as an aristocratic or aesthetic sensibility. As the withdrawal of mind and body from the stimulation and excitement of the world, fatigue here is a pleasurable sensation, a luxurious respite from labor or travel. The *Journal des Goncourts* records this perception, perhaps as an anachronism or nostalgic sentiment: "Excessive work produces a not unpleasant dullness, a feeling in the head which prevents one from thinking of anything disagreeable, an incredible indifference to the pinpricks of life, a detachment from reality, a want of interest in the most important matters."[96]

The modern image of fatigue contrasts sharply with this idyll of a natural boundary from the strains of civilization. In the medical literature of the late nineteenth century fatigue appears—much like idleness—as an obstacle to work, as the horizon of forces or energies within the body. It is the negative imposition of physiology or psychology on the body or the will: "The holding back of any power from exercise is positively painful, so its passing into energy, is, were it only the removal of that painful repression, negatively pleasurable."[97] Another early medical writer on fatigue, Pierre Révilliod, grouped all fatigue disorders under the single rubric *ponose* and described the experience of fatigue in this way: "The head is weighty, the spirit becomes lazy, thought, memory and the will all languish, and the apathy that results seizes control of the body's legislative and executive power."[98]

Clearly, fatigue was perceived as *both* a physical and a moral disorder—a sign of weakness and the absence of will. Maurice Keim, author of one of the earliest French medical textbooks of fatigue, noted that "we flee it by instinct, it is responsible for our sloth and makes us desire inaction." Keim also attributed fatigue to the "sad passions" that accompanied the desperation and *ennui* of the Siege of Paris.[99] As the portal of moral decay, fatigue accounts for, in the words of the writer Joris Huysmans, that "diabolic perversion of will which affects, especially in matters of sensual aberration, the exhausted brains of sick folks." Here the image is that of *acedia,* since "nervous invalids expose fissures in the soul's envelope whereby the Spirit of Evil affects entrance."[100] The novelist Barbey d'Aurevilly, who frequently took his subjects from articles in medical journals, once remarked that tiredness and unhappiness decide almost everything in the lives of men.[101]

THE POETICS OF FATIGUE

A breakdown of body and mind, fatigue was increasingly identified as a "modern" disorder of overwhelming social and physical consequence. This perception appears frequently in the poetic literature of exhaustion, which arises almost simultaneously with the medical and scientific literature on fatigue. As early as 1857, Baudelaire drew a direct line from the destructive power of fatigue to the figure of the devil in his allegory, "Destruction":

> *And so, far from God's sight, he leads me on,*
> *Panting and broken with exhaustion*
> *Into the plains of Tedium profound.*[102]

Similarly, Schopenhauer's reflections in his essay, "Physiognomy" (1851), also invoked the figure of fatigue as a negative component of the will, in which "moral character is reflected."[103] But it was Huysmans, an admirer of Schopenhauer "beyond all reason," who gave significant literary expression to the discourse on exhaustion. His 1884 novel *A Rebours (Against the Grain)* documents through the collapse of its youngest scion, "a frail young man of thirty, anemic and nervous, with hollow cheeks," the decline of the noble line of the family of Des Esseintes.[104] *A Rebours* traces the course of Des Esseintes' decline from fatigue to exhaustion to *ennui,* the world-weariness that ultimately consumes him. His ascetic withdrawal from a life of debauchery and hypocrisy in urban Paris into aestheticized solitude is recorded in microscopic detail. Numbed to even the most sensual pleasures by his excesses, overpowered by his *ennui* and the disintegration of his will, he collapses: "This was the end; as if all possible delights of the flesh were exhausted he felt sated, worn out with weariness; his celebration of the senses fell into a lethargy, impotence was not far off."[105]

Huysmans' novel is a recasting of the *acedia* encountered earlier, though in a modern sensibility—the aestheticist parody of aestheticism. Des Esseintes retreats into a fully privatized and hermetic existence. His apartment is a monastic cell where order, method, and control over all stimuli are deployed against the source of his moral infirmity, his fatigue. Every detail of interior design in his Fontenay retreat is constructed with this fatigue in mind, producing a bizarre interior architecture of nervous exhaustion: The corridor of his study is padded and sealed; the floors and walls are covered with thick carpets and hangings to prevent sound from entering; heavy double doors seal out all odors; and the servants are made to wear felt shoes. The entire structure of the inner sanctum resembles a ship moored within the larger apartment, with "no view of the outside world" permitted apart from a "faint dim 'religious' light."[106] Finally, there is even a lengthy disquisition on the relationship between color and exhaustion:

> Last comes the class of persons of nervous organization and enfeebled vigor, whose sensual appetite craves highly seasoned dishes, men of a hectic, over-stimulated constitution. Their eyes almost invariably hanker after that most irritating and morbid of colours, with its artificial splendors and feverish acrid gleams—orange.[107]

Flight from fatigue gives birth to a radically *interiorized* subject, whose personality is shielded from all stimuli by the tomblike heaviness of late nineteenth-century bourgeois decor. The goal was to construct an environment entirely of unnatural objects, to create artful surrogates, which were considered superior to vulgar reality or "nature," and to protect the individual from contact with any form of overstimulation. This design did not, however, exclude all color. On the contrary, as Dolf Sternberger pointed out, the *fin-de-siècle* interior had to become an "inner orient" in which color wages war against the disruptive incursion of that "trouble-maker in the kingdom of harmony—white."[108] Even the demand for changes in style could be attributed to fatigue. In 1887 *The Aesthetics of Architecture,* a work by the German art critic Adolf Göller, attempted to account for the "psychological causes for the changes over time in our feeling for the beauty of ornament in architectonic styles," by what he described as the "psychological law of the fatigue of the feeling of form."[109] Vladimir Jankélévitch, author of a major philosophical study of literary *ennui*, called fatigue a "delicate Monster," which haunted all pessimists and could be kept at bay only by a future without risk or hazard, by a career of complete rest, and a daily existence without tension.[110] As Walter Benjamin once remarked, the nineteenth-century interior was not only the private citizen's universe, "it was also his casing."[111]

Émile Tardieu, in his 1903 study of the medical and literary aspects of *L'ennui*, explicitly declared that the modernist sensibility and subjectivism characteristic of his generation was actually rooted in the physiological infirmities of writers. Their afflictions alone, and not any cultural advance, he argued, explained the seemingly endless panoply of physical and psychic symptoms and bizarre associations of several great figures of French literature. Tardieu defined *ennui* as "a psychophysiological complex of infinite variety, incomparably polymorphous. Its most essential characteristic is pain, but pain of an eminently unstable sort, not even recognizable as such, endowed with an astonishing power of metamorphosis. Psychically protean, capable of transforming itself with the aid of diverse elements, born of everything and nothing, it reveals itself and disappears, reappearing again, without revealing its law to us, without allowing us sufficient time to penetrate its secret."[112]

For Tardieu mental and physical exhaustion had to be ranked first among the causes of *ennui*. He warned that to ignore this physical aspect risked giving undue credibility to the literary representation of illness, thereby diminishing the significance of its reality as fatigue. By unmasking *ennui*'s literary aura, Tardieu claimed to reveal the true physiological basis of the distintegration of the will so cleverly disguised by the great writers of the *fin de siècle*. Fatigue, not art, caused their excessive subjectivism; fatigue not aesthetics, explained their profound "disorder of ideas, mental torpor, irritability, generalized asthenia—the primitive elements of ennui" evident in their work. To "insist on the central fact of exhaustion" in the etiology of *ennui*, wrote Tardieu, was to remove from the literature of *ennui* "its poetic garbe."[113]

FATIGUE AND SOCIETY

Tardieu's revulsion at the artistic *ennui*, which turned fatigue into modern writing, was of a piece with an increasing number of authors who by the 1890s understood fatigue as nothing less than the wellspring of social disorder and moral decay. Fatigue represented the membrane between morally sacrosanct labor and the violent, irrational impulses that constantly threatened to disrupt social order. The psychologist Charles Féré, a student of the great French alienist Jean-Martin Charcot, devoted one of his many books to the cumulative effects of fatigue on crime and vice, observing that "one of the principal effects of nervous exhaustion is the incapacity for sustained effort."[114] For the laboring classes the threat posed by fatigue was potentially more serious than among the easily seduced literary youth. A social pathology of fatigue was most evident among the "industrial and commercial classes," Féré noted, but "the fatigued of every social order seek to struggle against exhaustion by diverse excitements: luxury of attire, of furnishings, of diet, the pleasures of the body and the spirit."[115] And the leading French psychologist and editor of the fiercely materialist *Revue Philosophique,* Théodule Ribot, classified all humanity into three divisions: the "superior active," characterized by a "superabundance of force," the "average" active, whose "capital of energy is so limited they are compelled to economize, and the "asthenics" *(asthéniques)* whose "repugnance toward all effort, idleness, apathy and inertia were extreme."[116]

In fatigue the physical horizon of the body's forces was identified with the moral horizon of the species; the moral infirmity of the populace was directly proportional to the debilitating effects of fatigue. Fatigue "seems to consume our noblest qualities—those which distinguish the brain of civilized from that of savage man," wrote Angelo Mosso, the Turin physiologist and Galileo of modern fatigue research. "When we are fatigued we can no longer govern ourselves, and our passions attain to such violence that we can no longer master them by reason."[117] Ribot considered fatigue as an "impairment of the will" manifested by apathy and, in its more extreme forms, by complete abandonment. He distinguished between those cases in which the intellect remains intact but cannot function and those in which inhibition is absent and "the impulse extends itself entirely to the profit of automatism."[118] At worst the will becomes so "disorganized" that the individual is no longer "master of himself" and permits all sorts of chimerical impulses to range freely.

Nineteenth-century positivism was characterized by a profound suspicion of subjectivity and by the search for scientific laws that could transcend the disorder and instability of mental and physical states. For the scientific materialists who pioneered the study of fatigue, the representations of the will, or, as we would say, consciousness, posed a persistent problem to which fatigue presented a solution. The illusions thrown up by the fatigued will, the splenetic character of fatigued individuals, and the religious, literary, and philosophical chimeras cast off by evidently fatigued brains were products of a single pathology. Elaborately described by Ribot, Tardieu, Mosso, and others, such representations could be contained only by the objective laws of exhaustion, which, if discovered, could render those subjective effects transparent and, perhaps, superfluous. The preoccupation with fatigue exemplifies a critical feature of nineteenth-century positivism: attributing an objective basis to highly subjective states. But positivism also created a framework for knowledge, for norms, and for models of human nature that redefined the body and its external limits. Fatigue was central to the project of transforming the subject into the data of an objective and natural process. At the same time, all ideas not sanctioned by materialism and science were attributed to the negative effects of fatigue. The truth of the irrational subject was nothing less than the truth of its primordial fatigue.

The discovery and diagnosis of fatigue generated a proliferation of efforts to chart its course, find its cure, or at least modify its effects. Such efforts were joined not only by physiologists, but by social hygienists, engineers, psychologists, and social reformers for whom fatigue represented the threshold of human limitation; the barrier that society should strive to bring under the control of medicine, technology, and politics. Underlying the anxiety and hostility that surrounded fatigue was the utopian ideal of transcending it. The result would be not only a vast release of the latent energies of society but a productivist civilization, resistant to moral decay and disorder. Behind the scientific and philosophical treatises on fatigue lurked the daydream of the late nineteenth-century middle classes—a body without fatigue.

CHAPTER TWO

Transcendental Materialism:
The Primacy of
Arbeitskraft (Labor Power)

AN IMMENSE RESERVOIR OF ENERGY

THE single-mindedness that accompanied the medical, literary, and philosophical obsession with fatigue in the second half of the nineteenth century can be explained by a profound change in the way science thought about nature. Fatigue or exhaustion is linked to its opposing concept, energy, or *Kraft*, which after 1840 emerged as the dominant trope of scientific materialism. Nineteenth-century physics discovered energy as the universal force present in all matter, capable of converting itself into innumerable forms, yet inalterable and constant. The accompanying discovery that energy or heat was the source of all mechanical work led, according to the author of the first law of thermodynamics, Sir William Thomson (Lord Kelvin) "to the greatest reform that physical science had experienced since the days of Newton."[1]

The law of energy conservation accorded the concept of energy, or *Kraft*, undisputed primacy in the explanation of the natural world. Physics became the model of the sciences: the physicochemical basis of nature and the grounding of a materialist faith. At the root of that faith was a model of nature drawn from the technical instruments that could harness its powers. In his 1824 *Reflections on the Motive Power of Heat*, Sadi Carnot spoke of the steam engine as a "universal motor," which could be substituted for animal power, waterfalls, and air currents."[2]

The discovery of energy as the quintessential element of all experience, both organic and inorganic, made society and nature virtually indistinguishable. Society was assimilated to an image of nature powered by protean energy, perpetually renewed, indestructible, and infinitely malleable. The pioneers of energy conservation viewed the transformation of mechanical energy into heat, and subsequently, the transformation of all natural forces as manifestations of a single *Kraft*.

To be sure, the language of *Kraft* can be traced to numerous sources in the German intellectual tradition from Jacob Böhme, to Fichte, Herder, and Hegel.[3] In the later part of the nineteenth century, however, energy ceased to be used metaphorically as an "idea" immanent to natural phenomena in the sense that Romantic *Naturphilosophie* employed it. Recourse to "vital substance," "incalculable fluida," "irritation," or any other "imponderabilia of nature" was no longer required to explain motion.[4] As Thomas Kuhn has shown, between 1837 and 1847 several theories simultaneously emerged that attempted to give scientific credibility to the idea of a world of phenomena "as manifesting but a single 'force,' one which could appear in electrical, thermal, dynamical, and many other forms, but which never, in all its transformations, can be created or destroyed."[5]

Thermodynamics conceived of nature as a vast machine capable of producing mechanical work or, as von Helmholtz called it, "labor power." Initially a measurement of the force of machines, "labor power" became after the discovery of energy conservation the basis of all matter and motion in the physical world. Both a magnitude and a measure, labor was the universal fact of nature assimilated to, and exemplified by, the laboring body in the social world. For physiologists armed with the principles of thermodynamics, the energy of the body was not merely analogous to other natural physical forces, it became one among them. The purpose of nature was to yield "work," and as part of that equation, the body yielded the work of the nerves, the muscles, and the organs, which were subject to the same laws of nature as any other machine. Especially striking in this reconceptualization of the body as a thermodynamic machine is that work became a universal concept, the conversion of energy into use. This recasting of work in terms of the language of energy, or *Kraft*, was decisive. Raised to a transcendental category of experience, work became a metaphor for a physiochemical exchange. "Cleansed" of all of its social and cultural dimensions and considered "exclusively as a form of energy conversion, work could be applied to nature, technology, and human labor without distinction.[6]

Thermodynamics decisively altered the concept of labor, at once modernizing it according to the precepts of industrial technology and naturalizing it in accord with new laws of physics. The primacy of labor in the theory of property—prominent in the work of seventeenth- and eighteenth-century philosophers and economists like Locke and Smith—placed human work within a social division of labor that emphasized the identity of human industry and individual autonomy. Locke's famous "work of the hands and the labor of the body" expressed an already outmoded distinction between craft production and brute strength, between work as a creative act and alienated labor. This image of work, appropriate to a preindustrial era, still maintained the traditional distinction between labor as the source of property and selfhood, and labor as a burden. But a striking change occurred in the second half of the nineteenth century as this increasingly anachronistic vision of labor became superseded by the energeticist model of mechanical work. The work performed by any mechanism, from the fingers of the hand, to the gears of an engine, or the motion of the planets, was essentially the same. With this semantic shift in the meaning of "work," all labor was reduced to its physical properties, devoid of context and inherent purpose. Work was universalized.

The optimism in the image of a universe composed of unlimited power or *Kraft* announced by the law of conservation of energy was rapidly undercut by the pessimistic doctrine of the irreversibility of heat flow, or the "loss" of energy in conversion expressed in the second law of thermodynamics. In 1865 Rudolf Clausius adopted the term *entropy* to describe the irreversible decline or diffusion of energy that occurred in tandem with conservation. Thomson regarded the second law as confirmation of the biblical view of the universe's impermanence and as early as 1852 concluded that the earth's heat was decreasing, thus limiting the optimum conditions for, and duration of, human habitation.[7] Thomson's hypothesis challenged the predominant view of those nineteenth-century geologists, who following Charles Lyell regarded the history of the earth as a slow-moving process. Instead Thomson claimed that the geologist's calculations had radically exaggerated the earth's age and that the planet was cooling down over a much shorter time span. Countering the limitless energy of the conservation theory, the second law of thermodynamics raised the prospect of the cataclysmic "heat death" of the universe, the gradual dissipation of energy into an icy, lifeless void.[8]

As Charles Gillispie, the historian of science, pointed out, both energy and entropy "are highly sophisticated and abstract representations of certain elementary experiences of the world, certain serious intuitions: Energy, of the intuition that there is an activity, a 'force' in things beyond matter and motion, that makes nature go," and entropy of the "complementary experience of water seeking its own level, of hot bodies cooling, of springs untensing."[9] That these two perceptions were paradoxical as analogies of nature did not escape the attention of contemporaries. But they were unable to resolve the contradiction, perhaps because the paradox expressed through the complex metaphors of nature the deeper hopes and anxieties of an industrial civilization in its birth pangs.

The obsession with fatigue in nineteenth-century thought was not merely a sign of the "real" weariness of individuals in industrial society, but of the negative aspect of the body conceived as a thermodynamic machine capable of conserving and deploying energy. The body's fatigue was, as Helmholtz pointed out, a particular instance of entropy. Resistance to work was no longer located in the soul's impurities, but in the physical properties of fatigue. As the factory system and the machine consumed its first generations, the moral appeal against the regression of idleness lost much of its power. The new physics and the physiological theories that emanated from it dissolved the old moral categories by absorbing them into the scaffolding of a universe built out of labor power. Fatigue thus emerged at the threshold of the body's economy of energies; with its own internal laws of energy and motion, it was the corporal horizon of a mechanical universe. In the work of the body the cosmos was represented in microcosm. Jules Amar, an industrial ergonomist and fatigue expert, explained in his *Le Moteur humain* (1913) that the working body was a human motor, which functioned according to the principles of economy and regulation:

> Fatigue can be defined as the effects which limits the duration of work. In the case of living motors, man and animals, the fatigue either decreases the intensity of the muscular effort or reduces the contraction of the muscle. . . . The result of fatigue is a lessened aptitude for work.[10]

DEMATERIALIZED MATERIALISM

Gaston Bachelard once described nineteenth-century scientific materialism as a "dematerialized materialism," a materialism embodied in the primacy of energy: "The very fact that energy changes matter results in a peculiar shift of scientific language from metaphor to abstraction."[11] The new scientific materialism was predicated on a single, indestructible, and invisible *Kraft*, which could be perceived only in terms of its effects—in the material form of different kinds of mechanical work. Although energy was the source of all motion and matter, the materiality of the physical universe—energy—was nowhere to be encountered except in the manifest consequences of its enormous labor power. Materialism became, in a word, "transcendental."

Nineteenth-century materialism rejected any distinction between the laws of inorganic and organic nature. Nature was essentially the narrative of an inalterable demiurge whose powers could be harnessed and transformed but never depleted. If the cosmos could be subsumed under the universal laws of energy, society too was subordinate to natural laws of development that favored productivity, performance, and progress. In this sense energetics was well suited to a society on the crest of an astonishing industrial revolution at the end of the nineteenth century. In Germany Hermann von Helmholtz, Emil Du Bois-Reymond, Ernst Haeckel, and Wilhelm Ostwald each represented different aspects of German scientific thought characterized by materialism in the service of industrial modernity. They adopted the industrial machine as their model of the universe, and the universe in return became the source of unlimited power for industrializing society. During the late-nineteenth century, Germany was, as the Marxist philosopher Ernst Bloch explained, the classical land of "non-synchronicity," of both extraordinary rapid economic and social progress, and virulent resistance to modernity. If the chorus of antiliberal and antimodernist voices from Schopenhauer to Spengler was one familiar pole of that dialectic, the scientific materialists were surely the other, manifesting boundless optimism regarding the potential harmony of nature and industry. Indeed, the emphasis on the German tradition of antirationalism and antimodernity, highlighted by its centrality in National Socialist ideology, has all but obscured the significance, especially in intellectual circles, of this materialist modernity in Imperial Germany.

The emergence of scientific materialism in Germany was closely linked to the ill-starred revolution of 1848 and the subsequent defeat of liberal ideas in public life after its eclipse. To the extent that liberalism survived the political defeat of 1848, the traditions of German constitutionalism were largely preserved in its scientific, rather than political culture. In the absence of a liberal polity, German science became the most fertile terrain for the antireligious, antiautocratic, and democratic ideals that the post-1848 era extinguished in the public sphere. From its inception the first generation of scientific materialists (Carl Vogt, Jacob Moleschott, Johannes Müller, Ludwig Büchner, and Ludwig Feuerbach) tried to emancipate science and philosophy from recourse to any explanation that took refuge in metaphysics or departed from the sensuous and godless theory of the unity of matter and motion.[12] Everett Mendelsohn has argued that the materialist project in both French and

German physiology was in many ways the product of a self-conscious link between the post-1848 vision of a liberal and reformed polity and the reform of science and medicine.[13] The biologist Jacob Moleschott articulated the credo of the new scientific materialism when he wrote that "the materialists profess the unity of energy and matter, of the spirit, of the body, of God and the world."[14]

The early pioneers of *Kraft* were well served by the assiduous scientific popularizer, Ludwig Büchner, who used the new materialism as "a battering ram directed against the crudest conceptions of religious tradition."[15] Büchner's *Kraft und Stoff* (1855) became the "bible of the new materialism," an attempt to rally post-1848 German liberals under the banner of the unity of matter and motion. For Büchner, the unity of force and matter provided "an indestructible foundation" for a view of nature that would "decisively ban every kind of supernaturalism and idealism from the explanation of natural events."[16] Though Büchner was unaware of the doctrine of conservation of energy when he published the first edition of his *Kraft und Stoff*, he incorporated it into the subsequent edition only two years later. In the first edition Büchner had clearly located the materialist faith of his generation in the inseparability of matter and *Kraft*, expressed in the slogan: "No force without matter—no matter without force."[17] In short, the unifying principle of *Kraft* was the philosophical core of scientific materialism, even before it received its scientific pedigree. It was, to use Georges Canguilhem's distinction, a scientific ideology on the way to becoming a science.

Though more circumspect than Büchner and the generation of scientific materialists, the scientific generation trained by Johannes Müller in Berlin, which included Carl Ludwig, Rudolf Virchow, Helmholtz, and Du Bois-Reymond, rallied to materialism as both a scientific and philosphic doctrine: "We four imagined that we should constitute physiology on a chemical-physical foundation and give it equal scientific rank with physics."[18] Unlike the first generation of scientific materialists, who identified with liberalism and assailed religious and metaphysical doctrines, the "physiological reductionists" (as they have been called) refrained from linking science to politics or religion. These physiologists found in the principle of *Kraft*, or energy, the key to the natural laws of both organic and inorganic nature. The equivalence of different forms of energy and the infinite capacity for the conversion of these forms revealed that matter—whether in the form of nature, technology, or the body—could neither be separated from motion nor divorced from the energy that moved matter throughout the universe.

Even if science did not have to parade its ideological assumptions, materialism embodied a critical and antidogmatic ethos in its rejection of all theological and metaphysical predicates. *Kraft* thus described the transcendental force omnipresent in nature and implicitly discredited any nonmechanical explanations. In Germany first and then later in France, materialism took up the banner of enlightened republicanism by other means.

FROM THE HUMAN MACHINE TO THE HUMAN MOTOR

Scientific materialism also transformed the image of the body into that of a working machine, distinct from the animal and human machines envisioned by Hobbes and in Descartes' portrait of the human machine. The metaphor of the machine in physiology has roots in Aristotle's thought and was well established by the time Vesalius described the human organism as a "factory" in the mid-sixteenth century. Descartes' famous *Description du corps humain* (1648) explicitly distinguished the system of mechanical relations that accounts for the body's movements from the older Aristotelian image of a machine whose organs respond to the commands of the soul, as the foot soldier obeys a general. As Canguilhem recognized, however, for Descartes the moving power of the animal machine was innate—a mimesis of the power invested in it by the creator, the "maximum artificer."[19] This distinction separates Descartes' image from the most important and best known of the mechanical representations of the body, Julien Offray de La Mettrie's famous eighteenth-century treatise *L'homme machine* (1747). For La Mettrie, the human body was essentially a watchspring with unique self-winding properties. As the quintessential *perpetuum mobile,* or "self-moving machine," the human body created an enormous dilemma for replicating such motion by human artifice. This problem entertained the creative fantasies of generations of inventors until 1775, when the French Academy of Sciences officially refused to consider any further solutions to this famous conundrum.[20] La Mettrie, Condillac, and other French Enlightenment materialists shared the view that a principle of motion was inherent in all matter—for example, that each fiber of matter possessed "innate force"—which they attributed to "irritability."

The idea of self-moving power was the secret passion of eighteenth-century materialist physiologists such as F. Le Cat and Jacques Vaucanson, who constructed elaborate mechanical models of human and animal machines in the firm conviction that it was possible to replicate the mechanical qualities of human beings and animals. These "animal machines"—for example, Vaucanson's famous "duck," which ate and digested its food, or the flute player of the same artist, which produced exquisite sounds and moved all its fingers correctly—relied on an external source of motivating power. The technical fictions called "automata" were sophisticated machines designed to illustrate a biomechanical mode of explanation: a similitude between nature and the power of reason to mirror the processes of nature.[21] Voltaire recognized this purpose when he spoke of Vaucanson's "plan to create an automatic figure whose motions will be an imitation of all animal operations, such as the circulation of the blood, respiration, digestion, the movement of muscles, tendons, nerves and so forth."[22] By the end of the eighteenth century, physiological vitalists like Paul-Josef Barthez delighted in denigrating the automata by pointing out that "life" had eluded them and that their inventors had resorted to occult theories or "not yet invented machines" to explain why they could not breathe "life" into these artifices. The great automata provided a showcase for the analogic thinking of the age but disappointed in their apparent lack of "self motive power."[23]

With the invention of the steam and internal combustion engines, however, the analogy of the human or animal machine began to take on a modern countenance. As the philosopher Michel Serres has noted, the eighteenth-century machine was a product of the Newtonian universe with its multiplicity of forces, disparate sources of motion, and reversible mechanism. By contrast, the nineteenth-century machine, modeled on the thermodynamic engine, was a "motor," the servant of a powerful nature conceived as a reservoir of motivating power. The machine was capable of work only when powered by some external source, whereas the motor was regulated by internal, dynamic principles, converting fuel into heat, and heat into mechanical work.[24] The body, the steam engine, and the cosmos were thus connected by a single and unbroken chain of energy.

CONSERVATION OF ENERGY

In July 1847, at age twenty-six, Hermann von Helmholtz, still a physician at the military hospital in Potsdam, delivered an "epoch-making" lecture to the Physical Society of Berlin entitled "Die Erhaltung der Kraft." Remarkable for its conciseness and theoretical sophistication, Helmholtz's lecture assured the "triumph" of the principle of energy conservation and "diverted" science "into an entirely new channel."[25] According to the distinguished nineteenth-century scientist, Gustav

Robert Kirchoff, Helmholtz's formulation was the "most important contribution to natural science made in our era."[26]

Helmholtz's early presentation of energy conservation had already been anticipated to some extent by Sadi Carnot, Julius Robert Mayer, and James Prescott Joule. But his formulation offered a less technical, more philosophically coherent account of the epistemological significance of the new materialism that emerged from Müller's laboratory than any of its other protagonists had yet provided. In so doing, Helmholtz also elaborated a major tenet of modern critical positivism: a complete separation of scientific explanations of causality from metaphysics and a reliance on mathematics as the *sine qua non* of scientific proof.[27] Helmholtz maintained that matter and force could not be conceptually disentangled: "It is evident that the concepts of matter and force cannot be applied separately to nature. Pure matter would be indifferent to the rest of nature, since it would never produce any changes in it or in our sense organs; a pure force would be something which exists and yet does not exist, for that which exists we call matter."[28] In contrast to the dogmatic materialism of Moleschott or Büchner, Helmholtz was careful to distinguish between scientific knowledge, which admitted certain abstract concepts in order to gain indirectly knowledge of nature, and nature itself. Matter and energy were precisely such abstractions that attained conceptual coherence through their power to explain the effects produced in nature. But causation could not be explained: scientific knowledge could pursue its object until it "is forced by irrefutable limits beyond which it may not go." In short, "we can perceive matter only through its forces not in and of itself."[29] Helmholtz was perhaps closer to the earlier prophets of *Kraft* than he allowed, but he was also more insistent on demonstrable mathematical and experimental proof of the basic laws of energy.[30]

Helmholtz then proceeded by a series of "analogies," as he put it, to demonstrate the incontrovertible truth that it is impossible to create force out of nothing. Helmholtz thus subsumed the Newtonian laws of motion under the overarching theory of the conservation of energy, which held that there was a single, indestructible, and infinitely transformable energy basic to all nature. Motion presupposed the existence of this energy as a primordial cause *(Ursache)* that could explain the dynamic properties of matter. *Kraft* thus became an ontological concept basic to all matter and integral to the explanation of all mechanical causes.

It is significant that Helmholtz chose to illustrate the universal application of his theory by the futile search for a *perpetuum mobile,* which as his audience knew, had long been abandoned. According to his biographer, Helmholtz, who as a young boy had listened to the heated debates on the *perpetuum mobile* between his father and friends and as a young man had studied the eighteenth-century mathematicians who had wrestled with this problem, finally concluded that the idea of a self-moving mechanism was the phantasm *par excellence* of the French materialism of the previous century.[31] Indeed, if the French Academy ceased to offer a prize for the invention of a *perpetuum mobile*, it was because the *savants* had finally realized that to search for an intrinsic force to power machines was chimerical. The example of the *perpetuum mobile* demonstrated that energy is not something produced *de novo* but is simply communicated from the great store of the universe and transferred from one form to another. As Helmholtz observed in a popular lecture delivered some years later: "We humans cannot create any labor power for human purposes but can only appropriate it out of the great general storehouse of nature."[32] In all the natural and manmade uses to which energy is put, it cannot be generated without expenditure, and though a degree of "working force" may be lost in a particular conversion, it is never diminished in nature: "Nature as a whole possesses a store of energy which cannot in any wise be added to or subtracted from: the quantity of energy in inorganic nature is as eternal and unalterable as the quantity of matter."[33]

As Kuhn pointed out in a classic essay, energy conservation is a remarkable example of "simultaneous discovery," a complex story of more than a dozen prominent scientists often working in ignorance of one another's efforts but all arriving at similar conceptions of the equivalence of force in the 1840s.[34] Kuhn also identified three "external" influences, or events, that contributed to the discovery of energy conservation: the invention of the steam engine, the philosophic impact of *Naturphilosophie*, and the French engineering tradition of the early nineteenth century.

The steam engine demonstrated, albeit in a special case, the significance of heat for the production of work and provided the crucial illustration of the principle of conversion of force. As Charles Gillispie remarked, "Conservation of Energy lurked everywhere latent in the steam."[35] Late eighteenth-century mechanics provided the quantitative theorem "force times distance," which permitted French engineers to calculate the power of machines in motion. It is out of this tradition, especially the work of Sadi Carnot, Kuhn argues, that the modern concept of "work" emerged as the equivalent of the energy originally consumed.[36] Indeed, Helmholtz's examples of the mechanical equivalents of electricity, magnetism, and animal heat further illustrated this principle in each domain of nature.

Naturphilosophie, especially important in the work of Mayer and Helmholtz, drew on the idealist philosophy of Schelling and Hegel and posited the idea of an *Urkraft,* or *vis viva,* which contained the secret of life and energy in the universe. In an 1892 essay entitled "Goethe's Anticipation of Subsequent Philosophic Ideas," Helmholtz speculated on how "the germ of his insight into the constancy of the total amount of energy was already present in the eighteenth century and on how Goethe could well have been familiar with it."[37] In Helmholtz's version of energy conservation the technology of the industrial revolution and German idealist philosophy of nature were united in a single ontological principle.

Drawing on the engineering tradition of Navier, Coriolis, Poncelet, and others, who referred to "travail" or "puissance du travail" as a standard measure for the energy yield of machines in the 1830s, "Helmholtz first used the terms *Arbeitskraft, bewegende Kraft, mechanische Arbeit* and *Arbeit* for his fundamental measurable force."[38] Helmholtz thus extended a particular usage of "work" modeled on the machine to a general principle of nature. The engineering tradition was rederived by the pioneers of energy conservation, who needed the concept of work to compute successfully conversion of energy, a discovery that became "the most decisive contribution to energy conservation made by the nineteenth-century concern with engines."[39] Universalized as the demiurge present in all nature, the concept of labor power redefined the principle of motion in the universe in terms of its power to "perform work."[40]

As Helmholtz tirelessly pointed out: energy was a transcendental principle, "the groundwork of all our thoughts and acts."[41] This principle was not merely a homology; knowledge of nature was also transformed by the principle of *Kraft:* the forces of nature enter our consciousness only through their use and conversion into work energy. Nature thus became a vast cistern of Protean energy awaiting its conversion into work.

If the social implications of the conservation of energy were not immediately apparent, the image of inexhaustible natural energy and labor power *(Arbeitskraft)* underscored the optimistic faith of science in the productive potential of the age. The identification of nature with energy harmonized with the dynamic belief in the productive powers of newly harnessed sources of energy: electricity, electromagnetism, internal combustion, steam, and the new technologies of the age—the railroad and the factory. Implicit in the theory was also an image of nature as a productive force, capable of providing the unlimited and inalterable universal motive power of animal, human, and mechanical "motors." Energy conservation essentially confirmed the unity of all physical forces in the language of physics. It was a dramatic departure from the Newtonian universe where the laws of mechanics refer to distinct forces available in nature. Nature was a "force beyond matter and motion" powering the "work" of the cosmos. Energy conservation was "productivist" insofar as it placed the metaphor of the machine at the center of scientific explanation and the energy of the universe in the service of an order dedicated to the production of work. Although Helmholtz's reevaluation of *Kraft* has been frequently discussed in connection with energy conservation, his parallel discovery of *Arbeitskraft* has received far less attention. Yet this concept transformed the concept of labor in nineteenth-century Europe, a transformation that Helmholtz himself furthered through the dissemination of his popular scientific lectures.

A UNIVERSE OF *ARBEITSKRAFT:* HELMHOLTZ'S POPULAR SCIENTIFIC LECTURES

In 1849 Helmholtz became Ordinarius Professor of physiology and general pathology at the University of Königsberg. In 1855 he moved to Heidelberg and subsequently to Bonn (1858) and Berlin (1871). In addition to his major contributions to the psychology of sense perception, optics (*Handbuch der physiologischen Optik* (1855), and acoustics, Helmholtz also became "one of the most eloquent preachers of the gospel of energy."[42] Through his popular lectures on science, the idea of energy conservation as a transcendental principle—transcendental materialism—was elaborated in a self-consciously literary style. According to Helmholtz, the promotion of popular science was significant in Germany not simply because "the natural sciences have come to be a powerful influence on the social, industrial, and political life of civilized nations," but because it challenged what he called the "preeminence of language in European culture."[43] Whereas England had always been receptive to popular science, with its lecture societies and lending libraries devoted to promoting scientific themes, Germany lagged behind because of its classical humanistic tradition of *Bildung* with its emphasis on poetry and philosophy. Helmholtz's popular lectures, therefore, were more than illustrations of his scientific principles. They also affirmed the supreme place of science in the hierarchy of reason: through the example of energy conservation Helmholtz elaborated the implications of his transcendental materialism as a kind of enlightened *Naturphilosophie* for everyman.

In these writings, even more than in his specifically scientific papers, we see Helmholtz's powerful image of nature defined as *Kraft*, an image he cast entirely in terms of the power to produce work. It could be argued, in fact, that Helmholtz's lectures primarily demonstrate the fundamental significance of *Kraft* as a metaphor for labor in industrial society.[44] Helmholtz also points to the conceptual value of *Arbeitskraft*—the labor power that is produced by *Kraft* and the only means through which energy can be perceived. Kuhn's discovery that Helmholtz was the first German thinker to employ consistently the term *Arbeitskraft* to connote measurable force, can, I think, be shown to have had major social as well as scientific implications.[45] As a physicist and physiologist, he has long been credited as author of the most substantive version of the law of conservation of energy, but he should also be credited as a major contributor to social thought: the elaboration of the modern concept of labor power as the quantitative equivalent of work produced, regardless of the source of the energy transformed. Helmholtz was the first to demonstrate explicitly the equivalence between natural, inorganic, and social conceptions of labor power.

On 7 February 1854 Helmholtz delivered his lecture "Über die Wechselwirkung der Naturkräfte" ("On the Interaction of Natural Forces") in Königsberg.[46] It began with a memorable description of the great mechanical toys, or automata, that reproduced the movements of animals and humans. As anthropomorphized technology, they represented for Helmholtz the apotheosis of artisanal mechanics and the illusions of the *perpetuum mobile.* He described in detail Vaucanson's marvelous duck with its digestive tract and his skillful flute player. But most striking were his evocations of "the writing boy" of the elder, and "the pianoforte player" of the younger Jacquet-Droz, which followed its hands with its eyes while performing and bowed courteously to the audience at the conclusion of the perfectly rendered piece.[47] Helmholtz also remarked sympathetically on the fate of the younger Jacquet-Droz and his creation, both of whom were held by the Spanish inquisition and were suspected of being practitioners of "the black art." It would be incomprehensible, Helmholtz added, for the authors of these remarkable figures to have spent as much time, effort, and talent on them, "if they had not hoped in solemn earnest to solve a great problem." Did not these inventors envision a body without fatigue, without discontent, and without aversion to work? "Though these artists may not have hoped to breathe into the creature of their ingenuity a soul gifted with moral perfection," he speculated, "still there were many who would be willing to dispense with the moral qualities of their servants, if at the same time their immoral qualities could also be eliminated, and to achieve instead of the mutability of flesh and bones, the regularity of a machine and the durability of brass and steel."[48]

For our age of advanced robots and artificial intelligence, Helmholtz's fantasy may appear quaint. But, as he reminds us, the dream of the builders of automata was ultimately premised on redemption from labor. The automata were an anticipation and an idealization of the machine, an ideal that Helmholtz now claimed was realized by the industrial age: "We no longer seek to build machines which shall fulfill the thousand services required of *one* man, but desire on the contrary, that a machine shall perform one service, and shall occupy in doing it the place of a thousand men."[49] The great automata embodied mimetic illusion of the epoch as had "the philosopher's stone" in the seventeenth and eighteenth centuries: In the "endeavor to construct a *perpetuum mobile*" by these artificial beings, the chimera of a body that "creates energy out of itself" was singularly evident.[50]

In his 1847 essay, "Conservation of Energy," Helmholtz used the example of the impossibility of perpetual motion as the touchstone of his theory of energy. His subsequent popular writings reveal his awareness of the social vision implicit in the idea of self-moving power. At the root of perpetual motion was yet another dream—that of perpetual idleness. "Perpetual motion was to produce labor power *(Arbeitskraft)* inexhaustibly without corresponding consumption, that is to say, out of nothing. Work however is money. Here emerges the practical problem which clever people of all centuries have pursued in the most diverse ways, namely to create money out of nothing. The comparison with the philosopher's stone sought by the ancient alchemists is complete."[51] The *perpetuum mobile* was the phantasmagoria of a society dedicated to making work superfluous: the pervasive moral criticism of those who resisted work was accompanied by the illusory search for an alchemy of work without struggle.

Helmholtz's 1854 lecture can be read as a series of general illustrations of the ubiquity of energy in all forms of work. But it is also a meditation on the illusion of perpetual motion and the necessity of "work" made manifest by the discovery of labor power. Even the elimination of "labor" by the machine could never dispense with the requirement for labor power. By extending the idea of energy to all aspects of inorganic *and* organic experience, the theory of conservation of energy permitted the old phantom of *perpetuum mobile* to be exorcised. The invention of steam power dispelled the illusion and "forced us to ask: is it not similar with human beings?"[52] In place of the illusion of perpetual motion emerged a new image of a society powered by natural forces, of labor perpetually renewed, of the unity of society and nature in the process of mechanical—or industrial—production, and ultimately of the universe as a vast productive machine. Little wonder that Helmholtz held the builders of automata in high esteem. Their failure occurred because their attempt to replicate "life" in the machine was too narrowly conceived, their mechanical talents far beyond the scientific understanding of the time. The more ambitious and more daring invention of nineteenth-century physics was to construct out of nature a theoretical "automata": a universe modeled on a single powerful dynamo, fulfilling the dream of perpetual motion by eliminating its *raison d'être*.

The discovery of the laws of *Kraft* raised the concept of work to the dignity of a universal principle of nature, irrespective of the "moral perfection" of servants or of any other workers. The shift from mechanical "forces" to the language of *Kraft* also eliminated the need for a spiritual understanding of labor; the work ethic was eclipsed by a quantitative economy of energy. "Hence, in a mechanical sense," he notes, the idea of work has become "identical with that of the expenditure of force." At this point, Helmholtz claimed, "we have arrived at the concept of the driving force *(Triebkraft)* or labor power *(Arbeitskraft)* of machines, and will have much to do with it in the future."[53]

In a cycle of lectures on the conservation of energy delivered in 1862–63, Helmholtz acknowledged that his original model for the concept of work as the universal measure of *Kraft* was human labor: "The concept of work for machines or natural processes is taken from the comparison with the work performance of human beings, and is . . . comprehensible through a comparison with human labor." Helmholtz thus established the equivalence of human, animal, and inorganic mechanical work, applicable to all motion, irrespective of intelligence, skill, design, or any other extraneous circumstance. Work was the transcendental principle of nature and society—the pure productive power of *Kraft*.[54]

For Helmholtz the value of human labor was determined more by the force expended than by the skill involved, which is a product of "time and trouble." A machine, even one that executes work adeptly, can always be produced in quantity; human skill is valuable only where it cannot be replaced by machines. In physical or mental labor, however, the quantitative side of work, or "yield," is entirely comparable to that of a machine. The concept of "quantity of work" *(Arbeitsgrösse)* in machines considers only the expenditure of force, since most machines are built to exceed the work of humans and animals. Therefore, Helmholtz concluded, "In a mechanical sense the idea of work has become identical with the expenditure of energy."[55]

As the output of a single artisan or machine could be measured and evaluated in terms of cost and yield of production, all natural forces could be gauged and quantified. Helmholtz provided the following example. If a machine, with a certain expenditure of force, lifts the hammer a foot in height, the same amount of force must be expended to raise it another foot. In the steam engine, mechanical work is developed out of heat produced by combustion, but heat may also be generated by the force of a skilled blacksmith pounding iron. Work can be the product either of man, nature, or a machine.

The only essential difference between natural forces, machines, and human beings is in the utility and efficiency of the energy transferred: "The external work of man is of the most varied kind as regards the force or ease, the form and rapidity, of the motions used on it, and the kind of work produced."[56] Helmholtz gives the example of a blacksmith whose arms wield a powerful hammer, of a violinist whose hands produce delicate sounds, and of the lacemaker whose fingers work with threads barely visible to the naked eye. But he reminds us, all acquire the force that moves them in precisely the same way, with the same organs, and according to the same laws. And, it is the same with machines: "They are used for the most diversified arrangements. We produce by their agency an infinite variety of movements, with the most various degrees of force and rapidity, from powerful steam-hammers and rolling-mills, where gigantic masses of iron are cut and shaped like butter, to spinning and weaving-frames, the work of which rivals that of the spider."[57]

Helmholtz makes no distinction between the industrial dynamo and the preindustrial blacksmith, lacemaker, or violinist because energy remains entirely indifferent to its social use. Nature may demonstrate no interest in how society employs its gifts, but it is best served by utility or economy of design insofar as the most efficient social use of energy is one in which the greatest amount of labor power is generated with the least amount of waste. The only difference between nature and society is the greater variety of instruments available for production in the latter: "The modern mechanism has the richest choice of means of transferring the motion of one set of rolling wheels to another with greater or less velocity; of changing the rotating motion of wheels into the up-and-down motion of the piston-rod, of the shuttle, of falling hammers and stamps. . . . Hence this extraordinarily rich utility of machines for so extremely varied branches of industry." Despite these differences, "They all need a *driving force* which sets and keeps them in motion, just as the works of the human hand all need the moving force of the muscles."[58] The distinction is entirely in the quantity of expenditure and economy of force, not in kind. In short, "The sum of all the forces capable of work in the totality of nature remain eternal and unchanged throughout all variations. All change in nature amounts to this, that the labor power can change its form and locality without its quantity being changed."[59]

The new technology of the industrial age thus produced a new image of the body whose "origins lie in labor power." The body is not simply analogous to, but essentially identical with a thermodynamic machine: "The animal body therefore does not differ from the steam-engine as regards the manner in which it obtains heat and force, but does differ from it in the purpose for, and manner in which the force gained is employed."[60] Helmholtz did not demote the living creature to the machine; he transposed the character of an energy-converting machine to the body, indeed, to the universe. The metaphor of the machine rather than the machine itself—the automata—is anthropomorphized. In the naturalization of labor power as *Kraft*, the energy of the violinist's body and the energy output of the industrial mill are identical: both represent labor power.

THE FIRST BOURGEOIS PHILOSOPHER OF LABOR POWER

Unwittingly perhaps, Helmholtz was the first great bourgeois philosopher of labor power precisely because in his essays on *Kraft* he does not distinguish between natural, mechanical, or human labor power. For him, all expenditure of energy produced work, and conversely all work involved the consumption of energy. His conception of labor power reveals no self-moving power, no social labor that is not at the same time a natural force. With Helmholtz work was reduced to a quantitative phenomenon subject to a system of mathematical equivalents.

An optimist by nature, Helmholtz did not devote much attention to the second law of thermodynamics either in his scientific papers or in his popular lectures. In his 1847 memoir he was unconcerned with the problem of the direction of energy flow, as he was satisfied with the abstract equivalence of the convertibility of different forms of energy. Later, however, with the discoveries of Thomson and Clausius that energy is lost when heat is transferred to a cooler body, he acknowledged that in actual conversions a certain amount of energy may be used up, or lost, but not in nature as a whole, which retains its effective power, eternal and immutable. In his later lectures Helmholtz observed that the second law of thermodynamics indicated that all energy ultimately "dissipated"—its flow was essentially irreversible and tended toward inertia. This view gave rise, especially after 1860 to predictions of the "heat death of the universe," the great cosmological exhaustion implicit in the teleology of nature. Helmholtz accepted the notion that in theory the dissipation of energy into mechanical work would eventually extinguish all natural processes: "The universe from that time forward would be condemned to a state of eternal rest."[61] This idea reached an even wider public through its adoption into philosophy: Herbert Spencer, for example, assumed that evolution was furthered by the propensity of all civilizations to decay, while Nietzsche, resisting Thomson's "final-state" hypothesis, posited his famous doctrine of recurrence in contradistinction to it.[62]

For Helmholtz, even the profoundly pessimistic prognosis of a world ultimately condemned to inertia, or exhaustion, was in the grand scheme of things inconsequential. Through the movements of the planets and nebulae, he maintained that the universe might also replenish itself. Thus, "We find in the mechanical forces to which we refer such a rich source of heat and light, that there is no necessity whatever to take refuge in the simplistic idea of a store of these forces originally existing."[63] Reassuringly, by the 1870s Helmholtz retreated from the more apocalyptic conclusions of entropy. The unidirectional time-flow thesis did not take into account that the universe is endless in time and space, he claimed, a true *perpetuum mobile*, capable of replenishing itself. As pure labor power, the unending cosmos alone exhibits a "perfect economy," where the *Kraft* constantly lost in transfers of force comes to rest.[64] In this protean universe, modeled on human labor and the energy of the machine, plants or organic nature assume their place in the great chain of energy, which Helmholtz described as the "enormous treasure of constantly new and changing meteorological, climactic, geological, and organic processes of the earth." Helmholtz's cosmogeny, which he once compared to the "old myths of mankind and the forebodings of poetic fantasy," finds in the modern universe a new myth of the singular, constant and eternal reproduction of *Arbeitskraft*.

Michel Serres has eloquently described how the discovery of thermodynamics "shook the traditional world and shaped the one in which we now work."[65] With the laws of thermodynamics a new order of modernity supplanted the Newtonian universe of fixed mechanical relations. If the Newtonian order admitted only a stable and homogenous order of natural forces and relations, thermodynamics conjured up images of extraordinary power as well as decline, death, and with Ludwig Boltzmann's statistical interpretation of entropy, disorder and chaos.[66] This complacent world of diverse "forces," matter and motion, was consumed by a new image of nature whose topos was "conflagration"—a cosmos of fire, heat, and work. As Serres envisions it through the paintings of Turner, "Matter is no longer left in the prison of diagram. Fire dissolves it, makes it vibrate, tremble, oscillate, makes it explode into clouds."[67] The result is a material world far more random, arbitrary, and ephemeral. Only energy is constant and protean: its myriad effects are but transient states of heat briefly transformed into the work required for the universe, for motion, and for life, then soon lost or dissipated.

This paradoxical relationship between energy and entropy is at the core of the nineteenth-century revolution in modernity: on the one side is a stable and productivist universe of original and indestructible force, on the other an irreversible system of decline and deterioration. Progress and decline were incorporated in the new physics of the age as antitheses, coexisting as it were in the material laws of the universe. Physical nature could be comprehended as work without waste, perpetual motion without recourse to chimerical ideas, while the second law of thermodynamics channeled this irrepressible energy along a steadily eroding historical course leading inevitably to dissipation. The powerful and protean world of work, production, and performance is set against the decrescent order of fatigue, exhaustion, and decline.

This collision between a stable order of work and life, and a world condemned to the forces of disorder and dissipation would later be recast and given expression by Freud (whom Serres says "aligns himself with these findings"). The metapsychology of the early Freud was influenced by his teacher, the physiologist Ernst Brücke, whom the Helmholtzians called "our ambassador to the East," and was profoundly shaped by the energy model of conservation and discharge. The "work" of the mental apparatus is directed toward reducing internal and external tension and excitement. Freud's writing is suffused with the language of energetics: it juxtaposes, to use the philosopher Paul Ricoeur's terms, "two universes of discourse," that of force (or energy) and that of meaning, so that "meaning relations are entangled with force relations."[68] In his later works, especially *Beyond the Pleasure Principle* (1920), Freud introduced the idea that the human organism resists the excess expenditure of energy and strives toward the elimination of tension, not unlike the principle of inertia.[69] In its commitment to the energy model of the psyche, Freud's theory may represent the apotheosis of nine-teenth-century modernity as well as the beginning of its twentieth-century abandonment and dissolution.

ANIMAL MACHINES

Modern physiological mechanism was well established in the rationalist physiology of Descartes, Leibniz, and La Mettrie. Leibniz, for example, pointed out that "the machines of nature, that is, living bodies, are even in the smallest of their parts, machines *ad infinitum.* "[70] The machines constructed by engineers were considered imperfect realizations of these biological "machines," which the body and the universe personi-fied. The machine opened up not only the secrets of the body's struc-ture and function, but also the processes of thought and being, which could be "constructed" with the logical necessity, regularity, and pur-posefulness of machines. Indeed, machines were analogous to perfect order. By the end of the eighteenth century ever-more sophisticated images of the "animal machine" were articulated by physiologists, like Cabanis, who maintained that medicine could "teach us to recognize the laws of the living machine, the regular functioning of its sensibilities in the state of health, as well as the changes that disease can bring about in it."[71] It was the French chemist Antoine Lavoisier, however, who was the first to see characteristics of the "hydro-pneumatic fire machine"—a steam engine—in the animal body. Lavoisier's animal machine adopted a modern guise: more important than its mechanical function was its capacity for self-regulation (respiration, digestion, transpiration), which permitted it to produce and deploy force, in short, its motor function.[72]

The Committee for Public Safety permanently terminated Lavoisier's researches with a mechanical device invented by Dr. Guillo-tine. But other obstacles blocked the acceptance of his view of the body as an internal combustion machine. During the first half of the nine-teenth century, Lavoisier was nearly forgotten, as the social conserva-tism of the epoch favored a more "romantic biology" in which the mechanists lost ground to vitalists like the physiologists Xavier Bichat or François Broussais, who argued that the principles of "life" could not be supported by reductionist arguments.[73]

Georges Canguilhem has emphasized that during the Enlighten-ment, mechanism drew its programmatic strength from its rejection of animism and from its claim to situate "life" within the general laws of the universe. The vitalists, however, argued that life was more than the sum of its parts and that "vital principles could not be explained away by the simplistic reduction of the "unknown to the known."[74] After the Restoration, "autonomous life" became the slogan of numerous French physiologists who claimed, as did François Magendie, that "physiology is at the moment precisely at the point where the physical sciences were before Newton: it awaits only that a genius of the first order come to discover the laws of vital force in the same way that Newton made known the laws of attraction."[75] Only in the late 1840s did the debates on the sources and character of animal heat renew the impetus to the metaphor of the machine.[76] Though Justus von Liebig, for example, never abandoned the language of vital forces, in his *Animal Chemistry* (1842) he drew on the metaphor of the furnace to describe the "metamorphosis" of nutrition and oxygen, and his student, Julius Robert Mayer, used the metaphor of combustion in his early publications on the mechanical equivalence of heat.[77]

In the late 1840s the impact of Helmholtz's and the other discoveries of scientific materialism were registered in physiology as well as in physics. The ideas of vitality, or "life force," began to be superceded by an image of the body drawn from physics and the indissoluble unity of matter and motion in *Kraft*. Matter and *Kraft* are indissoluble because they "complete and presuppose each other," noted Emil Du Bois-Reymond, Helmholtz's friend and confidant, in a famous critique of the "life force" *(Lebenskraft)* delivered a year after his colleague's elaboration of "Conservation of Energy."[78] For the students of Johannes Müller and Carl Ludwig, physiology adopted the new precepts of physical reductionism as a rebellion against vitalism and against the religious and authoritarian principles which, they believed, were implicit in vitalism.[79] For Helmholtz and for the other representatives of the mechanical school of physiology, the essential point was succinctly made by Carl Ludwig in an 1848 letter to Du Bois-Reymond. Physiology, he said, "had to be founded on the physics and chemistry of the organism."[80] The search for the dynamic *Kraft* of the body paralleled discoveries in the fields of light, heat, magnetism, and electricity. Du Bois-Reymond, whose essay "Über die Lebenskraft" (1848) led the polemical struggle to establish the doctrine of energy in physiology, illustrated the new principle when he argued that energy could not exist apart from matter: in the body it had to be sought in "those forces with which the substances of the organism are supplied." Du Bois-Reymond further distinguished between the old and the new conceptions of corporal thermodynamics: "A life force should not be assumed when there are only life forces."[81] The physiological reductionists thus hoped to demonstrate the identity of the organism with the physical laws of the universe. The traditional separation of organic and inorganic nature was in their view, "completely arbitrary," a residue of the theological antinomy between God and the world, or body and soul. One day, Du Bois-Reymond noted, physiology would completely dissolve into organic physics and chemistry. The true "kernel of the method," he concluded, lay in bringing "natural appearance under the mathematical principles of causality."[82]

Interpreted through the dynamic language of *Kraft*, the body appeared as a field of forces, energies, and labor power. The metaphor of the machine underwent a change from that of a clockwork composed of diverse parts to that of modern motor modeled on a steam engine or electric-powered technology. By defining energy, or motion, as the essence of matter, the long-accepted dualism of matter and force was overthrown. Natural philosophy's idealist image of "Lebenskraft," *vis viva*, or "life force," was discredited, and with it, the barrier to the parallel with inorganic machines removed.[83]

This new physiological orientation also partook of the natural sciences' reorientation toward laboratory research and "pure" science, as opposed to the observational and taxonomic systems of earlier generations. What emerged was a conception of the body as a field of forces to be investigated and measured by medical technologies designed for that purpose—for example, Carl Ludwig's *kymograph* (1847), which measured blood pressure, and subsequently, Helmholtz's *myograph* (1849), which gauged the force and duration of a nerve's impulse.[84] In his *Untersuchungen über thierische Elektricität*, published in 1848, Du Bois-Reymond demonstrated the existence of independent electrical properties in the muscles and the nerves and postulated the presence of a "general muscular current" to replace the theory of nerve substance and fluid.[85]

By the 1870s and 1880s the problem of conservation of energy in the human machine was constantly discussed and debated by physiologists in the scientific journals and reviews on both sides of the Rhine. The terms of the debate were summed up by the French physiologist M. A. Herzen in 1887 when he said that "the living organism is a machine capable of producing heat, and it is a question of knowing if that machine is subordinated to thermodynamic universal equivalence."[86] Physiologists were concerned with the chemistry of metabolism, especially whether the energy required for work conformed to the general principles of energy transformation in the universe.[87] Claude Bernard attempted to find a rapprochement between mechanism and vitalism in the 1860s, arguing that biology ought to borrow the experimental method from physics and chemistry but retain its own special concepts, phenomena, and laws.[88] But until Max Rubner demonstrated conclusively in 1894 that "the exclusive source of heat in warm-blooded animals is to be sought in the liberation of forces from the energy supply of the nutritive materials," it was impossible to speak of a decisive shift from a general theory of *Stoffwechsel,* or metabolism, to the modern idea of *Kraftwechsel,* or energy conversion.[89]

The "fetishism of *Kraft,*" as Ernst Mach once called it, was not limited to physiology. In Germany, Gustav Theodor Fechner introduced the principle of energy conservation into psychology, arguing that mental systems essentially paralleled physiological processes and that they sought stasis and equilibrium. In the 1890s, his student Wilhelm Wundt criticized Fechner's lack of an experimental basis for his theories but stressed that the "will, the senses, the associations and apperception all rigorously follow the principles of energy conservation."[90] In France, Théodule Ribot also tried to establish a similar foundation for the psychology of the will, which he called "the highest force which nature has yet developed—the last consummate blossom of all her marvellous works."[91] For Ribot, as for Helmholtz, the body was aware of itself only through representations produced by its physiological states, a phenomenon he termed *synesthesia.*[92] In this view, the function of the ego was to coordinate conflicting impulses, or sensations, which derived either from society via education or from the physiology of nervous impulse. This "complex psycho-physiological mechanism, in which alone resides the power to act or to restrain," gives central importance to the psychophysical force of the will, as mental pathology was symptomatic of an enfeebled or fatigued will.[93] In the 1870s this point was gradually extended to a series of psychophysical phenomena, so that "those qualities of the nerves of the senses, are all different from one another, the nerves of each sense each have their own energy and their own particular quality."[94] According to Charles François-Franck, a physiologist trained by the French Helmholtzian Etienne-Jules Marey, "There are general laws which preside over the apportionment of the nervous activity in the different points of the system as there are mechanical laws governing the circulation of the blood in the vascular system."[95]

The social consequences of this new physiology and psychology of energy were far reaching. In physiology the focus on muscular force oriented scientific research toward the problem of work and ultimately led to a science of fatigue. In psychology, it led to the emergence of neurasthenia, or moral and psychic exhaustion, as the central diagnosis of mental pathology. Throughout the nineteenth century, the search for the physiological and psychological manifestations of thermodynamic processes confirmed the primacy of energy conservation as a social principle. Moreover, some writers argued that the course of progress was a by-product of the increasingly complex conversions of energy that characterized the natural and social world:

> The transformation of the mineral elements into organic elements, the transformation of cosmic movements into increasingly complex organic movements, the transformation of protoplasm into organized life elements, increasingly complex associations growing out of these elementary organic individualities into more and more perfect ones: these are the fundamental bases of the *law of progress by evolution* which governs the life of the elements and the organisms, as it also governs human societies.[96]

Despite the unmitigated optimism of a synthesis between evolution and thermodynamics, the principle of entropy still lurked. The crucial distinction between the great automata and chimerical *perpetuum mobile* and the new image of the body as a thermodynamic machine remained evident in the body's stubborn inability to sustain work indefinitely. In the 1880s and 1890s even the physiologists who argued for thermodynamic equivalence were becoming dissatisfied with the human machine's imperfection as compared with "real" motors: "The muscle is an imperfect machine . . . a machine for work that is unilateral," wrote Herzen; "the bicep can flex the forearm above the arm, and that is all; its action is exhausted by this work."[97] Physiological work could not be sustained without the omnipresent consequence of fatigue.

Expanding on the analogy between the body and inorganic nature in his 1863 lecture on conservation of energy, Helmholtz noted that the "expenditure of energy can be understood in terms of an analogy between human labor and that of machines. The greater the exertion and the longer it lasts, the more the arm is tired and the more the store of its moving power is for the time being exhausted. We shall see that this characteristic of being exhausted by work is also encountered in the moving forces of inorganic nature."[98] With the discovery of fatigue as the dystopia of universal labor power, the search for a thermodynamics of society began in earnest. As entropy revealed the loss of energy involved in any transfer of force, so fatigue revealed the loss of energy in the conversion of *Kraft* to socially useful production. As energy was the transcendental, "objective force" in nature, fatigue became the objective nemesis of a society founded on labor power. The metaphor of *Kraft*, which had, as we have seen, its origins in the concept of labor power and industrial technology, was subsequently adopted by physiologists and social thinkers who saw in the new principle of energy conservation implications for the conservation of energy in society as well as in nature.

CHAPTER THREE

~~~~~~~~~~~~~~~

The Political Economy of
Labor Power

THE SOCIAL IMPLICATIONS
OF ENERGY CONSERVATION

WITH Helmholtz's image of the cosmos as labor power, the work of the body became at once the material agent, the site, and the model for the transformation of nature into society. Political economy and social philosophy could not long evade the implications of this new image of nature. The Helmholtzian cosmos was a cosmos at work—nature conceived as a productive force. Consequently, if physicists could measure the amount of energy converted into work in each instance, they could also measure social energy and perhaps even conserve it.

Nineteenth-century thought did not discern as great a distance between discoveries in nature and their application to society as contemporary thought does. As Darwin revealed the mechanisms that governed the transformations of natural history, and perhaps human populations as well, so too, thermodynamics revealed those that governed nature and the deployment of labor power. And, as Darwin's writings left open the social implications of evolutionary doctrine—giving credence to conservative, liberal, as well as socialist versions of "Darwinism"—thermodynamics permitted an equally broad range of interpretations from progressive social reform to the more apocalyptic conclusions of Nietzsche's image of history. The most important nineteenth-century thinker to absorb the insights of thermodynamics was Marx, whose later work was influenced and perhaps even decisively shaped by the new image of work as "labor power."

By the 1850s, the implications of energy conservation were becoming clear to political economists: the nation that most efficiently used and conserved the existing supply of the world's energy—including both labor power and technology—would also win the race for industrial supremacy.[1] Particularly on the Continent, the discovery of energy conservation underscored that a scientific approach to human labor might lead to a more efficient use of a nation's productive capacity in competing with the achievements of English steam and industrial power. To be sure, Adam Smith referred to "the productive powers of labour" in the *Inquiry into the Nature and Causes of the Wealth of Nations* (1776), but the equivalence of labor and natural power was not a generally accepted concept in the vocabulary of political economy, even by midcentury.

The French economist and mathematician Baron Charles Dupin, who attempted to calculate the power of human labor as a "motive power" as early as 1829, was perhaps the first thinker to recognize the economic significance of Charles Augustin Coulomb's mathematical theories of force for the study of human labor. Dupin was primarily interested in enhancing the speed and duration of troops during wartime, but he drew attention to the fact that "in many professions workers consume a diet inadequate to replace the daily loss of force."[2] French political economists, like Pelligrino Rossi, followed suit during the 1840s and employed the engineering term *puissance du travail* to signify the measure of physiological capacity for work. Through Sadi Carnot's *Reflections on the Motive Power of Heat* (1824), they could comprehend labor power as a natural force in man, though perhaps not yet as a universal measure of converted force.[3] When Dupin published his six-volume *Force productive des nations concurrentes* (1851), surveying the "Création" and "mise en jeu" of the productive forces of all nations for the 1851 London exposition—the festival of progress—he was no longer alone in his belief that an accounting of the national and global supply of productive forces was paramount.[4]

Two years later the German economist Hermann Henri Gossen published a little-noticed, odd work entitled *Die Entwicklung der Gesetze des menschlichen Verkehrs und der daraus fliessenden Regeln für menschliches Handeln* (1853). Gossen foresaw the economic fruits to be harvested from the conservation of social energy, since society was composed of the energy of its members. For Gossen, the "totality of commodities over which a person disposes constitutes his economic energy, his wealth [*vermögen*]," which devolves to the benefit of the social whole.[5] Several years later the marginal utility theorist Léon Walras, who adopted some of Gossen's ideas, credited him with recognizing that the energy doctrine pointed to the necessity of balancing the needs of the individual and those of the state. The function of the state, Gossen argued, was to secure general conditions of existence, as individuals in turn contributed to society through their energy-producing capacities, that is, their labor power. From a social viewpoint the utility of any commodity could be measured by what he called its intensity, or energetic value, which was an expression of the labor power it had absorbed.[6] Since the goal of the state was to further the "free choice" of activity and the "pleasure" *(Genuss)* of each individual, it had to remove the two greatest barriers to the maximization of energy, which Gossen identified as the private ownership of land and the generalized lack of capital.

Gossen's work remained a "secret tip" for specialists who for the most part did not share his economic hedonism or his socialistic views on nationalization.[7] However, by the 1870s his general proposition that energy conservation was applicable to economics because of the centrality of labor power in determining value became common among conservative proponents of *national economy*—the doctrine that a nation should aggressively pursue its economic self-interest—in Wilhelmian Germany. Carl Neumann, one such economist, proposed that all economic life could be expressed by an exchange of energy—"the transfer of energy of physical bodies." Neumann considered the "energy of the self as the capital of the body" and held that insofar as the body contained such an energetic "component," it could be considered a commodity.[8] Furthermore, he claimed that in the act of consumption, energy passes from the commodity to the owner, culminating in satisfaction of need, which was the final economic form of the transfer of energy. These economists added, however, that the devastating effects of entropy could not be ignored: the gradual diminution of force over time presaged inevitable economic and social decline. For the proponents of energy conservation in national economy, both economic intensity and energy efficiency were constantly thwarted by the negative effects of entropy: "All commodities are directed to maximize their utility through exchange. The human being must constantly increase his intensity or economic energy, since in the course of nature all intensities diminish."[9]

As a consequence of this increasingly pessimistic view of the social implications of thermodynamics, some social theorists began to question whether the productivity of the nation was being optimally used and whether the regulation or even reduction of the overall expenditure of energy might not increase the yield if the nation's energy supply were better conserved. It was not, however, until the last decade of the century that we could clearly see the social effects of these ideas take shape in several different contexts, especially in initiatives to shorten the working day, to reduce the risk of accidents, and to increase the general health and well-being of the producing classes.

THE MARRIAGE OF MARX AND HELMHOLTZ

The social implications of the energy doctrine were not exclusively reserved for conservative proponents of a national economy. In the late 1850s Karl Marx was influenced by the new physics, which altered his general theory of labor and profoundly affected the thrust of his social theory. He began to see a quantitative equivalence between the "natural" productive force of labor and the productive forces of industry and technology. In his earlier works, especially the *Economic and Philosophical Manuscripts of 1844,* Marx considered labor the quintessentially human activity defining social being and offering humanity the way to reappropriate and regain its essential attributes. In his mature works Marx viewed labor as a burdensome necessity and as a constraint on freedom and self-realization. Reduced to one form of energy among others, labor became a means, not an end in itself.

In the late 1850s Marx substituted the term "labor power," *(Arbeitskraft)* for labor, a move that he and Engels consistently acknowledged as a major, if not primary, conceptual discovery. As Engels frequently pointed out, Marx distinguished labor power from the concrete dimensions of work in largely energeticist terms: "the sum total of the physical and mental faculties which exist in the living person of a human being and which he puts into motion when he produces useful values."[10] Paradoxically, because Engels has long been identified with a particularly "scientistic" reading of Marx, it has not generally been acknowledged—even by present-day defenders of Marx's materialism—that in Marx's mature work *Capital,* the discussion of labor power is explicitly drawn from the doctrines of scientific materialism, a development that decisively modified a basic precept of Marx's argument.[11]

Until the late 1850s, Marx viewed labor in Hegelian terms, as man's self-creation through history and, above all, as the uniquely creative impulse that would ultimately be restored from its alienated condition once private property was abolished. In *Capital,* however, this image is explicitly jettisoned for a concept of labor that no longer represented a distinct path to, or mode of, "realization" of the human essence. Henceforth Marx regarded labor as a negative restraint on freedom. Only through the expansion of the productive forces and the gradual supersession of labor by technology would emancipation from the time-bound circumstance of labor be possible. As "the everlasting nature-imposed condition of human existence," labor is "independent of every form of that existence."[12] If the mature Marx remained a utopian, his image of the "realm of freedom" is not concerned with improving or restructuring labor, but with reducing the need for labor; his positive image is directed toward the expanded potential of society to fulfill needs (which are conceived as social, not natural).[13] The point is to eliminate the amount of labor time required by society to the greatest extent possible.

By the early 1860s, Marx thus shifted his focus from the emancipation of mankind *through* labor to emancipation *from* productive labor by an even greater productivity. He became a "productivist," because he no longer considered labor a paradigm of human activity and because, in harmony with the new physics, he perceived labor as but one (central) mode of converting energy into work. In Marx's *Capital* labor power appears as a force equivalent to other forces of production. To use the philosopher Agnes Heller's formulation, Marx turned from a "paradigm of work" to a "paradigm of production."[14]

In the mature Marx, "labor power," very much in accord with Helmholtz's sense, becomes the engine of historical development. The language of energy appears frequently in passages from *Capital:*

> The more the productivity of labor increases, the more the working day can be shortened, and the more the working day is shortened, the more the intensity of labor can increase. From the point of view of society the productivity of labor also grows when economies are made in its use. This implies not only economizing on the means of production, but also avoiding all useless labor.[15]

The emphasis here is clearly on reducing the working time made possible by the simultaneous expansion of technology and the elimination of wasted energy. *Capital* consistently envisions human freedom in terms of "increased productivity and greater intensity of labor," which shortens socially necessary labor time.[16] Freedom is redefined as the rationalization of nature under the law of energy: "Freedom in this field can only consist in socialized man, the associated producers, rationally regulating their interchange with Nature, bringing it under common control, instead of being ruled by it as by the blind forces of Nature; and *achieving this with the least expenditure of energy* [my italics] and under conditions most favorable to, and worthy of, their human nature."[17]

Marx's most eloquent praise of capitalism is reserved for its revolution in production, just as his most hostile remarks are reserved, not simply for its unjust and exploitive aspects, but for its "anarchic system of competition," which wastes the energies of society: "the most outrageous squandering of labor power and of the social means of production."[18] The goal of a reorganized society is to eliminate wasteful energy, to abolish fatigue, and to release human beings from the burden of labor by the productive forces of technology.[19]

Marx characterizes industrial capitalism as a system designed to produce fatigue: "Factory work exhausts the nervous system to the uttermost, at the same time, it does away with the many sided play of the muscles and confiscates every atom of freedom, both in bodily and intellectual activity."[20] It requires the mechanization of the body in conformity with the demands of industry: "the main difficulty [of the industrial factory lies] in training human beings to renounce their desultory habits of work, and to identify themselves with the unvarying regularity of the automation."[21] The scandal of capitalism is that it depends on surplus labor and thus is unable to eliminate the exhaustion it produces.

THE SOCIAL PHYSIOLOGY OF LABOR POWER

The mature Marx considered his most significant achievement the discovery of labor power. The reasons for this pride of accomplishment are not, however, self-evident. What is this labor power, which Marx defines in several distinct ways? Labor power, Engels notes, "is, in our present-day capitalist society, a commodity like every other commodity, but a very peculiar commodity. It has, namely, the peculiarity of being a *value-creating force* [my italics], the source of value, and moreover, when properly treated, the source of more value than it possesses itself."[22] Labor power is therefore both social *and* physiological, historically specific *and* at the same time a form of universal energy, or *Kraft*. Labor power represents the quantitative aspect of labor under capitalism. First as an abstract, universal measure of labor and a "magnitude of value," it is clearly a social phenomenon as "the labor-time socially necessary for its production."[23] But it is equally a physiological concept, "devoid of all social and historical elements."[24] Both a social and a physiological magnitude, it is a measure of value and a measure of energy.

To confound generations of commentators further, Marx used the concept of labor power interchangeably to mean both of these alternatives simultaneously. Indeed, the great divide between "Eastern" ("scientific") and "Western" (Hegelian) Marxists runs directly along this fault line. Whereas the Eastern Marxists, even before 1917, denied the qualitative difference between the natural and the social sciences, Western Marxists from Georg Lukács to Jürgen Habermas have generally rejected the positivist implications of scientific Marxism. For the Eastern Marxists, labor power is primarily "a physiological concept, an ideal concept, and in the last analysis a concept which can be reduced to mechanical work."[25] Marx himself gave credence to the scientific version of the argument insofar as he always emphasized the physiological component of labor power as "the natural material transformed in the human organism."[26] For the worker, Marx adds characteristically, labor power is a commodity "which exists only in his living body."[27] Marx's definition of labor power as "mechanical work" brings the "aggregate of those mental and physical capabilities existing in the physical form, the living personality, of a human being," into harmony with the conversion of energy to use.[28] A close reading of Marx's *Capital* reveals that he always underscored this physiological model of labor power, though, as "Western" critics of the reductionist interpretation maintain, without ever denying the social character of the process.

Labor power for Marx becomes a reality only through its exhaustion: "In the course of this activity, i.e., labor, a definite quantity of human muscle, nerve, brain, etc., is expended, and these things have to be replaced. Since more is expended, more must be received." Marx emphasized that the sources of the energy expended in work had to be perpetually renewed, so that "the value of labor power can be resolved into the value of a definite quantity of the means of subsistence."[29]

As *Kraft* is also the sum of energy in the universe, *Arbeitskraft* is objectified in the mass of commodities present in society: "The total labor power of society, which is manifested in the values of the world of commodities, counts here as one homogenous mass of human labor power, although composed of innumerable individual units of labor power."[30] The concept of labor power here is clearly a quantitative measure of the expenditure of human energy in production. Marx stresses this point in *Capital:* "Tailoring and weaving, although they are qualitatively different productive activities, are both a productive expenditure of human brains, muscles, nerves, hands etc. . . . merely two different forms of the expenditure of human labor power."[31] Or more strikingly: "On the one hand, all labor is an expenditure of human labor power in the physiological sense, and it is in this quality of being equal, or abstract, human labor that it forms the value of commodities. On the other hand, labor is an expenditure of human labor power in a particular form and with a definite aim."[32] For Marx, labor power is both a social magnitude—as socially necessary labor time—and a natural force in the form of a specific set of energy equivalents located in the body.

THE EMERGENCE OF LABOR POWER IN MARX

As we have mentioned, the term *labor power (Arbeitskraft)* first appears in the second part of the *Grundrisse*, the draft of *Capital* written in the winter of 1857, where Marx distinguished between actual labor and labor as a potentiality that can be marketed and measured in units of time. Earlier Marx had spoken only of concrete versus "abstract labor."[33] But in the *Grundrisse* Marx defines labor power as a "specific, particular measure of force-expenditure [*Kraftäusserung*]."[34] Marx emphasizes the distinction between labor power and labor in this famous passage: "In exchange for his labor capacity as a fixed, available magnitude, he surrenders its *creative power*, like Esau his birthright for a mess of pottage."[35] This capacity for labor *(Arbeitsvermögen)* is above all a natural quality, since "the use value which he offers exists only as an ability, a capacity of his bodily existence; has no existence apart from that."[36] But in its new form, as a commodity, even this "living labor itself appears as *alien* vis-à-vis living labor capacity, whose labor it is, whose life expression [*Lebensäusserung*] it is, for it has been surrendered to capital in exchange for objectified labor, for the product of labor itself."[37] The "life expression" of the worker is turned into the "force-expenditure."

In the *Grundrisse*, then, labor power is distinguished from labor in energeticist terms. Here, however, we encounter a difficulty; we cannot proceed without understanding what Marx actually understood of the most recent developments in physics and physiology. In the *Grundrisse* Marx frequently shifts ground between an older, more available view of nature as composed of diverse forces, and the new idea of a single *Kraft* employed by Helmholtz and others who grasped the implications of thermodynamics. The *Grundrisse* combined this newer energeticist notion of labor power with an older, more obscure language of multiple vital "forces." We can thus speak of two narratives that are not yet disentangled: one emphasizes the "metabolic" element of man's exchange with nature; the second regards labor power as a conversion of force.

As Alfred Schmidt pointed out in his classic study of Marx's theory of nature, until he completed the preparatory work for the first volume of *Capital*, Marx conceived of the human relation to nature in terms of a metabolic exchange of matter, mediated through labor, which secured basic social needs and requirements.[38] Until 1857, Marx took as his model of nature the *metabolic exchange of substances and forces,* which reflected both the pantheism and the "metaphysical" materialism of his generation.[39] We recall, for example, that the early Marx viewed nature as man's "inorganic body," mediated through society as society is mediated through nature. The metabolism between man and nature is the framework for Marx's view of labor as the paradigm of nature and led him to characterize this relation as a process of exchange. But nature is also a mirror of labor, insofar as "the exchange of commodities is the process in which the social metabolism takes place." Labor is the model for all creative, life-generating, activity.[40]

After 1859 Marx gradually redefined labor from a metabolic exchange of substances between man and nature to a conversion of force. In the *Grundrisse,* and to some extent in *Capital* both conceptions still coexist. In the *Grundrisse,* Marx speaks of human needs as "necessaries with which to stoke the flame of living labor capacity, to protect it from being extinguished, to supply its vital processes with the necessary fuels."[41] But this language is not entirely consistent with other formulations of the same idea, for example, where Marx describes the body as a motor that both makes labor possible and limits what can be used up without being replenished, since "labor also is consumed by being employed, set into motion, and a certain amount of the worker's muscular force etc. is thus expended, so that he exhausts himself."[42]

In *Capital* Marx also refers to labor as "a process by which man, through his own actions, mediates, regulates and controls *the metabolism between himself and nature* [my italics]," but this can be read as metaphor for a process that is now understood entirely in energeticist terms.[43] The following passage reveals Marx's new understanding of the relation between man and nature: "He confronts the materials of nature *as a force of nature.* He sets in motion *the natural forces which belong to his own body* [my italics], his arms, legs, head and hands, in order to appropriate the materials of nature in a form adapted to his own needs."[44] Marx sees the laboring body as a natural force among others united by the universal equivalence of *Kraft.* Labor is no longer a creative or singularly human act, it is one kind of work aimed at the production of use values. In *Capital* Marx describes labor as intensifying and regulating energy expenditure. The process of metabolism is redefined as a constant recycling of forces. (Marx cites Lucretius, "nil posse creari de nihilo" [from nothing, nothing can be created], and in Volume 3 of *Capital* devotes an entire chapter to the problem of excretion and waste.) Labor is the natural necessity, which is not itself freedom, but its presupposition.[45]

Marx's philosophy of history is largely indifferent to the qualitative distinction between human labor power (in its social character as "capital") and any other "inorganic" productive force in nature or in technology. What distinguishes the worker and the machine is that "the worker consumes his provisions during pauses in the labor process, whereas the *machine* consumes what is essential to it *while it is still functioning.*"[46] In the notebooks to *Capital* Marx observes that laboring activity is distinct from *Arbeitskraft,* or labor power, only insofar as the latter actually *has* value. Whereas laboring activity is still a concrete exchange with the raw materials of nature, labor power conceives of the body as a natural force in itself. Since only labor power can be quantified and universalized in relation to all other forces and instruments of production, only labor power is potentially a productive force.[47]

Because Marx believed that "only labor which is directly *transformed into capital is productive,*" only the expansion of the productive power of labor can lead to reducing labor time, and ultimately to abolishing wage labor.[48] Technology is merely abstract labor operating with "indifference"—or put in another way: "Capitalist society, by generalizing labor and positing the basis for integral technological development, has brought about the preparation of what will abolish it."[49] Therefore only by expanding the forces of production can the historical limits of any social order be transcended. Since labor power is also the *via dolorosa* of history, its emancipation is predicated on transforming energy into capital and ultimately on transcending labor.

We have now reached the point where we can see precisely why, given his shared assumptions with scientific materialism, Marx's later energeticism eclipsed the idea of labor as metabolic exchange of substances and "creative" forces. Only with the discovery of *Arbeitskraft,* does the quantifiable magnitude of energy in a specific amount of labor time become fully equivalent to any other form of energy expended. Moreover, only with the concept of energy does the physiological aspect of labor become a constituent of labor power.

The sources for Marx's reception of the new thermodynamics can be documented, albeit with some circumspection.[50] Marx's first encounter with the ideas of Carnot, similar to that of Helmholtz's, derived from the French engineering tradition and centered on the theory of machines. In his notebooks to *Capital,* Marx excerpted the work of the political economist Pelligrino Rossi, who, as we have noted, used in the late 1830s and early 1840s the engineering term *"puissance du travail"* propagated by Navier, Coriolis, Poncelet, and other pioneers of hydraulics and mechanics. In *Capital* Marx makes use of Rossi in his discussion of the physiology of labor power, quoting him to the effect that "to conceive the capacity for labor *(puissance du travail)* in abstraction from the workers' means of subsistence during the production process is to conceive of a phantom."[51]

More important, Marx explains in the crucial section of *Capital* how the value of a day's labor power is calculated, and refers specifically to the energy equivalent of human labor power: "The value of a day's labor power is estimated, as will be remembered, on the basis of its normal average duration, or the normal duration of the life of a worker, and on the basis of the *appropriate normal standard conversion of living substances into motion as it applies to the nature of man"* [my italics]. Marx emphasizes the point by citing the following passage from a text by William Robert Grove, *On the Correlation of Physical Forces* (1867): "The amount of labor which a man had undergone in the course of 24 hours might be approximately arrived at by an examination of the chemical changes which had taken place in his body, changed forms in matter indicating the anterior exercise of dynamic force."[52] The experimental proof for this assertion was not established until almost three decades later, with the work of a German nutritionist and physiologist, Max Rubner. Nonetheless, Grove's book, according to Kuhn, presented the layman's view of energy conservation and was "one of the most effective and sought after popularizations of the new scientific law."[53]

For the mature Marx, necessity is characterized by the expenditure of energy, and freedom, by the liberation of energy from external constraint. The production process, even in the rare glimpse Marx permits us of a future in which man will only have to stand watch over a self-regulating complex of technology, remains within the realm of necessity. Only beyond the production of goods and the satisfaction of wants does that "true realm of freedom," which rests upon necessity, actually "blossom forth."[54] Labor power is the motor of history; its goal, however, is the replacement of men with machines.

For this reason Marx redefines freedom as existing apart from social labor. This freedom can only be gained as the consequence of capitalism's ever-greater economic and productive power. As the part of the day given over to material production is reduced, time expands for "the free intellectual and social activity of the individual."[55] The realm of freedom begins only when life dictated by necessity and material production ends. And even more important, human needs expand only as the technological means to fulfill those needs increase.[56] Both orthodox defenders of Marxian science and critics of his reduction of human experience to labor and the laws of history have largely missed Marx's own fundamental turn away from an exalted vision of labor as the activity that defines human nature to one that sees labor as a force of nature powering the productive engine of society.[57]

My recasting of the early-late-Marx distinction emphasizes that Marx's redefinition of human freedom from emancipation *through* labor to emancipation *from* labor was the consequence of his altered concept of nature—of his discovery of energy. From that point on, Marxism as a historical philosophy has been predicated on the intensification of production through the "least expenditure of energy." As is well known, in the *Grundrisse* Marx indulged his utopian vision of the dynamic triumph of machinery and technology over purely physical labor: "a motive force which moves itself."[58]

Marx also discovered the principle of entropy at work in capitalism: the unidirectional time flow of history produces the inevitable tendency of capitalism to decline as the productivity of labor power increases. By transforming labor into capital, capital amplifies its force: conversion of force is a conversion of magnitudes. The transfer is such that "whether the magnitude of the labor changes in extent or in intensity, there is always a corresponding change in the magnitude of the value created, independently of the nature of the article in which that value is embodied."[59] But because of the well-known tendency of capital to require greater and greater intensity to maintain or increase the profitability of labor in a competitive atmosphere, the conversion process yields proportionately less "value" for the capitalist, or as Marx expressed it, the rate of profit declines.

History exercises its sovereign "cunning" through the competitive "waste" of capital, which permits the expansion of technology ultimately to emancipate labor from the drudgery of physical work. When Marx denounces Adam Smith for conceiving of "Rest" as "the fitting state of things, and identical with 'liberty and happiness,' " he is expressing his opposition to idleness, not to productive leisure, which is freedom. This new form of human labor is "neither the direct human labor he [the worker] himself performs, nor the time during which he works, but rather the appropriation of his own general productive power, his understanding of nature and his mastery over it by virtue of his presence as a social body." Freedom only truly arrives when the worker "steps to the side of the production process instead of being its chief actor."[60] The song of emancipated labor can still be heard *sotto voce* in

Marx's *Capital,* but it is drowned out by the roar of expanding productive forces driving the *teleological* motor of historical progress.

In the later Marx, the distinction between social labor and nature is all but obliterated. Marx's phenomenology of labor power is not only contemporary with the Helmholtzian revolution in scientific perception; it is a direct consequence of it. By placing the weight of transformation on the "objective" expansion of the productive forces, Marx argues that the full development and unfolding of the productive potential of nature and technology is the organizing principle of society. The mature Marx held to a "paradigm of production" in which the distinction between the natural forces of production and the productive forces of society is no longer decisive. Marxism later built on this ambiguous legacy and adopted an ideology that both venerated productive over unproductive labor and waxed enthusiastically over the productive power of capital while anticipating its collapse.

HEGELIAN HELMHOLTZIANISM: ENGELS

The crucial step from social philosophy to a full-fledged "scientific materialism" was taken by Engels, who emphasized the natural scientific basis of Marx's theory.[61] Engels put Marxism on the pedestal of energy and transcendental materialism. His *Anti-Dühring* (1878) and posthumously published notebooks on science, *Dialectics of Nature* (1872–1882) are the canonical texts of the marriage of Marx and Helmholtz, though it would be more accurate to speak of a stormy marriage between Helmholtz and Hegel. For Engels, the law of conservation of energy became the fundamental principle of the materialist theory of both history and nature. Engels explicitly integrated Helmholtz into Hegel's philosophy of nature, in which matter exhibits the property of dialectical movement, with the result that dialectics became a theory of nature. Hegel then became a cudgel with which Engels beat Helmholtz for his undialectical positing of a single basic force inherent in all matter: "Such a primary force would be really not more than an empty abstraction, with as little content as the abstract thing in itself."[62]

Unlike Helmholtz who cautioned against transgressing beyond the legitimate boundaries of science, Engels had no qualms about presenting the doctrine of energy as the ground of a total philosophy of nature in which labor power is not simply a productive force, but a manifestation of "the true great basic process, knowledge of which comprises all knowledge of nature."[63] Engels' criticisms of eighteenth-century mechanical materialism are simply a replication of Helmholtz's own arguments about the impossibility of distinguishing matter and motion, or, for its lack of a theory of *Kraft*. But his interpretation of Helmholtz is much more; it derives all human activity, including human consciousness from the singular fact of energy: "The motion of matter is not merely crude mechanical motion, it is heat, light, electric and magnetic stress, chemical combination . . . and finally consciousness."[64]

For Engels, as for generations of "Eastern Marxists" who read him faithfully, Marxism was reduced to an ontological materialism rooted in the energetics of matter. But what Engels also accomplished was to simplify, and therefore to reveal the implications of, the doctrine of energy in Marx's theory of labor power. Engels believed that all events in history and nature were manifestations of the fundamental character of matter. Marxism as a science of the social world was built on physics and on the achievements of energeticism, which overcame the traditional materialism of Moleschott, Vogt, and Büchner. If muscular force is mechanical work, "merely transference" of energy, as Engels writes, then the distinction between the organic and inorganic forces of production is fully dissolved.[65]

With Engels, political economy became a subordinate science of labor power. It took its place among the other economies of energy: physiology, physics, and chemistry. Engels transfigured labor power into yet another "force" like heat, electricity, or steam: "That the quantity of motion (so-called energy) remains unaltered when it is transformed from kinetic energy (so-called mechanical force) into electricity, heat, potential energy etc., and *vice-versa*, no longer needs to be preached as something new."[66] Engels raised to the forefront "the indestructibility and the uncreatability of energy, which is one of the basic prerequisites of the materialistic conception of nature."[67] Political economy was thus harmonized with the doctrine of protean energy. Energy is the universal equivalent of the natural world, as money is the universal equivalent of the world of exchange.

By the last quarter of the nineteenth century the social implications of the doctrine of energy were becoming apparent. In national economy, as in the work of Marx, the body was the site where the natural force of labor power was converted into the energy to power the industrial dynamo. The working body was thus recast in the image of the Helmholtzian cosmos, as the cosmos itself became an industrialized automata. Yet it was a physiologist, rather than an economist, Etienne-Jules Marey, who should be credited with elaborating the social implications of this doctrine. Marey's investigations into the dynamic laws of the body in motion created a new science of human labor power based on thermodynamics. His importance as the founder of a European science of work and as the progenitor of a new concept of modernity—the modernity of the body cast as a social instrument—is the subject of the next chapter.

CHAPTER FOUR

~~~~~~~~~~~~

Time and Motion:
Etienne-Jules Marey and
the Mechanics of the Body

MAREY AND MODERNISM

CULTURAL modernity has generally been understood in terms of the intimate connection between the rapid industrialization of the second half of the nineteenth century and the corresponding disruption of traditional time-space perceptions in science, art, literature, and philosophy.[1] The visual and narrative field was dissolved by avant-garde writers and artists from Proust to Picasso while the natural sciences and mathematics were being transformed by thermodynamics and the emergence of non-Euclidean geometries. This widely shared sense of a crisis of space and time found cogent philosophical expression in Henri Bergson's assault on the "spatialized" time of "timekeeping" and measurement. When Edmund Husserl characterized Descartes' *Discourse on Method* as "absurd," he summed up the modernist protest against the hubris that "the whole world can be thought on the analogy of nature" and that time and space are objective truths of the natural world.[2]

Efforts to grasp the implications of this crisis of modernity have generated their own interpretive traditions. Two of these reactions are of particular interest for this study; the first concerns the relationship between aesthetic modernism to the discontents of society; the second focuses on the self-reflective and self-referential disposition of modernist works. The social and cultural significance of modernity has been the chief preoccupation of German social theorists from Max Weber and Georg Simmel to Georg Lukács and the Frankfurt theorists. In this tradition, the emphasis centers on the social costs of modernity conceived as the increasing colonization, or occupation, of social relations by abstract and quantitative relationships. Max Weber's concept of rationalization offered the most compelling contemporary explanation of this cultural predicament, sometimes elaborated with the added Marxist gloss that capitalism—specifically the "commodification" of social relations—was the key to the emergence of modernity. In this view, the formal rigidity and rule-bound modern bureaucracy; the division of labor and the rise of complex managerial strategies; and the impact of commercial culture reduced traditional ways of living and interacting to quantifiable relationships, subverting ethics, values, and universal norms. In this context of dislocation, aesthetic modernism is interpreted—especially by the Frankfurt School theorists (Theodor Adorno and Max Horkheimer)—largely as a sustained rebellion against rationalization and commodification, a revolt whose hope lay in the retrieval or preservation of an alternative conception of culture. T. W. Adorno, the most eloquent proponent of this view, claimed that artistic modernism's abstruseness, formal complexity, and distance from ordinary forms of experience or knowledge generated art's resistance to, and rupture from, the world of work and exchange. The inaccessibility of modernism, in other words, was the only means by which art could reclaim the lost autonomy of the individual ego from the constraints of society.[3] The mood of this critique was at once triumphal and defeatist; with the disastrous loss of social coherence, art represented the powerless victory of the unhappy consciousness of modernism over modernity.

The second tradition to confront the crisis of modernity is more militant and less melancholy. For poststructuralist critics all modernist literary and artistic works are essentially self-reflective and self-referential, mindful of only themselves and their formal, interior relations. Criticism therefore must adopt the pose of the modernist work and discard the nostalgic idea that some objective truth or intention is at stake. The key to this criticism is the role of language. However, in the poststructuralist rejection of art as a "counterpoint" to society, their critique of the hegemony of truth or meaning occurs within the framework of an "aestheticist" solution to the crisis—the liberation of language from meaning, art from representation. Instead of the classical contradictions of modernity (order versus instability; identity versus nonidentity, rebellion versus authority) being resolved, these contradictions become the inescapable intellectual conundrums through which modernity establishes, describes, and maintains itself.[4]

Although the Weberian-Frankfurt School thesis included the social dimension of modernity in its picture of the unfolding modernist crisis, it primarily sought to establish a dialectic between aesthetic modernism and social modernity. This dialectic of social rationalization and aesthetic resistance, however intuitively satisfying, was bound to a cultural view that is *reactive* to social events.[5] The larger question of how society and cultural modernism might be implicated in the same crisis of time-space perception remained vague and indeterminate. One reason for this lack of focus is that this perspective conceals a paradoxical attitude toward science. Nietzsche, Weber, and the Frankfurt School searched for a secure vantage point to criticize the inexorable and destructive extension of the "reason" of science and technology into society but conceded to science an ever greater power, taking refuge in the power-lessness of art. Paradoxically, by radically questioning the "uninformed dogmatism" of the scientific method, science became more opaque: its role in the constitution of modernity was never seriously investigated.

Yet, contemporary scholars of cultural modernity in the arts frequently point out that the dissolution of traditional modes of perception began with the profound epistemological crisis that first occurred within the natural sciences. This crisis elicited the critical skepticism of late nineteenth-century positivism (for example, Ernst Mach) as well as Nietzsche's unequivocal negation of the universality of reason and purposeful history.[6] "Modernism," writes the sociologist Anthony Giddens, "is neither only a protest against lost traditions, nor an endorsement of their dissolution, but in some degree an accurate expression of the emptying of time-space."[7] The breakup of the orderly continuum of time and space that invigorated the arts and characterized the late nineteenth-century crisis of perception also sprang from developments internal to the sciences, and not, as is often assumed, solely from a protest against the preeminence of scientific knowledge.

Because most commentators focus on only one side of the modern-ist predicament—that of rebellion—important confluences between developments in both late nineteenth-century science and in the arts have remained largely unexplored.[8] A case in point is the simultaneous impact of thermodynamics on the perception of the working body and on the representation of time and space in art and in philosophy. More-over, there emerged a social modernity, which is only now being discovered, whose concerns were similar to those of art and philosophy: applying new scientific modes of perception to social questions and bringing to bear a spirit of utopian and scientific "neutrality" that was opposed to the forces that rent nineteenth-century society along lines of class and ideology.

How social and cultural modernity emerged from a similar set of configurations can be more easily grasped by exploring one important instance where the emphasis on the economy of the working body converged with the crisis of time-space perception. In fact, a key convergence between social and cultural modernity can perhaps be located at this point where scientific materialism experienced its own inner crisis. Both Helmholtz and Marx conceived of the body as a field of forces capable of infinite transformation and conversion, simultaneously linking the cosmos to the body and to the productive order of work. This body mediated the laws of nature with the laws of production; it dissolved the anthropomorphic body as a distinct entity and made the industrial body subject to a sophisticated analytics of space and time. One consequence of that discovery was a shattering of perceptual certainties and the radical questioning of the nature of forms in the spatiotemporal world. It substituted an analytics of time and space for the *a-priori* certainties of traditional perspectives. But it also subjected the body to a series of objective, scientifically determined norms and economies of motion. The new analytics of nineteenth-century positivism thus fundamentally changed how the body was represented in both the social and the cultural domain.

A missing piece to the puzzle of how cultural and social modernity are connected can be found in the contribution of a unique but neglected figure in nineteenth-century intellectual history, a Parisian *savant* whose remarkable impact on the genesis of modernity registered in society and culture. Etienne-Jules Marey—physician, physiologist, pioneer of medical measurement, cardiologist, aviation pioneer, student of hydraulics, and photographic and cinema pioneer—remains an unexplored link between these disparate domains of modernity.[9] Indeed Marey's life work can be seen as an archimedean point at which social and cultural modernity intersected, illuminating the convergence of labor and culture in *fin-de-siècle* France.

For Marey, "movement" was the central fact of life. He devoted his career to charting and representing the movements of the "animal machine." Marey introduced a new language to describe the body at work, a language of time and motion. He prodigiously invented complex instruments of graphic inscription (by which he obtained tracings of heartbeat, pulse rate, gait, and the flapping of wings, among other movements), and later in his career, invented photographic apparatuses to record images of these movements. He believed that through these ingenious devices, the movements of the body could be analytically decomposed, broken down into its most discrete components. A science of the economy of the body could unlock the secrets of human labor power, emancipate working energy, and transform work. That Marey was the inspiration for Marcel Duchamp's *Nude Descending a Staircase,* for Bergson's *Creative Evolution,* and for the engineer Frank Gilbreth's (an associate of F. W. Taylor) photographic time-motion studies is more than a curiosity. In Marey's work the body was the focal point of the scientific dissolution of the space-time continuum. It thus became the provocation for Duchamp's images and Bergson's reflections on art and time perception, as well as the beginning of a new synthesis of science and labor. The Bauhaus theorist Siegfried Gideon paid tribute to this aspect of Marey's influence when he wrote that Marey was a "scientist who sees his objects with the sensibility of Mallarmé."[10]

AN ENGINEER OF LIFE

Born in the provincial town of Beaune (Côte d'Or), France in 1830, Marey was the son of Claude Marey, the "Commis" (steward) of a great wine house, Maison Bouchard Père et Fils, and Marie-Joséphine Bernard, a woman of "profound piety" and "superior intelligence." Marey's mother appreciated the rare qualities of her "dreamer son" and Marey's classmates, too, remembered him as a dreamy *collégien* who was "toujours absent."[11] By foregoing his own aspirations of becoming an engineer at the École Polytechnique, in 1849 Marey arrived in Paris to begin his medical training and to realize his father's dream of his becoming a physician.[12]

As a medical student, Marey proved both diligent and successful, his interests awakened by the great French physiologists who radiated in that epoch: Charles Robin, Charles Brown-Séquard, and above all Claude Bernard. In the late 1850s French medicine was being transformed from an empirical medicine founded on clinical observation to a new collaboration of clinical practice and experimental biology and physiology. Marey's teachers, like Philippe Ricord, a militant advocate of biological mechanism, were in the forefront of the growing campaign against vitalism, a campaign that had political as well as academic and intellectual implications.[13]

For young republican intellectuals, French medicine embodied the insurgent mood of anticlerical and positivist ideas in the late Empire. After the Prussian defeat of Austria in 1866, French liberal publicists drew the lesson that "Sadowa [the decisive battle] was a victory of superior education, not of Prussian institutions."[14] The obligatory scientific journey to the "Outre-Rhine" took on mythic proportions, and

French scholars often contrasted German accomplishments to sadly deficient domestic conditions. In contrast to French conservatives who tended to be militantly anti-German, French liberals were "Germanophile" in outlook. The Prussian victory of 1871 signified the advances of industry and progress, not merely military prowess, while Germany symbolized science, scholarship, and critical thought. Prussia became a synonym for innovation, dynamism, and the triumph of reason over clericalism, despotism, and deadening tradition.

Marey's 1859 thesis, *Recherches sur la circulation du sang à l'état physiologique et dans les maladies,* exemplified the new credo of physiological mechanism, stressing the role of elasticity in regulating circulation of the blood.[15] Throughout the late 1850s and early 1860s, Marey's research was devoted to demonstrating his conviction that "the laws which govern circulation are purely physical laws," a principle that guided his subsequent research on the heart, muscles, and neurological function.[16] Marey's appointment to the prestigious Collège de France in 1866 as successor to the eminent neurophysiologist Pierre Flourens (in the chair of "natural history of organized bodies") marked a shift toward public acceptance of the new Germanophile movement in French medicine. Auguste Chauveau, Marey's collaborator in his early years, recalled the "legendary penury" of French research institutions, to which the Collège was no exception. Marey frequently contrasted his cramped quarters and inadequate technical support with the "riches of instruments" developed by the Germans.[17] His chief supporter, the liberal minister of public instruction in the last years of the empire, Victor Duruy, was persuaded by a group of prominent physiologists— among them, Sigismond Jaccoud, Paul Joseph Lorain, Charles Adolphe Wurtz—to pursue aggressively the German model and introduce independent physiological laboratories in France.[18] Despite his complaints, Marey relished working at the Collège—the exciting center of the Germanophile movement, of materialism, and positivism—as opposed to the strongly Kantian Sorbonne, where metaphysics and idealism still reigned.

Marey's prodigious inventions, a plethora of fascinating devices designed to demonstrate the mechanical functions of the heart and vascular system already marked his early career. When presented to the Académie des Sciences in 1860, his invention of an improved instrument for measuring and recording the pulse, the *sphygmograph,* was considered a "revolution" in medical technology.[19] Marey's instant fame came from an incident that occurred when Napoleon III requested a demonstration of the sphygmograph at court. A few days after Marey's instrument detected irregularities in the pulse rate of a courtier, the unfortunate fellow was found dead in his bed, and Marey's prestige in royal and political circles was greatly enhanced.[20]

THE METAPHOR OF THE MOTOR

One of Marey's students later remarked that he was "never really a physiologist or a doctor in the usual sense of those terms, but above all, an engineer of life."[21] This comment captures the unique combination of invention and physiology in Marey's career. But it also describes the spirit of his work, which took as its leitmotif the metaphor of the machine for the functions of the animal organism. Although he did not discover this metaphor, Marey was among the first French physiologists to use it as the central metaphor to redefine the life sciences.[22]

The "animal machine," as Marey understood it, was not a complex of tools but an organized system, or motor, "endowed with direction." He was convinced that "the laws of mechanics are applicable to animated motors as well as to other machines," but, he added, "this truth has to be demonstrated."[23] For Marey, the "animal machine" was not simply an analogy, as it was for the "ancient authors" like Descartes. The study of the animal machine demanded both new instruments and new techniques whose application was firmly rooted in a mechanistic and materialistic theory. Marey, along with his German counterparts in the Helmholtzian school, argued that the unifying theory of energy conservation and the discovery of the principle of force in the universe overthrew the basic premise of vitalism—that is, a power inherent to life. As Marey often noted, "Modern engineers have created machines which are much more legitimately to be compared to animate motors."[24] The body's production of heat, for example, paralleled that of machines that create heat through combustion and the burning of carbon. By applying the principle of the conservation of energy in organized beings and by insisting on the equivalence between heat and mechanical work, physiologists had unlocked the key to the work of the muscles "as the furnace of our machines, where combustion is produced."[25] In their view, the body is a machine essentially capable of producing force and of maintaining an internal economy that deploys force for the benefit of the whole. It was the task of physiology to determine the laws of this machine. The "animal organism is no different from our machines," Marey concluded, "except for their greater efficiency."[26]

Marey's views were, it should be emphasized, hardly universally accepted among French physiologists, most of whom were still strongly attracted to those of Claude Bernard. Critics of the new mechanism found ample support in Bernard's well-publicized reaction, which he outlined in an official government-sponsored report on the state of physiology in France, published in 1867—the same year that Marey delivered his inaugural lectures at the Collège de France. Reserving some of his sharpest comments for the mechanists, Bernard argued that "there is one profound difference which separates the physiologist from the physicist. The physicist studies the instruments that he himself constructs," while the physiologist studies "living machines which he does not himself construct." The physiologist, unlike the physicist, "is obliged to interpret the workings of those living machines in an environment of errors and illusions" and, therefore, no analogy could serve as a complete explanation of physiological behavior.[27] Although he affirmed the physicochemical basis of life and acknowledged that this insight could sometimes reveal similar behaviors in the organic and inorganic world, Bernard maintained that the truth of biology could be revealed only in the living organism, not in a general view of nature, nor in physics and chemistry.[28] For Marey, on the contrary, "all that we know" in the sciences of life could be explained by the laws of physics and chemistry.[29]

Marey's inaugural lectures at the Collège, delivered in March 1867, explained the significance of motion for his concept of life: "Not long ago," he emphasized, "the origins of movement, that is to say, muscular action and nervous function were the most mysterious aspects of biology." Now, he claimed, "as a result of the contributions of the German school," they would be the "best understood."[30] To be sure, Marey oriented his investigations toward those physical phenomena most easily explained by mechanical laws: animal heat; the elasticity of the muscles; the circulatory system; optics; acoustics; and above all, locomotion. All of these were united, he held, by the centrality of energy for the deployment of force in motion.

For Marey, it was not until the second half of the nineteenth century that the life sciences could transcend a static view of the body: "We seem to have been traversing an immense gallery of mechanisms of greatly varied combinations . . . but everything here was mysterious in its immobility. The shift from organic structure to dynamics and the "interplay of organs" was a shift to mobility and to "motor function."[31]

Throughout the late 1860s and early 1870s we find evidence of Marey's preoccupation with Bernard and these controversies. For him the conflict in French physiology was a new and important stage in what Marey considered the eternal struggle in science between the mysticism of the vitalists and the rationalism of the mechanists; between the heirs of Plato and the heirs of Aristotle. He considered vital laws as "a fleeting hypothesis," made obsolete by the "general ensemble of physical or chemical laws."[32] "I cannot concede but two sorts of manifestations of life," he noted, "those that are intelligible to us, which are of a physical or chemical order, and those that are not."[33]

Marey's sometimes crude caricature of vitalism—and, of his protagonist Bernard—tell us more about his own Helmholtzian convictions than those of his opponents. Helmholtz's discoveries, according to Marey, had "changed the face of science," since energy, or force, became the unifying concept of all scientific explanation. If mechanics once provided seventeenth-century philosophers with a link between man and the cosmos, he claimed, the concept of energy truly reintegrated the organic and inorganic order of things. The unity of all material being *(transcendental materialism)* in energy conservation presupposed a total system of explanation: "The value of a theory depends on the number of facts which it embraces: that of the unity of physical forces tends to absorb them all. From the invisible atom to the celestial body lost in space, everything is subject to motion."[34]

BODIES IN MOTION

After 1860 Marey was increasingly concerned with charting the laws of motion of the body, "the rigorous determination of the intensity and duration of certain acts, of the form of different movements."[35] Marey's laboratory investigations of the locomotion of animals and birds in the late 1860s and early 1870s were predicated on the theory of energy conversion as the central fact of physiological life. Thermodynamics, he once said, was "the most remarkable theory of modern times," because it placed work or production at the center of its explanatory system.[36] This preoccupation with "work" as the common denominator of science and nature underlies all of Marey's concerns. It gives his physiology a decisively productivist cast of mind, making energy, economy, and output the central mechanisms and goal of life.

Marey cautioned, as did his predecessor Descartes, that animal machines remained distinct from man-made machines insofar as "animal motors cannot work incessantly, so that the work of the body only represents a fraction of what could be produced."[37] Yet, in that distinction the problem of how machines conserve energy—or dissipate it in fatigue or overwork—is central. First, the task of physiology is to determine the relative efficiency of the apparatus since "strict relations exist between the form of the organs and the character of the organs."[38]

Second, physiology maps the economy of function that each body contains; it describes the speed, energy, and duration that "are subjected to our will, and are regulated by it."[39] Motion had to be understood, not only in terms of the amount of energy used, but also in terms of how it is deployed in space and in time.

THE LANGUAGE OF PHYSIOLOGICAL TIME

A literary critic once remarked that Marcel Proust was the first to discover that "our body knows how to measure time."[40] In nineteenth-century science credit for this discovery belongs to Marey. His studies of motion are based on the problem of physiological time, a crucial concept of his mechanistic physiology. Marey, however, attributed the discovery of the internal physiological time, or duration present in any physiological process, to Helmholtz, who "had the boldness to undertake this measurement" in his pioneer investigation of the speed of the transference of a shock along the path of a nerve to the point of muscular contraction. This time, during which the muscle remains inactive until it acts on the impulse communicated to it by the nerve, is in Helmholtz's words, "temps perdu," time lost. This lost time, which consists of the relationship between duration and energy expenditure, is for Marey a basic component of the economy of the body.[41] The time expended in any activity, or reaction time, is a function of the body's internal laws of energy and motion. Marey's goal was to describe that economy of motion and to establish the laws of motion of the body. It is the search for these laws—"the regularity of phenomena and their rhythms"—that resulted in Marey's most important contribution to nineteenth-century physiology: the development of the first complex, sophisticated, and practicable instruments for measuring bodily function.[42] In the history of medical technology, he is best known as the pioneer of modern medical notation—graphic inscription—as a diagnostic tool.[43] Although he indefatigably promoted the standardization of measurement and the use of instruments of notation in diagnostic procedure, he did not consider the clinical aspect of his achievement as the most important. Rather, the triumph of medical notation, as his writings indicate, was a minor epistemological revolution in physiology, the triumph of technology over sense perception.

The discovery of physiological time connects the image of the animal machine to the tracing mechanisms developed by Marey to inscribe the body's motions on sheets of smoked paper. As Greenwich Time established an official standard of measurement for time around the world in 1884, Marey developed his techniques of physiological measurement in order to develop a standard for physiological time. For Marey, "All movement is the product of two factors: time and space; to know the movement of a body is to know the series of positions which it occupies in space during a series of successive instants."[44]

Marey's singular focus on graphic notation coincided with a growing scepticism of sense perception in physiology, ironically championed by both the Helmholtzian school and by Marey's adversary Bernard. Marey's philosophical differences with Bernard did not prevent him from sharing his view that knowledge could no longer be validated through direct observation. François Magendie and Claude Bernard both compared the dismal results of clinical medicine with modern physiology grounded in experimentation. Precisely because it possessed techniques to observe minute phenomena that "elude the senses," Bernard emphasized, medical progress could only derive from experimental physiology. Marey also argued that progress in science occurred despite, and often contrary to, sense perception and that new and more abstract "languages" of scientific explanation were necessary. His own solution was to advance the increasingly complex technologies of registration and measurement beyond the limits of the senses.[45]

This preoccupation with the technology of registration set Marey apart from an older positivist generation in nineteenth-century science. As he noted in his 1867 lectures, "Throughout the history of the human sciences this naïve positivism was based on such pieties as belief in incremental progress, hostility to prejudice of any kind, faith in the experimental method, and, of course, an absolute insistence on empirical verification."[46] Marey, however, preferred to see the development of science, not in the architectonic terms of a growing edifice with scientists providing new and sturdier building blocks with each new discovery, but in terms of archeology and linguistics. He criticized "that false method which produces systems," and disparaged constructivist approaches to science: "One pretends to *construct* a science, and one searches for the cornerstone which must support the entire edifice, [but] what guarantees that it is truly the base of the edifice?" "If a metaphor is absolutely necessary," he added, "I would be more inclined to compare the study of natural science with the work of archeologists who decipher inscriptions written in an unknown language *(une langue inconnue)*."[47]

The graphic method offered to decipher the language of duration within the space of the body as well as to map the body in space. It could establish an interior chronology of the body's rhythms and translate all changes in the activity of forces into a form that we might call the "natural language of the phenomena themselves, so much superior to all other modes of expression."[48] Marey implicitly rejected the idea of a single, homogenous time, applicable to all bodily functions. Each aspect of the body's rhythms was subject to its own discrete time, which could be traced. As Proust in literature or Bergson in philosophy, Marey was concerned with those dimensions of time and motion inaccessible to consciousness. "Not only are these instruments destined to replace the observer, and in his place, perform their function with an incontestable superiority; they will become irreplaceable in their domain. When the eye can no longer see, the ear cannot hear, or touch cannot feel, or even when our senses appear to deceive us, these instruments perform like a new sense with astonishing precision."[49] Graphic inscriptions were an attempt to express the body's most intimate movements with "a clarity that language does not possess."[50]

Marey considered his most enduring achievement the painstaking process of finding the variants of each physiological sign, of determining the conditions of each, and of tracing similarities between known and unknown inscriptions. This process constitutes the logic of Marey's scientific discovery. By deciphering the signs produced by the body's most intimate processes the graphic method "penetrates the intimate function of the organs, where life is expressed by an incessant movement."[51] The use of instruments to decode "nameless, nonacoustic languages . . . issuing from matter" is closer to the work of a linguist than a physician.[52] The physician's "gaze" seeks the invisible in the symptoms of disease; the physiologist, however, is confronted with signs that do not constitute a symptomology or even a pathology but can be perceived only by their duration and intensity. The graphic method, according to Marey, provides a "graphic expression of the most fleeting, most delicate, and most complex movements that no language could ever express."[53] These tracings reveal the "langue inconnue" of physiological time, the interior rhythms of the body.

Marey once compared physiology without the aid of graphic inscription to "geography without maps."[54] Without instruments, the physiologist is limited to the study of pathologies, or dead nature, with recourse to vivisection, which Marey resolutely opposed.[55] For the first time, the graphic method makes possible "the study *in situ* of living instruments." Not only could graphic inscription trace the most elusive movements of heart, muscles, or nerves, it could "capture" the lost time of the body: it could register duration, the temporal dimension of the body, infinitely divisible.[56]

Marey acknowledged that the graphic method did not originate with physiology, nor was he the first to use mechanical inscription in physiological research. Among his most important studies is a detailed history of graphic notation in which he traced its origins in the economics of trade and showed how in the early part of the century physicians had charted the course of an epidemic, how demographers had illustrated shifts in population, and how the Belgian social scientist Adolphe Lambert Quetelet invented social statistics by creating the ideal of "an average man."[57]

It was not simply the use of graphic representation, but the combination of this method with the technology of mechanical inscription that distinguished Marey's contribution to physiological measurement.[58] The first mechanical inscriptors originated in eighteenth-century meteorology, which registered changes in mercury columns in order to provide tracings of barometric pressure or temperature.[59] As early as 1807, the English physician Thomas Young developed a chronograph, which could register variations of impulse on a rotating cylinder, a device that Marey regarded as "the germ of all chronography."[60] In 1847 Carl Ludwig adapted Poiseuille's U-shaped mercury tube to register arterial pulsation. But Ludwig's pioneering *kymograph*, as he called it, could not be used on human subjects since it required direct insertion into the artery. A few years later Helmholtz produced a *myograph*, which could make tracings of a frog's muscle contractions.[61]

In 1858 Marey began his experiments, based on these models. His efforts resulted first in the more accurate *myograph* and, in 1860, the *sphygmograph*, which could measure pulse rate without being surgically attached to the arm. In 1863 he and his collaborator Chauveau succeeded in making tracings over longer distances, using air tubes, and two years later they constructed the first *cardiograph*, which could make tracings—forerunner of the cardiogram—of the human heart without an implantation.[62] A plethora of graphic inscriptors rapidly followed: a *pneumograph* for respiration, a *thermograph* for heat, and an increasingly sophisticated *myograph* for the nerves and muscles.[63] The myograph was enhanced by introducing an electric current, first through direct attachment to the exposed nerve of the frog and later through clips *(pince myograph)*. As a result, Marey could transmit a signal directly to the registering device and obtain consistently accurate tracings.[64]

Marey's inscriptors were not simply improvements over Ludwig's and Helmholtz's earlier models. They could be used in the laboratory and as clinical aids and were soon widely applied in medicine. The use of two interdependent lever drums (tambours) amplified each impulse into a broad sweep of the scriptor and gave "permanent expression to the subtle and fugitive phenomena of life."[65] Marey provided physiologists with instruments "whose refinement the limited senses of an observer could never attain, with absolute precision, without relaxation or without flagging."[66] His devices traced "that relation of space and time which is the essence of movement."[67]

Mechanical inscription combined observation and representation in a single image. With the graphic method the practitioner's "gaze" or touch was replaced by a technical apparatus that could make visible minute changes over time—"les infinement petits du temps." Observation was thus removed from the subjectivity of the physician and transferred to the machine's unerring eye. The technology of registration not only made possible the notation of any bodily act at the instant it occurred but could also chart its subsequent rhythm, frequency, and alteration over time. A kind of automatic writing, it united the body's own signs (pulse, heart rate, gait, the flapping of wings) with a language of technical representation. The mechanical inscription was "a new sense endowed with amazing precision."[68]

These remarks underscore the course that Marey would pursue in subsequent decades, expanding the scope of his investigations from the physiology of the heart, nerves, and muscles to the problem of locomotion, and ultimately to all forms of movement. Eventually, as we shall see, with the aid of chronophotography Marey's studies could document the decomposition of time and motion.[69]

MOTIONLESS BODIES DO NOT EXIST

The single theme that can be distilled from all of Marey's writing is that the body is a theater of motion. In his 1886 address to the Association Française pour l'Avancement des Sciences Marey noted that "in the body of a living being movement can be observed throughout: the circulation of the blood, the beating of the heart and the arteries; the lungs filling and emptying of air."[70] In his new laboratory established at the Collège de France in 1868, Marey turned from the problems of cardiology, circulation, and nervous function to the movement of bodies in general—to what he called "dynamic" physiology. His successes with graphic inscription permitted new studies of flight, human and equestrian gait, and even aquatic locomotion. The new subject matter was accompanied by a fascinating panoply of inventions. "The role of the physiologist," he remarked, "was to imagine all sorts of artifices to render these diverse movements 'graspable' and to rigorously determine their character."[71]

Marey, like many other inventors of his generation, was fascinated with the problem of flight. He frequently quoted the maxim, "The bird flies, but man will fly."[72] After centuries of debate, the question of how flight occurred remained a perplexing problem in the 1880s. The controversy over flight intensified during the course of the century, fueled by the hope that its resolution would one day permit that earthbound animal, man, to gain wings. By the first half of the nineteenth century, most ornithologists rejected the theories of the seventeenth-century Neapolitan physiologist, Giovanni Borelli, who had claimed that the force required by flying animals was far greater than that expended by animals in terrestrial locomotion. Instead, they proposed that the insect or bird relied on its respiratory canals, which permitted ascension as in a *montgolfiere* (hot-air balloon).[73]

In 1868 Marey applied his graphic method to the flight of birds and insects in order to "render to Borelli the justice due him."[74] To study insect flight, Marey adapted his tracing device to the flutter of the insect's wings, so that the tip of the wing simply came in direct contact with the tracing paper. By simultaneously recording the frequency of a tuning fork's vibrations on the same cylinder, the frequency of the wing movements could be calculated. Marey also put small amounts of gold spangle on the tip of the wing and held it before a dark background in direct sunlight. The result was a new, more complex conception of the trajectory of the wing—two figure eights and a dual motion from left to right and right to left during each complete wing cycle.[75] From these studies Marey concluded that "the movements executed during flight by the insect are sustained by an elevation and depression of the wing."[76] Marey found no evidence to support Borelli's view that birds or insects possessed a superlative strength. The rapidity of motion against the resistance of the air produced flight.

Perhaps more difficult was the measurement of a bird during flight. The frequency of the strokes of the wing, or the elevation and depression of the wing, could hardly be traced with a stationary bird; and a complex apparatus involving both an "electromagnetic switch" attached to telegraph wires and, ultimately, a myograph attached to the bird's thoracic muscle, presented severe technical problems. Nevertheless, Marey persevered in creating a new apparatus: "The bird flies in a space fifteen meters square and eight meters high. The registering apparatus being placed in the center of the room."[77] This experiment involved a drum (tambour) and lever attached to the wing and a "corset" that allowed freedom of movement. Because the instruments tended to weigh the bird down, he "chose a large bird to carry it, strong full grown buzzards."[78] Marey described the result: "After a great many fruitless attempts and changes in the fragile apparatus, which broke at almost every flight of the bird, we succeeded in obtaining satisfactory results."[79]

Marey did not limit these experiments to graphic notation alone. One extraordinary invention, completed in 1868, was a mechanical insect capable of simulating flight. It was designed to replicate the movements he had graphically inscribed, offering a kind of proof: "I have also realized the mechanical conditions which I have described through a schematic apparatus *(appareil schématique)*. I constructed a mechanical insect."[80] Marey's "automata" was the prototype for a series of mechanical birds and insects constructed by his co-workers in the 1870s, including a model airplane, designed by Victor Tatin, which reputedly flew for a few seconds in 1879.[81] Marey's *Le Vol des oiseaux,* published in 1890, and Louis Mouillard's *L'empire de l'air: Essai d'ornithologie appliquée à l'aviation* (1881) were the most important textbooks of the ornithological school of aviation, testimonials to a decade that for Marey "already hailed the coming of human locomotion through air."[82] It can only be added that the characteristically American discovery of flight depended not at all on these profound works.

Although Marey's fancy was with flight, his fame came from terrestrial locomotion, especially his studies of the gait of the horse. To understand the significance of the controversy over *les allures du cheval,* we should first recall the acrimonious debates throughout the early 1870s among equestrian experts on the movements of the horse. Even among the experts there was little agreement, not only over the famous question of whether all four hooves left the ground at the same instant, but about the "rhythm, duration and intensity" of the horse's gait. In his first paper on the subject, presented to the Académie des Sciences in 1874, Marey commented on the "discord that reigned between diverse authors on the succession of movements that characterize the position of the horse."[83]

In 1872 Marey attached inscriptors to the horse's hooves while the rider held the portable inscriptive apparatus.[84] As a result, Marey was able to provide a new and convincing answer to the riddle of the horse's gait: in gallop the horse supports itself on one, then on three, then on two, and again on one hoof, with periodic intervals when all four are in the air simultaneously. Contradicting all previous hypotheses, this discovery aroused the attention of a wide circle of equestrian experts, artists, and military tacticians.[85] Marey also designed a new apparatus for measuring human walking and running, consisting of a portable inscriptor and an experimental shoe. In addition, attached to the runner's head was an apparatus that registered the bounce of the entire body at various speeds.[86]

Marey summarized these studies in *La Machine animale* (1873), a work unique in its broad focus on the diverse forms of motion, and in its pioneering micrological analysis of bodies in time and space. By the 1870s Marey's interests led in two directions—utilitarian and aesthetic—both of which remained linked in Marey's mind. Each of these interests reached beyond the laboratory and physiological experimentation. As early as 1878 Marey published a long article in the popular Parisian scientific magazine *La Nature,* which emphasized the artistic and the social implications of his discoveries. The problem of the horse's gait, for example, was not simply of interest to equestrian specialists; it raised the broader question of the efficient deployment of the body's energies: "Whether we employ the horse, the donkey, the camel, or the deer, the same problem is always posed: how to gain the greatest possible advantage from the animal and to spare it the most fatigue and suffering."[87] To this end, he unveiled yet another inscription device, the portable *odograph,* which could measure the speed of the gait of a man or an animal, or the speed of a cart or wagon. The odograph was the first instrument to measure work, though it found little practical use until several decades later.[88]

In a further installment of his series on equestrian motion, Marey turned to the problem of artistic representation. Asking how the chronographic study of the horse's gait could contribute to a more "truthful way of representing the horse," Marey pointed to two benefits of the motion study: "The cavalryman tries to distinguish the different attitudes of the horse in order to correct those that he finds to be deficient by educating the horse; the artist, in attempting to represent the horse, tried to find the best means to translate its attitudes with the greatest fidelity, to best express its force, suppleness and grace."[89] While the study of the efficiency of animal motors led ultimately to a science of work, the problem of the artistic rendering of motion led to an analytic approach to representation, and ultimately to Marey's photographic motion studies. For Marey, the determination of beauty and the determination of utility were inseparable. Both relied on the accurate representation of a moving body. With graphic inscription artists could avail themselves of the new techniques and learn the exact movements of each gait: "which position we find each of the limbs, either raised or placed on the ground, at each instant."[90]

Marey's hippological investigations brought him into contact with the Anglo-American photographer Eadweard Muybridge, who had made his reputation in the American West.[91] As early as 1872, under the auspices of California Governor Leland Stanford, Muybridge used a primitive apparatus to photograph the horse in motion. In the spring of 1878 Muybridge resumed his efforts after a half-decade hiatus, in part because of the appearance of the English edition of Marey's *Animal Mechanism,* which Stanford (incorrectly) believed contained inaccuracies about the horse's gait at gallop.[92] Muybridge published a series of six new photographs under the title, *The Horse in Motion,* and during the following year he completed the studies that appeared in *Attitudes of Animals in Motion.*[93] Marey responded to these developments by writing to his friend, Gaston Tissandier, *La Nature*'s editor, that his "enthusiasm is overflowing" and that his research had already turned in the same direction:

> I am impressed with Mr. Muybridge's photographs published in the issue before last of *La Nature* [they were actually drawings of the photographs, since reproduction of photographs in illustrated magazines was still quite rare, A.R.] Could you put me in touch with the author? I would like his assistance in the solution of certain problems of physiology too difficult to resolve by other methods. For instance, on the question of birds in flight, I have devised a gun-like kind of photography [*"fusil photographique"*] for seizing birds in an attitude, or better, in a series of attitudes, which impart the successive phases of the wing's movement.[94]

Marey also suggested that Muybridge could place silhouettes on a revolving wheel to create the impression of motion—the principle of the zoetrope, based on a popular children's toy of the late nineteenth century. These photographs would, he believed, "create a revolution" for artists, "since one could furnish them with true attitudes of movement."[95] In a letter also published in *La Nature,* Muybridge acknowledged Marey's influence on his work and announced his plans to extend his experiments to "all the imaginable postures of athletes, horses, oxen, dogs and other animals in movement."[96]

Marey immediately recognized the superiority of Muybridge's photographs over his "incomplete" graphic notations and the drawings based on them. "The image of a horse caught at 1/500 of a second, provided, even at the very rapid strides, the real position of the animal almost as distinctly as if it had been standing still."[97]

In September 1881 Muybridge arrived in Paris, where he became an instant celebrity. Appropriately, his first public appearance was at Marey's new house, 11 Boulevard Delessert, on the Trocadéro, where an illustrious group of "foreign and French savants" gathered for a demonstration. As the Paris *Globe* commented: "The attraction of the evening consisted of the curious experiments of Mr. Muybridge, an American [sic], in photographing the movement of animated beings."[98] With

Marey providing translation and commentary, Muybridge projected his zoetropic images, creating the illusion of movement. Among those present were von Helmholtz, Brown-Sequard, the famous photographer Nadar, Duhousset, Tissandier, and several scientists attending the International Electrical Congress in Paris. The impact of the photographs was described by the Globe's writer: "In that diabolical parade, that infernal chase, the deer ran after the dogs, the steer pursued the deer, and even pigs showed the mad pretensions of grace and speed."[99] Marey later recalled with pride that Muybridge's demonstration was the first successful attempt to synthesize motion: "It was Muybridge, to my knowledge, who was the first to do so. He placed the instantaneous photographs of a running horse end to end and projected that series along a sort of phenakistoscope [a child's optical toy consisting of a spinning disc with figures reflected by mirrors]. This occurred in my home in 1882 [sic]."[100]

Shortly thereafter, on November 26, the famous painter Anton Meissonier, known for his panoramic historical tableaus and an acknowledged expert on the horse, held an even larger soirée attended by two hundred luminaries of the French art world. The assembled guests at Meissonier's studio were hardly disappointed, though Meissonier was patently less than thrilled with their approval for Muybridge's techniques, which convincingly demonstrated that the great painter had erred. The writer Alexandre Dumas, for example, remarked that one of Muybridge's California landscapes was "marvelously arranged," to which Meissonier stiffly retorted: "Nature composes . . . well."[101]

Muybridge's appearance in Paris caused an overnight sensation. His photographs were hailed as conclusive proof that all four hooves left the ground during gallop, and his zoetropic device signaled the first dramatic example of the photographic synthesis of movement.[102] Though his initial reaction was to explain that the "apparatus saw falsely," even Meisonnier henceforth "based all of the scenes in which the horse appears on chronophotographic documents."[103]

Marey's decision to abandon the graphic method for photography, the most significant turn in his career since his decision to abandon the clinic for the laboratory, dates from Muybridge's visit. At that time Marey already had several reservations about Muybridge's achievement.[104] Muybridge, he said, had never been able to obtain more than a "silhouette of a horse."[105] He also saw a more serious problem. With Muybridge's system of using multiple cameras in tandem, triggered by the animal passing over a trip-wire, the element of time could not be controlled. Less than a year after Muybridge arrived in Paris, a more sober assessment of his work appeared in Marey's writings about his most passionate interest, the flight of the bird. Marey pointed out that unlike the photographs of the horse, Muybridge's images of the bird did not serially reproduce flight, but consisted only of random images:

> Last September M. Muybridge arrived in Paris carrying a rich collection of instantaneous photographs which not only portrayed the horse in different poses, but also man engaging in different exercises: running, jumping, duelling, wrestling etc. In M. Muybridge's collection were also some photos of the bird in flight, but these were not representative of successive attitudes, as they were for the man or the horse. They were images analogous to those previously obtained by M. [Louis] Cailletet [a French physicist (1823–1913)] some years ago, and showed the wings of the bird sometimes in a unique position, sometimes elevated, sometimes lowered or in some intermediate phase. These photographs were nevertheless very interesting: they verified what the graphic method permitted me to grasp in regard to the mechanism of flight. But above all they promised valuable information *if it were possible to obtain the images in series* [my italics], as M. Muybridge was able to do for man and the quadrupeds. I resolved to concentrate that Winter on realizing my old project of the photographic rifle *(le fusil photographique)*.[106]

These remarks signal a fundamental difference between Marey and Muybridge that would become even more apparent in years to come. Whereas Muybridge's interest centered almost exclusively on the decomposition of movement into phases, Marey wanted to determine the precise relationship between time and motion in the sequences. Muybridge's photographs, done with multiple photographic apparatuses and electromagnetic trip-wire shutters, could not determine the time elapsed between the sequence of images.[107] Muybridge, Marey said, could freeze the individual moments in series, but he could not integrate the crucial variable of time. His images were spatially distinct but temporally blurred. Muybridge had always included the significant disclaimer: "The intervals of time between each phase is an *average* of the intervals of time between all the phases, or an approximation."[108]

Unlike Muybridge, Marey pursued the integration of time into the dissection of motion by developing his own device, a "photographic rifle," developed in February 1882, with a single objective and fixed plate. As Muybridge had successfully achieved the decomposition of motion, after 1882 Marey pursued the more elusive decomposition of motion into time. He devoted himself almost entirely to this endeavor, which he called first "photochronography" *(photo-chronographie)* and later "chronophotography" *(chronophotographie)*.

CHRONOPHOTOGRAPHY:
THE MICROSCOPE OF TIME

Marey's work was motivated by a utilitarian image of science guiding politics and even economics. "Scientific questions," he wrote, "are intimately connected to economic problems, it might even be said they dominate them."[109] Throughout the 1870s and 1880s Marey drafted numerous appeals to the French government for support, stressing the potential benefits of the motion study for military training and for the conservation of labor power. He claimed that this knowledge would put an end to fruitless debates. Recruits would not "be condemned to certain military exercises"; "the optimum use of the horse" in cavalry exercises could be determined "if we knew under what conditions the maximum speed, force, or labor which the living being can furnish may be obtained."[110]

Finding his research facilities at the Collège cramped, Marey frequently traveled about Paris to find suitable outdoor facilities to conduct his experiments. To the astonishment of onlookers, he was seen with his bizarre apparatuses "in the neighborhood of the grounds of the Luxembourg" or conducting his equestrian experiments in the riding schools of Paris. Such efforts sometimes came to nothing if his delicate instruments were damaged en route. From that time on, Marey recalled, "I had one desire, that of finding a spacious ground where I might unite a workshop, a laboratory, and an experimental field."[111]

On several occasions Marey appealed to the French government whose lack of commitment to research he said "inhibited the application of physiological studies to 'practical life.'"[112] In 1878 he came close to realizing his dream. After the Universal Exposition on the Champs de Mar closed, the minister of war General Farre, perceived the potential military uses of Marey's equestrian studies and offered him the grounds where part of the exposition stood. As it turned out, however, these grounds were destined to house, not Marey's vision of an outdoor laboratory, but another more famous symbol of modernity, the Eiffel Tower.[113]

Marey persisted, convinced of the potential of his research for "the development and utilization of muscular energy; in education and military gymnastics, manual trades, sports and so on."[114] Three years after the Champs de Mar disappointment, Marey's appeals finally brought results. In August 1881, at the request of M. Hérédia, president of the Paris Municipal Council, the city gave him a spacious ground on the periphery of Paris, at the Parc de Princes, near the porte d'Auteuil.[115] In late December he received 22,000 francs for equipment and research in gymnastic education, which subsequently became an annual subvention.[116] One of the first projects undertaken by the new facility was a revision of the existing "manual of physical exercises" issued by the Ministry of Public Instruction.[117]

Marey continued to pursue instantaneous photography, combining the basic precepts of Muybridge with techniques developed almost a decade earlier by the famous astronomer Pierre Jules Janssen (1824–1907), director of the Paris astrophysical laboratory and president of the Paris photographic society.[118] In December 1874, in order to photograph the positions of the planet Venus during its transit close to the earth Janssen invented an "astronomical revolver." Janssen combined a photographic apparatus with a telescope to produce seventeen images on a single glass plate.[119] As a photographic enthusiast, Janssen wrote in 1876 of the potential benefits of his revolver for motion photography:

> The characteristic of the revolver is that it affords an automatic means of taking a series of photographs of the most variable and rapid phenomena in a sequence as rapid as may be desired, and thus opens up for investigation some of the most interesting problems in the physiology and mechanics of walking, flying and various other animal movements.[120]

Janssen also understood that his revolver made possible the integration of the element of time in the production of precise images of motion. His revolver accomplished the opposite of what the phenakistoscope or zoetrope achieved by synthesizing rapidly moving images passing before the eye. "The photographic revolver gives, on the contrary, an analytical representation of movement in series in its elementary phases."[121] In the winter of 1881 to 1882 Marey wrote to Janssen, asking him to provide the exact specifications of his photographic revolver, which he eagerly did.[122] Marey found that Janssen's revolver was more than adequate for tracking the slow movement of the planets, but for his purposes the "extreme rapidity of movement" and the "shaking of the apparatus" distorted the image. Marey then conceived of constructing a variation on Janssen's idea "in the form of a hunting rifle," which could easily provide a dozen images in a second, with each individual image requiring no more than 1/700th of a second.[123] Marey's *fusil photographique*, which he produced sometime in the spring of 1882, could trace the flight of a bird, recording the successive attitudes of the wings. It consisted of a long barrel with a camera lens, a smaller, round, or oxygonal plate, and rotating disc mechanism similar to Janssen's, with twelve windows capable of twelve revolutions per second.[124]

The photographic rifle seemed to provide the solution to many of the problems Marey encountered in Muybridge's work: it offered greater clarity, portability, and faster exposure time. Above all, it provided "a measure of duration," which Marey estimated at one-twelfth of a second between each exposure. But it did not prove as practicable as Marey initially hoped. Its images were fuzzy and sparse and because it was handheld, it could not measure the distance traversed by the target. The goal of achieving, on a single fixed plate, a series of successive images representing the different positions of a body in measurable space and in relation to "a series of known" instants, remained elusive.[125]

Marey returned to Paris from Naples in the spring of 1882, eager to pursue his chronophotography at the newly completed Physiological Station where "space, open air, and unobstructed light are indispensable for the study of living creatures." The station fulfilled Marey's hope of emancipating physiology from the constricting atmosphere of the laboratory.[126] It boasted two concentric horizontal circular tracks (the longer one 500 meters)—one for the horse and one for the human studies. The two tracks were bounded by a series of telegraph poles that when passed emitted a signal communicated to an odograph housed in one of the buildings. The device allowed the speed of the unencumbered runner to be measured accurately. At the center of the track was a tower from which the speed of the runner or horse could be regulated.[127]

Abandoning the *fusil* for a stationary camera with a stable fixed plate, Marey began a new series of experiments with human and animal locomotion. The new camera consisted of a disc with a window, creating a series of images on a fixed plate.[128] This device, which "produced rapid exposures at regular intervals of time," was housed on a unique railway spur, for which Marey constructed a wagon containing the apparatus, *la chambre noir roulante.*[129] The photographer rode inside the small wagon, viewing the scene through a red glass while communicating with the assistant through a loudspeaker. At the other end of the spur was a dark screen *(écran noir)*—actually a three-sided camera obscura painted black, with a velvet curtain across the rear wall—which approximated what Marey called "absolute blackness."[130] "A man, entirely dressed in white and vividly illuminated by the sun, walks, runs, or jumps in front of the photographic apparatus, supplied with a shutter device rotating more or less rapidly, which takes his image at nearer or further distances."[131] As the man in white moves across the screen, the rotating disc records the successive phases of his movements and registers the precise intervals between them on the plate.

The first photographs of the runner were published in *La Nature* in July 1882. Marey soon added the device that would become his signature. In order to provide an image of time within the photographic image, he attached a clocklike device with a luminescent dial (chronometric dial) to the front of the camera obscura. Marey's subsequent images can be identified by the small wheel with radii of equal angles and black and white spatial markers at the base of the dark screen, distinctive signs of his chronophotography. The chronometer signified the complete integration of time into the study of motion.[132] "By thus multiplying the images at very short intervals," he wrote, "we can obtain, with perfect authenticity, the successive phases of locomotion."[133]

By the end of 1882, Marey pursued the main objective of chronophotography. Whereas Muybridge decomposed motion into space, Marey analyzed motion "by means of a series of instantaneous photographs taken at very short and equal intervals of time."[134] Any speed, any trajectory, or any body could be traced in this manner. Chronophotography, as Marey remarked a few years later, made it possible to know "the complete law of movement of a point in space."

Despite his success with the fixed plate, Marey remained dissatisfied with his earliest images.[135] Time and space still seemed to be in conflict. A rapidly rotating disc with increasingly short exposure times could encompass an increasing number of images, but often resulted in confusingly superimposed images, especially in slower movements. This was less of a problem with rapid motion, where relatively long exposure times were able to distinguish different phases of motion, but with slower movements the disc produced a confusing number of images.[136] For example, in the case of a jumper and a hurdle, the image became obscure—superimposed—as the jumper slowed to a halt in the final phase.[137]

In 1883 Marey found a solution in what he called "partial," or "geometric," photography, a method which consisted of "reducing the complexity of the images in order to repeat them at extremely short intervals."[138] By means of a specially designed suit with silver stripes, highlighting only the parts of the body that are of interest and suppressing the others, Marey created a remarkable series of abstract photographs. The lines that designated the parts of the torso permitted him to "multiply the number of images" and indicate the phases of the movements more accurately:

A man dressed completely in black, and consequently invisible upon the dead-black background, wears certain bright points and lines, strips of silver lace attached to his clothes along the axes of his limbs. When this man, so rigged, passes in front of the apparatus, photographs will result that will be accurate diagrams to scale, showing without confusion the postures of upper and lower arms, thighs and lower legs, and feet at each instant, as well as the oscillations of the head and of the hips.[139]

The more lines used to express the movement of the limbs, the easier it was to increase the number of images and create more complex representations. With these photographs, Marey achieved an extraordinary economy of representation—the reduction of the body to a "geometric" pattern of lines in space along a line of time. In Marey's images the body became a trace on a glass surface, a trajectory of decomposed movement.

Throughout the 1880s Marey improved his chronophotographic techniques, expanding the range of his subjects, improving his ornithological studies, and even experimenting with "four-dimensional forms," for example, conoids, spheroids, and the like. In 1886 to 1887 he produced the first of several model "syntheses" of flight in the form of wax or bronze sculptures of the successive phases of a gull's movements.[140] Another major turning point in chronophotography came in 1887 when Marey first began to experiment with a moveable film "ribbon," leading to the offshoot of chronophotography, which we now call cinematography.[141] By then Marey had achieved what artists, physiologists, and equestrian experts had unsuccessfully tried to accomplish for years, perhaps centuries—an intricate and mathematically precise line drawing of the human or animal stride. As we shall see, these images, which had a profound impact on philosophy, on the art world, and on the study of the economy of labor power, were his most influential achievement. Marey concluded that "motion was only the relation of time to space."[142]

TIME AND MOTION

The question of what constituted the "true" nature of time and space was already hotly debated in the Parisian intellectual reviews in the 1880s, often finding its way into the popular press.[143] These discussions not only had profound consequences for science, but also for the decomposition of the spatiotemporal field in art, literature, and philosophy.[144] Defenders of the traditional Kantian view of time and space as stable, innate, *a-priori* concepts increasingly found themselves on the defensive as new ideas of "curved space," "four dimensionality," and higher orders of space perception gained currency. The advocates of the new theories of space, which at one point even included Marey's friend and idol Helmholtz, argued that such alternative conceptions were theoretically possible, though Helmholtz believed they could not be conclusively demonstrated or represented. Helmholtz's physiological investigations, especially his important contribution to optics, convinced him that our sensations are in no sense reflected "images" of external objects. Rather they are signs that have an analogic relation to the external world but do not directly correspond to the objective laws of nature, as there were clearly phenomena that eluded the senses. Hyperspace was not open to empirical verification.[145]

Thus, time and space were not absolute categories that existed apart from the perceptual apparatus and were perhaps even a limitation of the three-dimensional world of the senses. A leading proponent of the new space in France, the mathematician Henri Poincaré wrote as early as 1887 that though the competing Euclidean or non-Euclidean hypotheses could not be judged absolutely true or false, both were merely conventions, and as such, four-dimensional space could not be definitively ruled out. Any conception of space was purely relative; major alterations in the universe could occur without being either measured or perceived.[146] These ideas resonated as a result of intense debates in such critical journals as the *Revue Philosophique* and the *Revue de Metaphysique et de Morale,* which assumed neopositivist and neo-Kantian viewpoints respectively. As Linda Dallrymple Henderson shows in her study of this perceptual revolution, "The philosophical impact of non-Euclidean geometry in the nineteenth century was far greater than simply its initial challenge to Kant. It substantially shook the foundations of mathematics and science, branches of learning that for two thousand years had depended on the truth of Euclid's axioms. As a result, optimistic belief in man's ability to acquire absolute truth gradually gave way during the later nineteenth century to a recognition of the relativity of knowledge."[147]

Though Marey refers to these debates only obliquely, some of his experiments with objects in motion in the early 1890s were clearly a response to them—an effort to demonstrate that at least some propositions of speculative geometry could be represented through motion photography: "By means of threads stretched between metal armatures, one can show how the successive positions of a straight line can produce cylinders, cones, conoids, and hyperboloids by revolution," These figures do not correspond to our "normal" idea of three-dimensional space, he added, since "in reality we are dealing with hypothetical figures which find no counterpart in nature."[148] Marey's chronophotographs also challenged the notion of a static spatialization of time and motion by showing how the trajectory of any object was dependent on the ability of an apparatus to "decompose" it spatially and temporally. More important, his work showed how distinct the mental processes of perception could be from the "objective" laws of objects in motion. As Henderson points out, "Coming full circle from its early days as a tool of the empiricist positivist Helmholtz, non-Euclidean geometry contributed substantially to the demise of traditional positivism."[149] By 1900 these ideas had become common parlance in the world of science and philosophy.

Along with those who theorized about the geometries of the new space, were others who argued that time was a "fourth dimension," reaching its popular apotheosis in H. G. Wells' *The Time Machine* (1895). In 1885, the philosopher J.-M. Guyau published an article on *"L'évolution de l'idée de temps dans la conscience,"* in which he, too, rigorously investigated the idea of time. He asked whether "in sum, everything is present in consciousness, if the image of the past is a sort of illusion, if the future in turn, is a simple projection of our present activity, how do we come to form and organize the idea of time?"[150] Time, he concluded, was an "essential factor of progress, an ordering of experience, yet ultimately a matter of perspective."[151]

A more radical approach was that of Henri Bergson, whose *Essai sur les donées de la conscience* (1889), was written from 1884 to 1886, only a few years after Marey's chronophotographs became well known. Bergson argued that in organizing the facts of succession in our conscious life, we create "for them a fourth dimension of space which we call homogenous time"—the "connecting link" between space and what he called real time, or "duration."[152] Remarkably, Marey's role in these debates has been consistently neglected, though much of Bergson's writing on time is a commentary on Marey who is often invoked, though rarely mentioned by name.[153] When Bergson chose to illustrate his discussion of homogenous time in *Matière et mémoire* (1896), he took Marey's chronophotographs as his image of how we construct an image of time: "In just the same way as the multitudinous successive positions of a runner are contracted into a single symbolic attitude which our eyes perceive, which art reproduces, and which becomes for us all the image of man running. The glance which falls at any moment on the things about us only takes in the effects of multiplicity of inner repetitions and evolutions."[154] As opposed to the fragmented and immobilized time displayed by the chronophotographs, Bergson argued, our memory "solidifies into sensible qualities the continuous flow of things." What we perceive as movement is only "discontinuity" that is spatially "fixed" by consciousness. Bergson's view of the spatialization of time often refers to biology, and the final chapter of his *L'évolution créatrice* (1907) is entitled *"Le mécanisme cinématographique de la pensée et l'illusion mécanistique."*[155]

Bergson and Marey were colleagues at the Collège de France, and in 1902, the year before Bergson delivered his famous lectures on time consciousness, they participated in a group for the investigation of parapsychological phenomena. The "Groupe d'Études de Phénomènes Psychiques," met to "explore that region, situated at the intersection of psychology, biology and physics, where we encounter manifestations of forces not yet defined."[156] They also conducted what Bergson called "rigorous experiments," using Marey's inscriptors to test certain well-known Parisian mediums. Finally, for Bergson, Marey's famous comparison of the chronophotographs of the horse and the "perfect rendition of the gallop in the Parthenon frieze" by Phidias serves as the crucial distinction between chronophotography and time consciousness in Bergson's *L'évolution créatice.*[157]

Phidias had long been the object of Marey's fascination. In his 1878 article on the horse in art, he noted with with undisguised admiration how in the "great age of Greek art, Phidias had indeed correctly portrayed a galloping steed in his Parthenon frieze." Marey confessed that in his astonishment he had at first believed that "in the epoch of Phidias the science of the horse's gait was already possessed by artists." But, after having examined the entire frieze, he was convinced that this had been "a happy coincidence."[158] Bergson, however, had another purpose in underscoring the frieze/chronophotograph comparison. For him, its significance was not that Marey had proven Phidias "correct" or that Phidias had "anticipated" the principles of chronophotography, or graphic notation. Rather, Marey's instantaneous photographs revealed for the first time the essential difference between art and chronophotography, between truth and illusion in the representation of time, and illuminated for Bergson the critical feature of modern time consciousness:

> It is the same cinematographical mechanism in both cases [Phidias and Marey] but reaches a precision in the second that it cannot achieve in the first. Of the gallop of a horse our eye perceives chiefly, a characteristic, essential or rather schematic attitude, a form that appears to radiate over a whole period and so fill up a time of gallop. It is this attitude that sculpture has fixed on the frieze of the Parthenon. But instantaneous pho-

tography isolates any moment; it puts them all in the same rank, and thus the gallop of a horse spreads out, into as many successive attitudes as it wishes, instead of massing into a single attitude, which is supposed to flash out in a privileged moment, and illuminate a whole period.[159]

Phidias's gallop "freezes" the true image of time, the single moment of illumination; it also captures the passage of time into duration. In a series of lectures on "the history of the idea of time" delivered at the Collège de France during the winter semester from 1902 to 1903, Bergson further elaborated his view of the crisis in the modern conception of time. In contrast to time as duration, objective time consciousness is always expressed as space; its infinite and irreparable discontinuity is mirrored in chronophotography. Time is a "sign par excellence," which permits us to grasp relationships spatially. Its mode of perception is analytic in the sense of decomposition into units expressed in spatial terms. Knowledge of external reality is predicated on the recomposition of different perspectives, based on signs and concepts that permit us to assimilate experience into "a language with which we are already familiar" *(une langue déja connue).*[160]

Viewed externally, movement is "a displacement, a trajectory," but when viewed subjectively, it appears as experience, a simple state of the soul, something indivisible and original. Thus, objective knowledge of motion can never be more than "a relation of multiple similarities," a series of points in space. Spatialized time, whether as points on a line or a sphere, is essentially a futile attempt to capture the lost "mobility" of our experience of time—as in Bergson's famous "durée."[161]

The symptoms of the crisis in time consciousness were most evident in the linguistic confusion that surrounds time: "Duration is always expressed in extensity. The terms that designate time are borrowed from the language of space. When we evoke time, it is space that responds to the appeal."[162] Only by radically dissociating the pure and immediate intuition of time, or duration, from the spatial forms in which we envelop it, can the experience of time be redeemed. The clock mystifies time by objectifying it as order.

Marey's chronophotographs mirrored for Bergson the ordinary mode of grasping motion or external reality as "natural," for example, as "moments" ordered spatially. As in the chronophotographs, Bergson noted that "instead of attaching ourselves to the inner becoming of things, we place ourselves outside them in order to recompose their becoming artificially. We take snapshots, as it were, of the passing reality."[163] The instantaneous photograph was for Marey no more than a superior kind of observation, an instant captured by the speed of a shutter. But, for Bergson, chronophotography became a metaphor of our objective knowledge of consciousness. The memory of the eye is an invention, like time itself, an artificial image. We "bring back continuity" by producing stable images of bodies from multiple objects in time, a single image in space. In short, "The mechanism of our ordinary knowledge is of a cinematographical kind."[164]

Marey's images thus provide a visual parable of the crisis of space-time perception: instantaneous photography is the concrete "illusion" of objective time consciousness. Even historical time is illusory. As Paul Valéry noted, these photographic images engender "a state of mind that is curiously anti-historical, that is to say, an acute perception of the completely immediate substance of our images of 'the past' and of our inalienable freedom to modify them as easily as we can conceive of them."[165] Marey thus confirmed Bergson's diagnosis of the crisis of all perceptual systems: objective time was infinitely divisible; moments of experience were organized spatially, that is, the reduction of quality to quantity.

This brief detour through Bergson's philosophical critique of spatialized time brings us back to yet another crucial dimension of Marey's chronophotography. Bergson went further than the prevailing skepticism of his contemporaries, such as Guyau or Georges Lechalas, by rooting the "crisis of the real" within the scientific worldview.[166] He saw that the problems posed by the infinite spatialization of time, and conversely, of the temporalization of space, were endemic to a perceptual system entrenched in quantification and abstraction. Whereas time is in its essence "mobility," the measurement of time in spatial terms undercuts all other possible conceptualizations because of its "immobility." The emptying of all time into space—as points on a trajectory—is for Bergson the subordination of human experience to an objective and external measure. For the ancients it was the privileged or salient moments that mattered. For moderns there are no such moments. Decomposed time, scientific time, cinematographic time are identical, opposed to "real time, regarded as a flux . . . as the very mobility of being [which] escapes the hold of scientific knowledge."[167] With these arguments Bergson demonstrated, without mentioning him by name, that the most fundamental aspects of Marey's method—the decomposition of motion into spatialized time and the discovery of the *langue inconnue* of the body's forces—were expressive of a century of positivism in crisis.

Bergson's encounter with chronophotography provided him with a metaphor for the modern experience of time, to which he contrasted his own sense of time as experience. The ancients "composed" the visual and temporal field; the modern sciences "decomposed" it. Nevertheless, Bergson's view can also be characterized as one polarity of classical modernist criticism—the attempt to redeem the totality of experience at the expense of scientific rationality. The opposite pole is represented by Marey, who identifies with the scientific project of decomposing the absolutes of time-space. Walter Benjamin's famous evocation of the true image of the past, "which flashes up at the instant when it can be recognized and is never seen again," is similarly directed against the "homogenous, empty time" of progress.[168] At one end of the spectrum stands Phidias, at the other Marey.

Contemporaries clearly saw the modernist implications of chronophotography for other cultural domains. Marey's experiments with four-dimensionality made possible an early voyage beyond the three-dimensional universe and the depiction of new shapes. Poincaré called him "a veritable artist of the mechanics of life" and hailed the accessibility of his science for the Parisian public."[169] Among those fascinated with instantaneous photography was the painter Edgar Degas, whose dancers were also "motion studies."[170] It is likely that the painter Seurat, too, was influenced by Marey's photographs.[171] Charles Henry, the founder of scientific, psychophysical aesthetics— and a bicycle enthusiast—was an admirer of Marey (and is reputed to have been the man in white in Marey's bicycle chronophotographs).[172] Henry's achievement, the analytic decomposition of visual sensation into light, color, and form, closely parallels Marey's decomposition of motion, and Henry's impact on the neo-impressionists via Paul Signac and Seurat is well documented.[173]

Apart from Bergson, there is the critic Paul Valéry, whose reflections on the aesthetic implications of the metaphysics of motion are unsurpassed. For Valéry, Muybridge's photographs "showed how inventive the eye is, or rather how much the sight elaborates the data . . . imposing continuity, connection and the system of change which we ground in the labels of *space, time, matter,* and *movement."* Valéry identified "the law of unconscious falsification," the idea that the mind provided intelligibility of movement through a "creative seeing by which the understanding filled the gaps of sense perception."[174] It was not their use of photography to depict reality more accurately, as Marey and Muybridge had assumed, that propelled them into the aesthetic revolution in Paris. Rather, it was their decomposition of the visual field, their experiments with the boundaries of perception, and their impact on the crisis of absolute time and space that was most important for the arts.

Marey's texts are embedded in the language of nineteenth-century science, but his images belong to the canon of twentieth-century art. The attempts to cross the threshold of dynamic physiology, to capture physiological time, to chart the unexplored terrain of motion within the human motor, and ultimately to represent the energy of the body in action—all led Marey to produce his remarkable chronophotographs, the first iconography of energeticism. Detached from their scientific context, such striking photographs as the walking man whose white stripes alone are captured provides a purely graphic archetype of motion, approaching the spatial sensibility of the modernist avant-garde. Asked if he was influenced by cinematography in a now famous interview with Pierre Cabanne in 1967, Marcel Duchamp replied that it was chronophotography that inspired the *Nude Descending a Staircase.*

P.C.: Chronophotography?
M.D.: Yes, I saw it in the illustration of a book by Marey, where he showed men who were fencing, or horses in gallop with a system of dotted lines delineating the different movements. That is how he explained the idea of an elementary parallelism. It is a bit pretentious and formulaic, but amusing. That is what gave me the idea for the execution of the nude descending a staircase.[175]

For the Italian Futurists, who also held Bergson in high esteem, Marey's images demonstrated "the non-reality of the motionless body."[176] In 1913 Anton Giulio Bragaglia published his *Fotodinamismo Futurista,* a book that can be counted among the earliest works of aesthetic chronophotography. Umberto Bocaccio, the great theorist of Futurism, credited Marey with the "unification of the concept of space, to which Cubism was limited, with that of time."[177] In rendering visible "movements that the human eye cannot perceive" and in converging with Bergson, with cubism, and with Futurism, Marey entered the vocabulary of modern art.

THE ECONOMY OF WORK

A less frequently recognized, but perhaps more important facet of Marey's precocious modernism is his view of the body as a system of economies of work. Marey's original purpose in conducting his experiments is worth recalling: to understand the thermodynamic laws of motion, to grasp the ineffable "lost time" of movement, and to discover the economies of objects moving through space. This aspect of his research, with its emphasis on the "animal machine" and the mechanical work of energy systems in the body, represented the utilitarian and social dimension of Marey's modernism.[178]

In his 1886 address to the Association Française pour l'Avancement des Sciences, Marey outlined the practical use of chronophotography: "As we regulate the use of machines in order to obtain a useful result with the least exertion of work, so man can regulate his movements . . . with the least fatigue possible."[179] The task of a physiology based on chronophotography and graphic notation is to determine the most efficient movement of the body. These comments are among the first succinct statements of modern "ergonomics," or the science of efficient movement. Marey located his science, not in the debates over time and space, but in a scientific approach to the conservation of energy as labor power. In *La Machine animale* Marey was already concerned with how the work of animated motors can be measured and made more efficient, and later he used the odograph along with chronophotography to calculate the "mechanical work expended in different movements."[180] In short, chronophotography made possible a science of fatigue and a rationalization of the body's movements—an economy of energy that led to a distinctly European science of work.

The first application of chronophotography to the study of physical labor was undertaken in Marey's laboratory in 1894 by Charles Fremont, an engineer concerned with establishing optimum work performance with specific tools. These studies were published in the popular Paris monthly, *Le Monde Moderne* in February 1895.[181] Fremont compared his chronophotographs of two (sometimes one) forgers hammering a red-hot iron on an anvil with earlier depictions of a forge in Mongé's eighteenth-century sketches and, of course, in Diderot's *Encyclopédie*, to demonstrate how "all the movements are false."[182] Unlike the highly aestheticized and sanitized preindustrial milieu shown in the *Encyclopédie*, Fremont's forgers perform before a dark field with only a chronometer visible in the foreground (to indicate the uniform intervals between blows).

These photographs, taken with a ribbon of film, were largely aesthetic. More important was a second series, which used a fixed plate (as in the photographic rifle) in which "all the successive poses occurring during a complete cycle of the striking of a hammer" are shown. These fifteen superimposed shots completely blur the body of the man, while only the hammer is visible. In order to capture the successive positions of the hammer and the hands, the chronophotographer must obliterate the worker. Only the essential corporality is retained, as in the geomet-

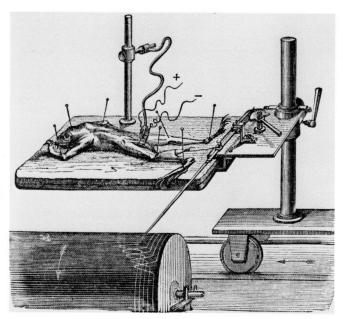

FIGURE 1: E. J. Marey, Simple Myograph and tracing mechanism with electric current, 1867. (E. J. Marey, *Animal Mechanism: A Treatise on Terrestrial and Aerial Locomotion* [New York, 1874].)

FIGURE 2: E. J. Marey, Apparatus for tracing and measuring wing movements during flight, Collège de France, 1874. (E. J. Marey, *Animal Mechanism: A Treatise on Terrestrial and Aerial Locomotion* [New York, 1874].)

FIGURE 3 RIGHT: E. J. Marey, Runner with portable registering apparatus, 1873. (E.J. Marey, *Animal Mechanism: A Treatise on Terrestrial and Aerial Locomotion* [New York, 1874].)

FIGURE 4:   E. J. Marey, Portable Odograph and cart, ca. 1887. (E. J. Marey, *Le Mouvement* [Paris, 1895], courtesy National Library of Medicine Washington, D.C.)

FIGURE 5 TOP: Janssen's Astronomical Revolver during the transit of Venus past the sun, 9 December 1874. (*La Nature*, 8 May 1875, courtesy New York Public Library Photographic Services.)

FIGURE 6 LEFT: E. J. Marey, Testing the Photographic Rifle at the Bay of Naples, 1881. (*La Nature*, 22 April 1882, courtesy New York Public Library Photographic Services.)

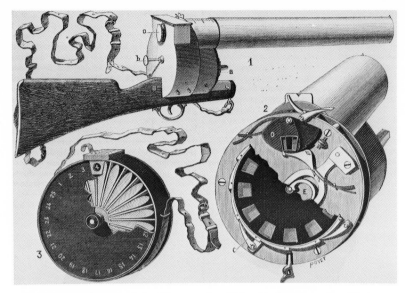

FIGURE 7: Mechanism of the Photographic Rifle: *(1)* the apparatus; *(2)* shutter mechanism and aperture; *(3)* twenty-five photosensitive plates. (*La Nature,* 22 April 1882, courtesy New York Public Library Photographic Services.)

FIGURE 8: Physiological Station with *écran noir,* railway spur, moving chamber, and track, 1883. (*La Nature,* 8 September 1883, courtesy New York Public Library Photographic Services.)

FIGURE 9:   Stationary Odograph and track at the Physiological Station, ca. 1887.
(E. J. Marey, *Le Mouvement* [Paris, 1895].)

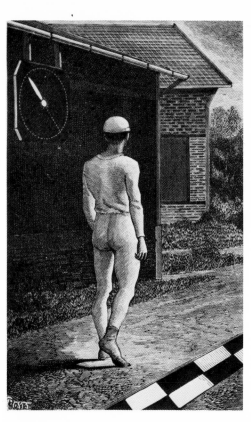

FIGURE 10 LEFT: E. J. Marey, Walker in white suit in front of *écran noir*, 1882. (E. J. Marey, *Développement de la méthode graphique par la emploi de la photographie* [Paris, 1884].)

FIGURE 11 BOTTOM: E. J. Marey, Chronophotograph of runner on fixed plate, 1883. (Deutsches Filmmuseum, Frankfurt am Main.)

FIGURE 12 TOP: E. J. Marey, Geometric Chrono-photograph of the man in the black suit, 1883. (Musée Marey, Beaune.)

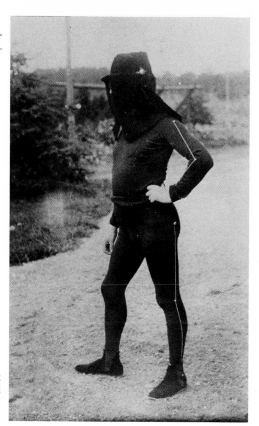

FIGURE 13 RIGHT: Georges Demeny in the black suit with striping and points for geometric or partial chronophotography, 1883. (Musée Marey, Beaune.)

FIGURE 14: E. J. Marey, Chronophotograph, pole vaulter, 1890/91. (Musée Marey, Beaune.)

ric or partial photographs of 1883, in which Marey discovered how to eliminate the superfluous parts of the body. The result is an extraordinary photograph, which concentrates the labor power involved in a specific task into its essential physiognomic elements while "the body of the worker is ignored." Fremont then translated the chronophotographs into a mechanical line drawing of the economy of movement during a "complete cycle."[183]

Fremont's chronophotographs initiated the study of diverse industrial occupations with the goal of discovering the optimum expenditure of energy by a micrological decomposition of a given action. Although similar in conception to the detailed studies of work in several American factories by the American engineer F. W. Taylor, they antedated the arrival of his works in Europe by almost two decades. These first photographic studies of time and motion also preceded by twenty years Frank B. Gilbreth's famous cyclographs completed in 1914.[184]

Taylor's goal was the maximization of output—productivity—irrespective of the physiological cost to the worker. As an engineer he considered the body as a "machine," which either operated efficiently or it did not. He did not consider, as did the physiologists concerned with the "human motor," how energy and fatigue might be optimally calculated for long-term use, rather than productivity, *per se.* Chronophotography demonstrated the potential for greater economy to be attained from "training," which Fremont believed could be developed according to studies of each work process. In the case of the forgers, for example, he noted with pride an almost innate tendency toward an economy of force, which could be logically explained by "the instinctual movement of the worker." A careful study of the diagrams revealed "the least amount of force exerted," and as a result, "the least amount of fatigue."[185]

The day would soon come, Marey confidently predicted, when similar studies would reveal "the laws that govern the mass of different tools," and more important, would result in a systematic elaboration of the economy of the working body—a physiognomy of labor power. Such a theory, he added, could be divided into four main areas: (1) heat and the production of energy; (2) the relationship of time and motion; (3) the suppression of shock, or elasticity; and (4) the rational deployment of energy. As Marey recognized after 1895, his forty years of effort could be subsumed under the economy of work, a concept that linked his earliest studies of muscle physiology and circulation to the animal mechanics of the 1870s and to the chronophotographs of the 1880s.[186]

His motion studies showed that in all animals locomotion could be characterized by the transformation of abrupt and disjointed movements into consistent motion. The central feature of all work—whether of humans or machines—was the suppression and transformation of irregular, inconsistent, and jarring shocks into regular and uniform activity.[187] Shock was at once productive and destructive. In the case of humans or animals doing physical labor, such as pulling a wagon or cart, there was the additional resistance of the added weight and of further shocks to the system. Like a machine's piston, the body possessed an elasticity that permitted the suppression of shock into regular effort. The muscles, for example, act to turn abrupt movements into dynamic work, though "the energy at our disposal is always circumscribed by real resistances."[188]

In Marey's last published writings the ideal of an "economy of motion" is central. He attempted to demonstrate how animal and human motors were naturally efficient and capable of improvement. If we could envision a future in which "all machines work silently with grace and waste very little energy," he predicted, "we could also see a more efficient deployment of energy in the manual trades."[189] Marey conceded, however, that an insurmountable difference existed between man and machine. What gives the human motor its most machinelike quality, its capacity to determine and regulate the deployment of energy, also prevents it from realizing a fully efficient mechanical economy of work. Even the most efficient of movements, determined by "geometric chronophotography" could not overcome the natural resistances of the body. Experimenting with military recruits, Marey was disappointed with the body's inability to sustain even the most thoroughly rationalized cadences. Inevitably, the wasteful aspect of human labor power came to the forefront. Entropy, or fatigue, rather than the maximization of effort eventually triumphs. The muscle "cannot surpass a certain maximum: the point arrives when resistances vanquish every possible effort; finally muscular acts occur without the production of any external work; the muscle itself becomes exhausted."[190] Fatigue, in short, appears as the subversive element in the human motor, the body's fifth column. In this juxtaposition of the economy of energy and its perpetual nemesis of fatigue, a science of work came into being.

As future laborers in Marey's garden were to realize, fatigue was the physiological limit of even the most perfectly executed work, the horizon of the metaphor of the human motor. Marey's techniques— graphic inscription and chronophotography—as well as his principles of analysis—the decomposition of motion into discrete units and the measurement of the body—became, so to speak, tools of the trade. The fecundity of his inventions, and the popularizations by which he disseminated his ideas, attracted the attention of French republican politicians, among both liberals and the left, who viewed science as a means of eliminating the traditional social obstacles to greater productivity and economic well-being.

Marey's accomplishment, to establish a science of the human motor, was more than a precocious aesthetic or a socially utilitarian modernism. It was also a modernist politics, the politics of a state devoted to maximizing the economy of the body. A Saint-Simonian ghost hovers over the science of work, urging the collaboration of progressive men of science, the open-mindedness of an entrepreneurial elite, and the guidance of an enlightened state. Marey's social modernism thus linked the spirit of analytic positivism to a new kind of productivism—the optimum deployment of all forces available to the nation.

CHAPTER FIVE

The Laws of the Human Motor

The human body has been compared, insofar as the source of movement is concerned, with a machine that functions by means of heat. We know that no machine ever creates energy. The most perfect motors can do no more than transform heat into movement.

—Fernand Lagrange
Physiologie des Exercises du Corps

SOCIAL HELMHOLTZIANISM

B Y the 1890s the achievements of German physics and physiology, as well as Marey's investigations of the human and animal machine, produced a program for investigating the physiology of labor power. The techniques pioneered by Marey, together with the efforts of a small, international avant-garde of European physicians and physiologists, were enlisted in the search for precise laws of muscles, nerves, and the deployment of energy in the human organism at work. Marey's laboratory served as both a training institute and a clearing house for the new energeticist science.

These physiologists translated the ideas of energetics into a program of social modernity that conceived of the working body as a system of economies of force and as the focal point for new techniques that could eliminate social conflict while ensuring productivity. Firm believers in the achievements of materialist physiology, and dedicated to social progress, these pioneers of the science of work were for the most part liberal republicans and in some cases socialists. They saw industrial physiology as the intersection of social policy and medicine. For these progressive *savants* "social hygiene" could be served by the new ideas of biological materialism and by the triumphs of the laboratory. Their hopes were predicated on a neutral and benevolent state that would implement their achievements and ensure the rational application of scientific solutions to the "social question."

"Social Helmholtzianism" thus created an image of work entirely unconnected to milieu, class, or political interests. It presented the body of the worker as a universal, degendered motor, whose specific and nonenergetic needs could be bracketed. Energy conservation became a social doctrine. The author of a major treatise on the physiology of work, André Liesse observed that "the human being . . . always remains a machine, but that machine directs itself, within the limits of the environment in which it performs, and the faculties it employs. It is, if we permit ourselves the expression, a self-propelling machine, powered by itself, with the aid of nutrition, the source of its energy and the will to conservation."[1] Armand Imbert, a Montpellier physiologist, an assiduous researcher of fatigue and a dedicated reformer sympathetic to the workers' movement, elaborated some of its most cherished ideals in his 1902 synthesis, *Mode de fonctionnement économique de l'organisme.* "It is a fascinating idea," wrote Imbert, "to think that our organism, as a machine which produces work, is constructed according to a general model and presents a mode of functioning in which every wasteful, or even less powerful expenditure of energy, is avoided."[2] Imbert's characterization underscores the productivist utilitarianism of the new science of the working body. Its goal was to discover efficient expenditures of labor power, to find "the degree of energy economy which we unconsciously realize in the functioning of our organism as a motor."[3]

By the later 1890s the problem of fatigue was identified as the paramount manifestation of the body's limits to produce work. Fatigue, European scientists and social reformers concluded, was a greater threat to the future of a modern industrialized nation than either rapacious capital or the atavistic worker lacking in discipline and time sense. "Any considerable amount of work," Imbert noted, "measured in duration or in terms of quantity in relation to a unit of time, involves fatigue, and it is in reality because of the preoccupation with reducing fatigue, that we regulate our functions."[4] The so-called natural desire to avoid work did not lead to an unproductive laborer; it was the body, undermined by excessive, irregular, poorly organized, and exhausting work. Consequently, a rational solution to the labor problem could be found in physiology, not politics. Labor power was seen as a purely physiological issue, a single instance of the more general problem of energy conservation and its transformation into work.

The new science of labor made possible the search for the quantitative laws of the muscles, the nerves, and the deployment of energy within the human organism. Though the physiology of work could not be an exact science, it nevertheless "confirmed and explained the basic law of all physical processes, the 'law of the least effort.'"[5] The *law of the least effort* (sometimes called the law of "least action"), to which Helmholtz devoted several theoretical studies in the mid-1880s, posited that masses in motion always exert the least amount of work in passing from potential to kinetic energy. Helmholtz claimed that this principle could be demonstrated in a variety of specific cases, including electromagnetic conversions.[6] The idea that nature essentially found the shortest means to achieve its ends could, the physiologists of work concluded, also permit the discovery of the most suitable external conditions for efficiently using those forces employed in labor. Moreover, the physiologists applied this idea to the body: "Following the principle of the conservation of energy, which has been indisputably verified on the muscles, the heat given off, and the mechanical work produced, will be found to be equivalent; that which the human motor gains, on the one hand, it loses on the other."[7]

For the pioneers of the new science of the laboring body, social Helmholtzianism envisioned a labor force that did not have to be inculcated with eternal truths about the importance of will, the sin of idleness, or the value of work: Moral exhortations, the idealization of the "true worker," and the reflections of political economists on the needs of the worker were obsolete. Instead, the physiology of labor power offered a neutral approach free of social conflict: in the future the industrial workplace would be modeled on the technology it served.

The emergence of a physiological approach to labor coincided with important changes in work during Europe's second industrial revolution. If the first industrial revolution was essentially a revolution in steam-driven technology, textile production, and the railroad, the second was a revolution of electric power, of steel and chemical production, and of the rise of industries producing heavy machinery. European industrial production soared after 1895: the industrial enterprise expanded accordingly, with increasing emphasis on planning, administration, and design of the plant.[8] By the end of the nineteenth century the patterns of industrial discipline typical of the first half of the century (paternalism, familialism, surveillance) also changed markedly. With a huge workforce, rapid turnover, and immigration, ideas of industrial edification were either obsolete or took a back seat to issues of time, work norms, and wages. The "new factory" of the second industrial revolution was rapidly eliminating craft production and was employing an unskilled workforce in a setting where new technologies increasingly determined the nature of work and the organization of the workplace. The moral discipline of work, which middle-class reformers attempted to inculcate among earlier generations of industrial laborers, was giving way to the discipline of the workplace, to conflicts over the tempo of work, and the risks of mechanized industry. Unions were increasingly challenging management over work norms and the right to determine them.[9] In an era when work was measured by time and motion rather than by desire or will, the prospect of a purely technical or scientific solution to what Michelle Perrot has called the "crisis of discipline" was particularly welcome.[10]

For this reason the apparent scientific neutrality of the medicalized discourse on labor power appealed to progressive reformers on both sides of the Rhine. Efforts to regulate scientifically the tempo of labor coincided with the first attempts by the governments to regulate labor conflict and create a "scientific labor policy." For the proponents of the new science of work, the efficiency of the human organism provided reformers with a "higher law" than the vicissitudes of class conflict and authoritarian labor policies. Nevertheless, only after the science of labor power had established itself as an acknowledged source of expertise, or social knowledge, did it intervene significantly in the political arena. As we shall see, industrial accident legislation, the length of the working day, the duration of military service, and the economics of the national deployment of the labor force were all eventually judged by the standards of research in fatigue.

We can distinguish three phases in the early development of the European science of work: (1) a theoretical development in the study of the economies of energy, heat, motion, and fatigue in the body (1867–1900); (2) the growth of a laboratory science of work and the collection of data on modern industrial work (1900–1910); and (3) the first interventions in the workplace and efforts to influence state policy on questions such as length of the working day and the causes of industrial accidents. This latter period also witnessed the first challenges posed by the American system of industrial management devised by F. W. Taylor. In each of these phases scientists attempted to reduce labor to a purely instrumental, or technical act, which lent itself to the rigors of physiological experiment and social science. Although in this chapter we will be concerned with the initial phase from the 1880s to roughly 1900, the economics of energy remains a motif of the science of the working body.

MUSCULAR THERMODYNAMICS

The central problem confronting the nineteenth-century physiologists who adopted the thermodynamic model of the human motor was the production of animal heat and the physiological processes that consumed and replenished the body's energy supply. The historian of biology William Coleman has emphasized that "during the nineteenth century, the overall chemical and physical relationships of the respiratory process—which subsume the question of organic heat production—were brought into close agreement with the principle of the conservation of energy."[11] The effort to demonstrate the principle that equivalent amounts of energy produce equivalent amounts of mechanical work—that the consumption and the expenditure of energy were equivalent—became a recurrent theme of late nineteenth-century physiology. In Germany and France, physiologists attempted to prove that Julius Robert Mayer had been right to claim in the 1840s, though without substantiation, that "the organism in its overall measurable relations with the external world—that world serving as source and drain for the organism's energy supply—was an energy conversion device, a machine no less than those scrutinized by mechanics and thermodynamics."[12]

After 1850 muscle physiology was a fertile ground for applying energy conservation to physiology.[13] As early as 1845 Mayer wrote that "the muscle is a tool, by means of which the transformation of force is effected, but it is not the substance which is transformed into the performance produced."[14] Mayer showed how oxidation was the ultimate source of energy for the organism's capacity to do mechanical work.[15] And Helmholtz, in one of his earliest essays, stated that research on animal metabolism confirmed that the materials supplied to the body by respiration and digestion "provide the entire sum of vital warmth during the successive stages of their combination."[16]

However, the question of how the muscles converted heat to work remained a source of persistent controversy. By the late 1840s developments in muscle physiology shifted to the "electro-motor" power of muscle tissue discovered by the Italian physiologist Carlo Matteucci and by Du Bois-Reymond. The latter's *Untersuchungen über thierische Elektricität* (1848) was a mechanist manifesto and a strongly worded polemic against vitalism.[17] In the 1850s Helmholtz's famous experiments measuring the duration of muscle spasms demonstrated that "the course of these inner transformations remained exactly the same, even when the muscle performed under different external conditions of load."[18] Marey's myograph (1867) also enhanced the popularity of studies of the intensity of muscle contraction under various conditions, providing some of the earliest tracings of muscle fatigue.[19]

In the second half of the century, German physiologists remained in the forefront to demonstrate the relevance of energy conservation to muscle physiology.[20] Adolf Fick, an early pioneer of muscle physiology, confirmed that the comparison between the action of the muscle and the steam engine was "apt and instructive [since] in both cases we are concerned with the effects of chemical forces through which the motion of certain masses and also heat are created."[21] Concerned with the chemical changes in a muscle as a consequence of fatigue, Fick also scrupulously investigated the chemistry of "muscle substance" during work: "The fact that chemical changes in the muscles take place is already indicated by certain general phenomena which are easy to observe on one's own body. Everyone knows that when a group of muscles works very energetically for a time, this does not continue with the same force despite the impulses of the will. This phenomenon, which is well known by the name of fatigue, proves irrefutably that the muscle undergoes an inner transformation through its work."[22]

Initially Fick closely followed the chemist Justus von Liebig's view that the decomposition of nitrogenous substances, especially protein, was the chief source of the muscle's work, a theory abandoned only in the last quarter of the century. Paralleling energy conservation, Fick claimed that the work of the muscles revealed that "the sum of potential and moving energy of the system is always of constant size, which cannot be altered by any positive or negative work of any force within the system."[23] Experimenting on themselves during a strenuous climb of the Faulhorn (literally lazy mount), Fick and his collaborator, the chemist Johannes Wislicenus, found that the protein used up in the effort was only a small fraction of the total calories consumed. These and other experiments confirmed that the analogy of the human motor could be applied to the physiology of nutrition as well: "A machine is built of iron. Yet, in work this iron is hardly consumed in significant quantity, while the burning of coal provided the work."[24] Coal represents, of course, carbohydrates, while the iron, or protein, was necessary in only limited amounts.

Fick's discovery, as well as the experimental work of a London chemist Edward Frankland, revealed that muscle protein (as Liebig had held) could not alone account for the work performed, though attention soon focused on the importance of intermediary metabolic processes, especially the metabolic exchange of foodstuffs and nutriments.[25] Carl von Voit, professor of physiology in Munich, who studied and collaborated with Liebig, undertook some interesting empirical studies of the food consumption habits of Bavarians during the 1870s. But von Voit's most famous work, with his co-worker Max von Pettenkofer, were attempts to measure the content of both gaseous and solid wastes, the energy-equivalents of food intake. In a series of well-known experiments conducted on dogs, Voit demonstrated that the animals converted nitrogenous substances into energy, but found that the dog's urea or fecal substance could not alone account for the production of heat. Using a respiration chamber—an apparatus that could control food intake as well as end-products and waste over time—they ascertained that the total daily input of carbon, hydrogen, nitrogen, and oxygen was present in the body's excretions and exhaled gases. Voit and Pettenkofer also made the important discovery that oxygen utilization differed according to the food consumed, influencing the body's efficiency. Although he refuted Liebig's theory of protein as the primary source of muscular energy, Voit's effort to discover an accurate and universally applicable measure of the value of foodstuffs in relation to energy output was hampered by his reliance on a traditional framework of the chemical theory of metabolic "substances," or *Stoffwechsel.* A mathematically reliable system of equivalence between the amount of potential energy ingested in the form of nutriments and the amount of energy produced, remained elusive.[26]

A more conclusive answer was provided by one of Voit's students, Ludwig Max Rubner, who jettisoned the idea of chemical substances and used calories to calculate the energetic content of each nutriment (fats, protein, starch) required to produce an equivalent amount of energy. Despite Voit's resistance, Rubner's theory of the substitutability of foodstuffs was ultimately confirmed, leading in 1894 to his establishing of the exact caloric values of all nutritive substances. Rubner had finally solved the problem of energetic equivalents, or *Kraftwechsel,* in human physiology:

> Calculation of the energy content of nutritional materials henceforth provides us with a legitimate expression of their capacity to provide for specific needs of the cells. Useful energy is a measure of their value, transformed energy a measure of the biological energy created. These are the foundations of an energetic conception of warm-blooded life.[27]

Rubner's experiments with a calorimeter (1889) assured the triumph of *Kraftwechsel* over *Stoffwechsel,* offering conclusive proof of the organism as a "heat machine."[28] After 1891, when Rubner became professor of hygiene in Berlin, he greatly expanded his "hygienic" approach to physiology, working not merely as an advisor to the military, but popularizing his ideas in a series of works on hygiene and nutrition.[29]

ELASTICITY AND EFFICIENCY:
AUGUSTE CHAUVEAU

The view that the human machine possessed a superlative efficiency was a key precept of the energeticist physiology. The German physiologist Emmanuel Munk observed that the human motor was "the most complete dynamic machine" as compared with the "machine" proper, which wastes nine-tenths of its heat in the conversion to force, whereas the body uses 40 percent of its chemical materials in the production of work.[30] The source of the body's greater efficiency was the "law of the least effort"—the internal economy of the conversion process within the organism.

In his 1867 course at the Collège de France, Marey also argued that "the chemical actions which take place in the organism are the cause of the production of heat in animals," and consequently "the animal organism is no different from our machines except by its more advantageous efficiency."[31]Demonstrative proof of the human machine's greater efficiency was the contribution of Marey's co-worker, Auguste Chauveau, whose prodigious work in the 1880s resulted in the "law of the least effort" in physiology. For Chauveau, "the contracted muscle is the result of a special and absolutely perfect elasticity of the muscles . . . adapted to the functional purposes envisaged and anticipated by muscular work."[32] His experiments on the flexors of the forearm demonstrated the "unconscious, but constant effort to reduce the total expenditure of energy to a minimum." The principle of efficiency is contained in the economy of work performed by muscles, Chauveau claimed, so that in its design and execution, the human motor always chooses "the most economic course."[33] Chauveau therefore concluded that a "mechanical optimum" *(optimum mécanique)* might be obtained if certain loads and speeds could be calculated for each kind of physical labor.[34]

These initial efforts to produce a thermodynamics of the working body did not always meet with universal acclaim. In France, the growing challenge of German physiological mechanism produced considerable controversy over how far the analogy of the motor could be pursued. Defenders of Claude Bernard's view that biology was sovereign in its own domain were skeptical, as were some physiologists sympathetic to materialism, like Gustave-Adolphe Hirn, who acknowledged the primacy of energy in explaining the universe but who remained dubious of Marey and Chauveau's overliteral use of the metaphor of the machine.

Hirn cautioned that the metaphor was being pushed too far: "The phenomena of energy does not in reality correspond to the reality of things, as the acts of an automata do not correspond to those of a living being."[35] He admonished the mechanists for their zeal, claiming that "the living motor was neither a thermal nor a caloric motor."[36] Nevertheless, defenders of the metaphor continued to assert, as did Chauveau, that "vital properties are nothing other than the aptitude to transform . . . *potential energy* or *forces of tension* . . . into a special mode of actualized energy."[37] And the leading defender of physiological materialism in France, Charles Richet, contended that even mental activity could be understood as a transformation of force, producing a chorus of rebuttals by those whose Cartesian sensibilities were profoundly injured by his refusal to accept the dualism of mind and body.[38] One writer argued that the increase in "crimes of blood" during rising temperatures proved that body heat or "external temperature was one of the elements that determined the forms of thought."[39]

Despite these controversies, by the 1890s the physiology of human and animal energetics established a more sophisticated understanding of the human motor's efficiency at work. Both Marey and Chauveau believed that the crucial elements in the science of work that emerged from their research included (1) the conversion of energy and the measurement of the chemical forces that undergo "metamorphosis" in work; (2) the problem of "physiological time," for example, in measuring the duration of an impulse sent from the nerves to the muscles; (3) the elasticity of the muscles and the phenomena of shock; (4) the problem of fatigue; and (5) the analysis of time and motion, the decomposition of the act of work into its smallest measurable units.

CARE AND FEEDING OF THE HUMAN MOTOR

Studies of the basic principles of muscle physiology underscored the critical role of diet and nutrition in creating labor power. As André

Liesse put it: "Human labor, considered from a physiological point of view, could be evaluated in terms of calories and kilogrammeters."[40] But which diet was appropriate for the optimal performance of the worker? Some German physiologists asked whether "our worker with his poor diet of potatoes can satisfy all his needs," pointing to chronic lack of protein in the national diet. "The iron of the steam engine is also used up over time [and] protein-rich nutrition gives the worker well-formed, powerful muscles."[41] The diet of the working man became a *cause célèbre* of physiologists whose search for the proper combination of nutritional elements was frequently linked to the first comparative nutritional studies of workers of different cities or nations. Von Voit, Rubner, and others found that fats and meats were necessary for reproducing the body's albumin supply, while starches and sugars were necessary for regulating and properly utilizing protein. How much of the required starch might be derived from different foodstuffs remained unresolved.

One common argument, for example, was that the English workers' diet of meat and wheat-flour bread could account for their superior productivity, whereas the German workers' diet of potatoes constituted an economic disadvantage. This controversy culminated in the famous "Brot versus Kartoffel" debate in the 1870s and 1880s, when von Voit calculated that the worker should consume roughly 70 percent of his carbohydrate diet in bread and the remaining 30 percent in "the form of potatoes and vegetables."[42]

The diet of the French worker caused concern among physiologists who tried to ascertain whether different kinds of work and different individuals demonstrated different metabolic rates. As early as 1858, Hirn questioned whether "the caloric sum, which, for example, produces chemical reactions in our bodies, is not different when we are at rest, and when we work, when we raise a weight, when we meet resistance." He also correlated the "age, sex, and temperament" (for example, "nervous and bilious, lymphatic, strong and robust") with three measurements: (1) calometric; (2) the oxygen inhaled and expelled; and (3) a dyanometric calculation of the work produced "when the individual functions as a motor."[43] In the 1860s, Jules-Auguste Béclard performed simple experiments on human subjects to measure the amount of heat produced in the body before, during, and immediately after physical work to determine the quantitative relationship between each of these phases of work.[44] Richet and Hirn attempted to ascertain whether "the machine is subject to the universal law of thermodyamic equivalence, that is, if it consumes heat in order to furnish positive work, or whether it also accumulates heat during rest or 'negative work.' "[45] The distinguished physician and political figure, Paul Bert, devoted a two-volume work entitled *La Machine humaine* (1867–68) to a study of the diet of English prisoners.[46]

By the 1890s ambitious experiments were frequently undertaken to provide data on the comparative caloric requirements of different social groups. Armand Gautier, a professor of physiology at the University of Paris and, in 1870, director of the first laboratory of chemical biology established in France, attempted to construct a profile of different social groups and different nationalities, according to their food consumption and chemical wastes, an ingenious effort at physiology "from below."[47] Gautier asked how could "the needs and losses of the energy of the man whose organs function normally" be measured, and how do they "vary according to this same activity?"[48] For Gautier, the laws of physiology ordained that in a healthy organism the alimentary needs of a normal adult man will "always be proportional to the expenditure of energy of which he is the center; and such is the principle of a new method which in its turn will give us the measure of those needs." Gautier believed his calculations were vastly superior to those using general statistics on the food consumption of large human communities or to those relying on the observation of individual cases.[49] In short, the dietary requirements of different groups could be scientifically forecasted by the comparative analysis of their excretory products.

Employing a *respiratory calorimeter,* similar to the one used by Rubner and further developed by the American physiologist W. O. Atwater, Gautier measured an experimental human subject who "eats, works and sleeps in the chamber for several days." The device established the quantities of heat lost, work accomplished, oxygen absorbed, and water, carbonic acid and excretory matter lost" in precise figures.[50] Gautier employed two broad groupings, work and rest, subsequently combining his own studies with those already accomplished by other physiologists to produce a comparative picture (table 5.1) of the daily consumption of nutrients by different social groups and different types of workers.

On the basis of these calculations, Gautier believed he could find the minimum number of calories necessary to power a worker or soldier at different tasks. Gautier distinguished between the nutritional requirements of work performance and the caloric requirements of a body concerned only with maintaining heat and other normal functions. As a result, physiologists could distinguish quantitatively between "useful work," or economic work "appreciable in terms of its results," and the total energy required by the body under any circumstances.[51] For a worker at rest 2604 calories in 24 hours might suffice, whereas a

TABLE 5.1

Daily Dietary Consumption According to Social Groups (in grams)

	Protein (Gr.)	Fats (Gr.)	Carbohydrates (Gr.)	Source
Alimentation at rest				
French bourgeois with moderate exercise	120	70	330	Gautier
Average Paris population	115	48	333	Gautier
English bourgeois with moderate exercise	92	72	352	Forster
German worker	137	72	352	Voit/Pettenkofer
Swedish soldier (peacetime)	130	40	530	Almen
Prisoners	87	22	305	Schüster
Silesian peasant	80	16	552	Meiner
Average	108	49	403	
Alimentation during work				
French worker (much work)	190	90	600	Gautier
English blacksmith (fatiguing work)	176	71	666	Playfair
Swedish worker	146	44	504	Hildesheim
French soldier (wartime)	192	40	651	Gautier
Swedish soldier (during a campaign)	146	59	557	Almen
Bavarian soldier	118	56	500	Voit
German soldier	130	40	550	Moleshcott
Average	150	60	563	

Source: André Liesse, *Le Travail: Aux point de vue scientifique, industriel, et social* (Paris: Guillaumin, 1899), p. 23.

worker at hard work requires 3556 calories, "an increase destined to furnish the supplement of energy necessary for the labor of a worker without excess."[52] For fatiguing work, an average of 3800 calories was required, but for "exceptionally severe work" 5000 utilizable calories might be required."[53] Other physiologists had found that in extremely demanding work, for example, among the Russian woodcutters of Astrakhan, the miners of Tomsk, or German brickmakers, more than 5000 calories daily was hardly extraordinary.[54] These figures calculated the relative nutritional values required for each task, for a given number of hours of work, and for a precise number of days and months.

Gautier emphasized that observing the worker in the working milieu rather than in the laboratory might yield more useful information about nutritional requirements. The actual workday demanded different calculations to determine the ratio between the food, the energy required, and the amount of work. To calculate the ratio of "utilizable amount of labor" to the actual work expended, Gautier studied the wine and spirit workers of Midi, in the South of France. Their primary task was to raise the level of the water or wine in large vats by means of a pump over the course of nine or ten hours. By breaking the work down to its component parts, Gautier estimated that the total work expended by a "good worker" laboring "to the borders of fatigue," was about 250,700 kilogrammeters and that studies of other laborers and mountain climbers also showed that "a good workman furnishes in a day of eight to ten hours from 260,000 to 280,000 kilogrammeters" of work.[55] This measure necessarily includes the energy expended in all the body's functions and does not, of course, entirely translate into *utilizable* work.

To his surprise, Gautier discovered that an actual work situation hardly measured up to his expected calculations. In fact, caloric consumption and work performance were more elastic than he had anticipated. In the case of the wine workers, the principle of entropy, or loss of energy, could be reaily demonstrated. Gautier calculated that only 25 to 65 percent was translated into useful work. Their total work expenditure usually required a considerable amount of "additional daily foods" (an average of 1779 calories over and above the daily amount required without work). He also calculated that although the additional calories should theoretically be sufficient for any work that might be required, in fact, less than a third of this potential energy was converted into "real and tangible work." He could only explain this discrepancy by hypothesizing that the "total energy dispensed by the human being during work" diminishes over time.[56] In other words, fatigue accounted for the extraordinary inefficiency of output that Gautier confronted in the Midi wine workers.

Gautier's simple caloric measurements could not adequately explain the complex physiology of energy production, nor could they account for the "loss" of caloric energy in the actual work. In contrast to the more optimistic Chauveau, Gautier suspected that "the human machine, from a muscular point of view, was capable of a rather weak efficiency when compared with the maintenance necessary."[57]

Gautier ultimately concluded that the wine workers were not economizing their energy output to full advantage. The "loss" of energy in transforming caloric intake to work performance could also be explained by the lack of economy of force exercised by the worker, by the differential course of fatigue, and by the inefficient use of time and motion during work. Gautier had clearly reached the point where the search for the laws of muscular thermodynamics became a science of fatigue. The principle of energy conservation had become a *fait social*, an empirical reality that could be precisely measured and quantified in terms of specific work performance. By the turn of the century, energy conservation became the fundamental fact not only of nature but of society as well.

THE LAWS OF FATIGUE: ANGELO MOSSO AND THE INVENTION OF THE ERGOGRAPH

The physiology of the muscles, of nutrition, and the empirical observations of Gautier—all led to the same conclusion: fatigue was the chief source of the body's resistance to work and lack of efficiency. Fatigue was identified as the natural barrier to the efficient use of the human motor. Before 1870 there were hardly any studies of fatigue, by 1890 the floodgates released an outpouring of literature on all aspects of fatigue. Already in 1867, Marey identified the central role of fatigue in his law of human effort: "A muscle is subordinate to two influences, one is reparative: nutrition; the other is exhausting: its motor function. The body's capacity to produce motion varies according to the one or the other influence affecting it."[58] Fatigue represented the corporal analogue of the second law of thermodynamics, diminishing the intensity of the energy converted in the working body, tending toward decline and eventually, inertia. A dynamic force, autonomous of the will, fatigue was subject only to the physiological laws of function. The science of fatigue could determine "the best conditions for production and indicate which person is most suited for the work accomplished, which tool is most appropriate for which organism, or which task; it makes possible a rational selection of persons for the diverse professions in conformity with their well-being and happiness. Each superfluous fatigue can be eliminated." The science of fatigue was "a *hygiene of efficiency.*"[59]

In the early 1870s, Hugo Kronecker, a German physiologist who had once worked with Marey in Paris, made tracings of the contractions of frogs' muscles over a period of time. Kronecker observed that in a fatigued muscle, the intensity of the contractions diminished with *regularity* until the organism was incapable of working.[60] He concluded that fatigue had its own dynamic laws—as heat is converted into work—revealing an economy of decline that was irreversible.

The author of the classical text of the new science of fatigue was Angelo Mosso of Turin, an Italian physiologist and educational reformer. His *La Fatica* (1891) instantly became a minor sensation, the result of a decade of laboratory investigation and a synthesis of German physiology and French technique. Mosso had studied with Jacob Moleschott and Carl Ludwig in the early 1870s, and in 1873 worked closely with Kronecker whom, he later noted, "first fired me with the desire of applying myself to the study of fatigue." He also absorbed the practical experimentalism of Marey (a debt he always acknowledged) with whom he had worked briefly in Paris in 1874. Combining a profound admiration for German physiology with a solid grasp of the inscription techniques pioneered by Ludwig, Helmholtz, and Marey, Mosso placed the study of fatigue within the canon of mechanical materialism.

Mosso considered the law of energy conservation to be "the greatest discovery of the last century." Its irrefutable principles were "the thread of *Ariadne* which guides us in our search of the unknown; by their means the most secret region of science becomes illuminated by a ray of light."[61] In 1884 Mosso invented the first efficient and accurate measure of fatigue, the *ergograph* (register of work). With this device he constructed "an instrument which would measure exactly the mechanical work of the muscles of man and the changes which, as the effect of fatigue, may be produced during the work of the muscles themselves." His ergograph consisted of two parts. First, Mosso built a supporting platform and metal glove that fixed the hand and forearm in such a way that the index and ring fingers could not be moved, while permitting the middle finger to remain free. Attached to the unhindered finger was a string and a weight that then set in motion a registering apparatus.[62] By raising the weight, the forearm muscles were isolated and quickly fatigued, establishing a tracing of their diminishing intensity.

With the aid of the ergograph Mosso produced hundreds of graphic representations of fatigue, or "fatigue curves," which plotted the rate of fatigue in different individuals and with different weights. Mosso's first studies, conducted on his own laboratory assistants, contrasted sharply with the uniform results of Kronecker's research in the muscles of frogs. Mosso's tracings showed remarkable variety on the "fatigue curves" of different subjects, including those of relatively equal strength, age, sex, or occupation. While some individuals maintained muscle contractions of relatively equal height and intensity, dropping off markedly once fatigue set in, others demonstrated a more gradual decline of muscle force under exactly the same conditions. These early studies seemed to indicate that "the way of living, the night's rest, the emotions, mental fatigue, exert an obvious influence upon the curve of fatigue."

But fatigue exhibited unpredictable and frequently perplexing behaviors. For example, the feeling of physical tiredness was often totally unrelated to the actual onset and course of fatigue. Fatigue also differed from individual to individual, displaying an astonishing variety of patterns. Muscle fatigue and brain, or mental, fatigue were sometimes mutually constitutive, but just as often they were mutually exclusive: "There are some people, robust so far as the development and energy of their muscles are concerned who are incapable of any intellectual work."[63] Mosso recounted his witnessing of physically fit soldiers being forced to take written examinations to prove they were literate: "I have often seen great strong men perspire until drops of sweat fell upon the paper. At Lecco I saw one faint during the examination, then, feeling better, demanded another trial; but on the threshold . . . he turned pale and fell into a fresh faint."[64]

These apparent differences only strengthened Mosso's unshakeable determination to demonstrate that fatigue's dynamic laws remained constant. His great purpose was to eliminate all subjective aspects from the study of fatigue, to find its discrete and therefore unchanging laws of motion. To resolve these quandaries, Mosso subjected all types of fatigue to intense observation and quantification. Paradoxically, the result of these studies, he argued, was to show that the "intimate and most characteristic feature of our individuality—the manner in which we fatigue" was also subordinate to the laws of nature. "If, every day at the same hours we were to make a series of contractions with the same weight, and in the same rhythm, we obtain tracings which all had the same outline, and thus we should convince ourselves of the constancy of individual fatigue."[65]

Experimenting on his assistants with electrodes attached to the arms, Mosso compared the involuntary spasms of their muscles produced by mild shocks to voluntary movements with strikingly similar results. His conclusion was that each person fatigues differently, but that each individual's fatigue curve displays the same regularity, and the same pattern, regardless of the causes of fatigue, and independent of the kind of work performed. Even intellectual fatigue, Mosso asserted, displayed the same regularities as physical fatigue. Those who fatigue gradually in physical labor, fatigue gradually in mental work; those who fatigue rapidly in mental work, fall prey to its effects more quickly in physical labor. Ergographic tracings during arduous intellectual exertion showed that mental fatigue visibly diminished the efficacy of muscular contraction.[66]

Moreover, Mosso contended that fatigue was a poison. In a famous experiment he injected the blood of a fatigued dog into the body of a rested one to show how the toxins immediately produced the characteristics of fatigue. Immediately after receiving the fateful injection of tired blood, the rested dog became exhausted. With this experiment Mosso claimed two decisive discoveries: First, that fatigue was an objective phenemenon with laws directly analogous to the laws of energy. Second, fatigue demonstrated a consistent "diminution of muscular force," which could be graphically represented and measured.[67] Mosso's "law of exhaustion" holds that the course of fatigue is always constant and independent of the work done. Fatigue obeys its own dictates—a fatigued body refuses to work until sufficient rest and nutrition replenishes its supply of energy. Mosso thus posited his *primary law of fatigue and of sensation*—namely, "that intensity is not at all proportional to the intensity of the external cause which produces them." Simply stated, the exhaustion of our bodies does not increase in a direct ratio to the work we do. Fatigue often begins to show its effects before the *potential* for work has ended. In this sense fatigue exhibits a protective dimension. It "saves us from the injury which lesser sensibility would involve from the organism."[68]

Mosso's breakthrough resonated far beyond his Turin laboratory. He believed that if fatigue could be carefully observed and studied under controlled conditions, then "the conservation of the internal energy of the muscles" could be enhanced.[69] The discovery of the laws of fatigue would lead directly, he hoped, to its more efficient control, if not to its ultimate conquest. The science of fatigue thus went beyond "muscular thermodynamics" to the direct investigation of the sources of resistance to work. The psychologist, mathematician, and aesthetic theorist Charles Henry called the 1894 French translation of Mosso's work "a new chapter in animal mechanics." The discovery of the "curves of exhaustion of different animated motors" permitted the calculation of fatigue in various activities: the movements of the body in the march, on the track, and so forth.[70] Marey's method of graphically representing motion—the decomposition of motion into micrological units over time—was applied, not to the body as a whole, nor even to the limbs and muscles, but to the phenomenon of fatigue. Henry's prescient remarks foretold how laboratory research in the "excellence of the animal motor," and the discovery of its law of motion, would soon be applied to actual situations.

THE SCIENCE OF ERGOGRAPHY

Within a year of the publication of Mosso's work in French, Fremont's sensational photographs of the forgers appeared in *Le Monde Moderne*.

By 1900 laboratories devoted to the experimental science of the "human motor" were established in almost all continental countries. Ergographic fatigue studies were taken of "the maximum utility of work under a variety of conditions." Mosso's Turin laboratory produced extensive research on the conditions governing the optimum of work intervals necessary for reducing physical and mental fatigue.[71] His original ergograph was improved and extended to ever-more complex forms of work, while overcoming the "artificial limits created by the fixation of the digits" by measuring the upper arms, legs, and torso.[72] Some of Mosso's students compared involuntary and voluntary muscle fatigue; some plotted the effect of fatigue on the neuromuscular system; others charted the relationship between fatigue curves and different types of intellectual work.[73]

Marey and Mosso's pioneering work soon spawned a second generation of field and laboratory researches, first in France and Belgium, and later in Germany. In Belgium, the philanthropist and social philosopher Ernest Solvay established an institute of energetic physiology bearing his name, Laboratoire d'Énergétique Solvay, which after 1902 was renamed the Institut de Physiologie de Bruxelles and was directed by the capable Polish-born and Paris-trained physiologist, Josefa Ioteyko.[74] Solvay was motivated by his conviction that "all the factors which directly or indirectly intervene in the organic phenomena relevant to the human being living in society have a physio-energetic value capable of determination by means of the same unit"—a belief that became a search to improve "social efficiency."[75]

In addition to the physiological institute, Solvay founded several other institutes, including an institute of sociology to realize his plan for developing a science of "physio- and psycho-sociological energetics." The physiological institute attempted to derive precise mathematical correlations of age, sex, and occupation with energy consumption; to compare the rates of plant and animal growth; and above all, to provide a mathematical basis for subsequent research in fatigue.[76] Solvay also combined physiology with the psychological studies of intelligence and mental fatigue pioneered by Alfred Binet, the French psychologist (and inventor of the intelligence test), to create a new social science based on translating the laws of energy conservation to social forces. Charles Henry's "measurement of intellectual and energetic nature according to the laws of statistical probability" in 1906 was a first effort to calculate social phenomena mathematically in accordance with energetics.[77]

After 1900 the journals of the German, French, and Belgian scientific academies overflowed with literature on every aspect of fatigue. Physiologists turned "their attention to analyzing fatigue, the manner in which it affects the output of nervous and muscular energy, the functions of circulation and respiration, and the production of organic poisons, or auto-intoxication."[78] Ergonomic studies investigated the influence of weight, rhythm, heat, cold, anemia, blood chemistry, and other factors on the fatigued body. Fatigue experts—for example, Ioteyko and Henry in Belgium, Jules Amar and Armand Imbert in France, Kronecker in Germany, and Mosso in Italy—engaged in the search for physiological laws of the body's economy and for ways of reducing the effects of fatigue. By defining the normal limits of this fatigue, they hoped, "over-exertion may be avoided with certainty."[79] Fatigue became the most important "criteria of expenditure of energy."[80]

By the first decade of the century, Ioteyko's innumerable studies of fatigue, combined with those of Mosso, Kronecker, Henry, Binet, Fernand Lagrange, and others, produced what might be called an "objective phenomenology" of fatigue and its representations. In 1904 Ioteyko provided the new science of fatigue with its first official name, *ergographie,* or "ergography" after Mosso's instrument. Ioteyko defined fatigue as "the diminution of effort as a function of time." In the same year she announced that "we have all of the elements necessary for the establishment of a mathematical law of effort and of fatigue," which she attempted to express mathematically in a series of papers devoted to the "fatigue quotient."[81]

According to Ioteyko, Mosso's great achievement was to plot the course of fatigue independent of the tiredness experienced. However, when Ioteyko tested her students at the University of Brussels and at the Solvay Institute between 1899 and 1900, she discovered that fatigue was not constant but highly idiosyncratic.[82] Despite individual differences in "fatigability," Ioteyko found that each fatigue curve revealed a distinct, though paradoxical pattern: muscles that were accustomed to work fatigued most rapidly at first but their capacity to work was rapidly restored as work continued. From this fact she concluded that "fatigued muscles worked more economically" and that "training" augments efficiency at the moment of fatigue, constituting "a self-regulating mechanism of fatigue."[83]

The notion that fatigue was largely self-regulating and demonstrated an economy of efficiency—not unlike the law of the least effort—also led to investigations of the varying *forms* of fatigue arising from different tasks or activities. If Mosso "was content with tracing individual fatigue characteristics," subsequent researchers found that the curves of one person did change markedly under altered conditions of work. For example, Ioteyko observed that fatigue was cumulative, "though not directly in proportion to the work done." An increased amount of work could be compensated for by slower pace or a proportional amount of rest so that "the accumulation of fatigue varied with the intervals of rest between the curves."[84] Fatigue thus demonstrated its own dynamic laws of motion, representing a local, self-regulating and economic character that could be traced ergographically. Above all, the laws of fatigue were distinct from the subjective feelings that accompanied different kinds of work, as "the sentiment of fatigue" was distinguished from its objective course, often lagging behind it.[85] Fatigue accumulated at a rate initially imperceptible to the individual, since the tiredness becomes conscious only at a later stage: "This central and conscious mechanism, intervenes late, appearing only when the peripheral mechanism is not sufficiently attended to. [Thus the feeling of fatigue] is an expression of a particular state of the muscles becoming conscious at a particular moment."[86]

The chemical basis of fatigue remained particularly perplexing. Was fatigue the result of chemical changes and the product of the nervous system, or were chemical changes in the muscles produced by fatigue? The problem had vexed Mosso, who expressed his belief in the former. Also puzzling was the physiochemical explanation for the tenacity of fatigue. Once the muscle became tired, only rest permitted it to recover its capacity. Chauveau believed that this phenomenon was a function of lessened muscular elasticity, whereas Zaccaria Treves (a student of Mosso) attributed these apparently intractable effects to an impairment of the spinal cord. Ioteyko, on the other hand, pointed to the weakened power in the synapses of the nerves.[87] By 1900 most researchers had concluded that the nervous system provoked the debilitating chemical reactions as fatigue intensified. But they also found that fatigue did not originate "centrally" in the nervous system nor in the chemistry of the blood, but "peripherally" in the muscles directly undergoing fatigue. In short, the course of fatigue might be modified if the conditions of fatigue could be altered.[88] Fatigue, noted the French hygienist Jules Amar, was "fundamentally an *intoxication;* if the brain and the muscles function in a disorderly fashion as a result of excessive effort or too great a rate of exertion, the blood is no longer able to cope with its task of purification. The waste products of this intense cellular activity accumulate; the blood loaded with toxic produces fatigue in any animal into whose veins it is injected."[89] The intoxication of fatigue could be eliminated only through the prophylactic of reduced work.

The early laboratory studies of fatigue were limited to tracings of specific, isolated muscles subjected to artificially induced stress. The impact of fatigue on the working body as a whole and the effects of different kinds of mechanical work under real conditions were still inaccessible to the fatigue curves. In 1903 a Parisian doctor, A.-M. Bloch, undertook the first general survey *(Enquête)* of fatigue. He asked workers in various strenuous occupations two questions: "When you work a lot, where do you feel fatigue; does the fatigue always occur in the same place?" At first the answers appeared paradoxical: The baker kneading dough all night complained of leg pains; the forger did not, as one might expect, complain of fatigue in his arms or shoulders, but instead, of difficulties in his back or kidneys; and the shoemaker, bending over his last, complained of abdomen pains. Bloch concluded that there was a rational explanation for these phenomena. Those groups of muscles "immobilized" by contractions were subject to extreme fatigue, while those that remained more active were spared "even in excessive work." Bloch immediately recognized the practical implications for education or military training: "One must exercise the auxiliary muscle groups . . . as often as possible."[90]

Fatigue could now be classified according to its "degree of intensity," its "pathological effect," or "conditions of origin." The intensity of fatigue ranged widely from the minutely observable "diminution of effort" perceptible only in the ergographic fatigue curves; to the "sentiment of fatigue"; to exhaustion; to the "fièvre du surmenage," which incapacitates the exhausted body; finally to "auto-intoxication," or death by excess fatigue toxins—as in the example of the famous Athenian (or contemporary) marathon runner who crosses the finish line and expires.[91]

Each degree of fatigue corresponded to a specific set of physiological and mental symptoms. Philippe Tissié noted the following distinctions: "lassitude," or weariness, which disappears after rest; "l'épuisement," or enervation, which decreases the capacity for recuperation and provokes symptoms such as rapid heartbeat or arterial tension; "le surmenage," or exhaustion, which impairs the appetite, suppresses sleep, causes hypertension; and finally, "le forcage," or extreme exhaustion, which constitutes "a serious illness" resulting in pathological psychic reactions, such as the "dissociation of the self."[92] Fernand Lagrange, a French physician and expert on clinical aspects of fatigue, claimed to have discovered an entirely new fatigue, "l'essouflément," or "breathlessness," which he described as "a malaise produced by the body as a result of a violent exercise or intense muscular effort," such as a last-minute dash for a train. According to Lagrange, fatigue was aggravated by "an excess of speed, an excess of intensity and excess of time."[93] Even dietary excess could produce fatigue, as "an excess of alimentary excitation fatigues the nerves of the digestive system."[94] Different types of fatigue were also classified according to their origins. "Active fatigue" was an effect of the muscles and of voluntary behavior, while "passive fatigue" resulted from modern stresses—for example, the effect of "railroad travel," of modern forms of communication. Finally there were sedentary, or "intellectual fatigue," "emotional fatigue," and the fatigue from extreme physical pain *(fatigue dolorifique)*.[95]

For the majority of these European researchers, it was axiomatic that the experience of fatigue was intimately connected to the demands of industrial society. Fatigue was a pathology of productive, routinized labor, of the intensified pace of life in the modern factory, and in society. Yet, Ioteyko proposed that the sensation of fatigue also be considered "a defense which protects us against the dangers of a work pursued to the extreme."[96] Biologically, fatigue "could be considered to be a generalized defense of the organism against excitations which are too intense or too prolonged." In this case, fatigue is an "immediate defense" like the "peripheral paralysis" of the overtired muscle. "The entire mechanism of fatigue is founded on the protection of the nervous centres from noxious excitations." Finally, fatigue is also a "consecutive defense" that prepares and accustoms the body to increased work and "renders the organism resistant to fatigue." She called this a "prophylactic fatigue, kinétophylactique, or esthophylactique, which safeguards movement" from overwork. Included in her definition of fatigue as protection is also *ennui*, "the sentiment with which we defend against monotonous work."[97]

Clearly, for Ioteyko fatigue had a normative dimension, which, like pain, protects us against the sufferings inflicted by work and by society. Fatigue thus revealed two faces of modernity. On the one side it was a defense, marking the limits of the body's ability to convert energy into work, a limit beyond which the human motor could not function. On the other, fatigue was the body's method of economizing its energy, acting as a regulator of the body's expenditure of energy. In order that science have a true knowledge of fatigue and its costs to modernity, it must determine, Ioteyko claimed, the individual constitution, the conditions of work, and the most economic way of accomplishing it. Not only useless or wasteful fatigue had to be eliminated, but the productive side of fatigue, the individual's fatigue quotient, also had to be directly acknowledged in the modern workplace. "It is not impossible," she believed, "within the impassable limits of the law of the conservation of energy, to communicate an activity to the human motor which will favor the liberation of one form of energy rather than another."[98] This knowledge of fatigue could also serve the interests of productivity, by regulating the expenditure of energy of the human motor, ensure its most economical working, and "guide the animal machine to adapt to the best conditions for work."[99]

A FATIGUE VACCINE?

The search for the laws of fatigue were accompanied by an equally intense search to discover its chemical properties. If the ergonomic approach might result in the optimal deployment of the body's energy economy, the chemical approach promised an even greater panacea: the discovery of a vaccine against fatigue. In 1904 a German physiologist at the University of Erlangen, Wilhelm Weichardt, called attention to his remarkable experiments with the blood chemistry of fatigued rats. By subjecting the rodents to strenuous physical exercise, he produced the symptoms of pathological fatigue—lowering of the body temperature and shortness of breath. By accelerating this excessive exercise, he artificially induced in the rats a kind of "narcosis," during which breathing gradually slowed to "a complete standstill."[100] However, when he permitted the rats a brief respite from this tortuous labor, they restored themselves "remarkably quickly." Weichardt could not resist the analogy to a brief thundershower, in which fatigue approached "like a dark cloud which quickly approaches, sinking ever more closely to the ground like a dead-tired mass." Weichardt's experiments showed that "in this pure, uncomplicated fatigue, substances grow in living organisms" and that if the fatigue does not cease, these substances rapidly accumulate, causing "stupor" and then death.[101]

Weichardt believed that the chief cause of death by exhaustion was the gradual "strangulation" of the animal's life-giving properties by a specific fatigue toxin. The increasing spasms evinced in fatigued muscles, as demonstrated by Marey and Mosso, and by Claude Bernard's investigations of the structure of fatigued muscles, further indicated to Weichardt that "the fatigue of the most different organs could be traced back largely to the accumulation of deleteriously acting metabolic products," which causes exhaustion if not counteracted by oxygen. Fatigue, in short, does not simply produce exhaustion but "exhaustion causes fatigue to an equal degree."[102]

Weichardt contended that these "poisonous" fatigue substances that quickly exhausted the body were the key to conquering fatigue. He acknowledged that other fatigue experts, like Ioteyko, were extremely

skeptical of discovering the chemical basis of the fatigue toxins.[103] However, he noted that developments in both immunology and blood-
serum analysis had made possible a breakthrough, not only in analyzing
the fatigue toxins, but in producing a chemical antidote. As Mosso had
already shown, a dog injected with the blood of an exhausted dog
becomes tired. Conversely, the production of antitoxin capable of resisting fatigue might be equally successful in reversing its effects. Weichardt concentrated his efforts on producing *in vitro* a fatigue substance "independent of the body of the animal"—in other words, a
fatigue vaccine.

Weichardt was unflagging in his efforts to eliminate fatigue by
first distilling its pure chemical essence and, then, by creating an immunizing substance. Not completely heartless, he occasionally regret-
ted his ruthlessness in pursuing the fatigue toxins in his experimental
rodents. It was "personally uncomfortable for me to conduct these
studies," he wrote, "especially since the animal must not be given
the slightest opportunity to recuperate," lest the fatigue toxins be sul-
lied. But it was the only way. He named the fatigue toxins that he
produced (from albumin) in his chemical laboratory "kenotoxins."
When injected with this material, the experimental rats displayed extreme fatigue but as Weichardt predicted, soon developed an "active
kenotoxin immunity" and resistance to fatigue.[104] He then experimentally produced a similar chemical that might be resistant to fatigue—an immunizing substance he called "antikenotoxin." Antikenotoxin was then injected into the bloodstream of the rodents,
suppressing the fatigue toxins and allowing them to outperform their
fatigued compatriots. Weichardt was euphoric: the possibilities for
human use were legion.

Weichardt's moment arrived when he prepared to inject human
subjects with his new substance. In 1903 he tested his antikenotoxin on
several groups, employing both the ergograph and the Griesbach *aesthesiometer,* a more sophisticated device that measured fatigue on the
surface of the skin. He also used a new method that Weichardt devised
for his own purposes. It consisted of a 2.5 kilogram dumbbell that the
subject held horizontally and moved from the front to the side with
outstretched arms to the accompaniment of a metronome. The subject
also lifted his right and, then, left foot to knee-level. This simple exercise
became "gradually more difficult, and suddenly the arms sank as a
consequence of the most extreme fatigue."[105] The number of seconds
that elapsed before this inevitable outcome was a more reliable indicator of the onset of fatigue than the ergograph, which tested only the

forearm muscles. Even small amounts of antikenotoxin subcutaneously injected in the subjects resulted in an increased capacity for the exercise. Exhaustion occurred significantly later than in those subjects who received a placebo.[106]

On 30 June 1909, armed with sprayers containing amounts of 1 percent antikenotoxin solution, Weichardt and an assistant appeared in a Berlin secondary school, where they sprayed a classroom with the chemical. The unsuspecting students were told that the sprayers contained materials to improve air quality. The room was sprayed in the morning; in the afternoon a series of mathematical exercises were provided to test fatigue by the Kraepelin error method (see chapter 6). The result was extraordinary. Although the students had already completed five hours of instruction, they performed their prescribed calculations with "considerable improvement." The speed of calculation increased by 50 percent; the number of errors were reduced; and some students, usually tired and sleepy afterwards, were now fresher than in the morning.[107]

German scientists immediately began experimenting with the vaccine by converting it to a gas and filling laboratories and classrooms with it. The enthusiasm that accompanied the discovery of the fatigue vaccine did not fully abate until 1914, when several influential physiologists concluded that Weichardt's claims had been exaggerated. First, the substance only partially combatted fatigue but in no way eliminated it, since it did not affect the "negative side" of fatigue—the exhaustion of the body's own materials in the effort.[108] Second, other researchers demonstrated that kenotoxin was not produced, as Weichardt believed, by fatigued muscles alone, but was frequently present in equal measure in relaxed muscles.[109] Finally, on the eve of the Great War, the heightened interest of German and Austrian military physicians in the benefits of a fatigue vaccine delivered a fatal blow to Weichardt's hopes to deliver mankind from fatigue. Experiments conducted by the Austro-Hungarian army concluded that the performance of those men injected or sprayed with the antikenotoxin did not much differ from those who received an ineffectual chemical.[110]

On this front, the battle against fatigue proved chimerical. But exposing the vaccine's illusory powers led to a more important discovery. In contrast to the antikenotoxin subjects, a control group injected with concentrated caffeine exhibited marked spurts in energy and productivity. This further intensified the search for other "nerve whips," or stimulants, which, like tea, coffee, or cocaine seemed to erase the signs of fatigue. In the long run, however, these too were inefficient, since they only either masked the real symptoms of fatigue until complete physical exhaustion set in, or were absorbed by the muscle toxins until they required ever-more dangerous amounts to provide more work.[111] With the failure of Weichardt's vaccine, it became evident that the battle against fatigue would not be won in the laboratory and that the war would have to be waged on other fronts.

CHAPTER SIX

Mental Fatigue, Neurasthenia, and Civilization

MENTAL FATIGUE

JUDGING from the extraordinary proliferation of studies of fatigue in the 1880s and 1890s, we can conclude that fatigue was consuming the energies of society. These studies confirmed the anxiety of many nineteenth-century liberals who believed that society could not withstand the demands of modernity, and who prophesied a general social decline not unlike the entropy revealed by the second law of thermodynamics. Social decline, as the psychologist and pedagogical expert Alfred Binet argued in his *Les Idées modernes sur les enfants,* was a direct consequence of "physical decline."[1]

For nineteenth-century physicians, physiologists, and reformers, physical fatigue did not exist in isolation from the will, from morality, and from the ensemble of social forces that comprised the nation. Philippe Tissié, the most prolific advocate of a national policy of hygienic resistance to fatigue in *fin-de-siècle* France, warned that "a nation, like a fatigued individual is always prepared to obey any master which imposes itself on it brutally and with force. Propelled by the sentiment of weakness, it requires a protector; incapable of sustained attention and will, it abdicates to suggestibility and snobism. No longer master of itself, it lacks freedom. Reason may triumph over passion, but in fatigue, it is passion that triumphs."[2] Exhausted by fatigue, the nation, like the individual, is abandoned to the vicissitudes of the will, the emotions, and the enemies of the productive order. Fatigue did not merely express the entropy that accompanied the conservation of energy for social use, but seemed to threaten modernity itself.

Nowhere was that order more threatened than in the physical and mental state of the nation's youth. By the mid-1880s in response to a widely perceived crisis of educational exhaustion in both France and Germany, public debate about the current state of mental fatigue had intensified. During the summer of 1887 the French Academy of Medicine was absorbed by the problem of mental exhaustion, especially among high school students. The crisis was set off by a "cry of alarm" raised by the physiologist Gustave Lagneau in April 1886 in a highly publicized article entitled "intellectual exhaustion and the sedentary life."[3] Surveying several decades of French educational reform, Lagneau challenged what he perceived as a chronic crisis of overwork and exhaustion. In both secular and in clerical institutions he claimed that "the youth of our lycées, seminaries, and colleges is surcharged with excessive demands to the great detriment of their body and spirit; the cultivation of the physical forces is nil or insufficient in all of our houses of education; the overdoing of the required examinations is as fatal for good education as it is for the good hygiene of the adolescent."[4] Lagneau warned officials entrusted with public instruction or rectors of the universities to act quickly against the "exaggerated expansion of knowledge" required of young people. Lagneau was certainly not alone in this view: "Numerous physicians are equally convinced of the growing morbidity of our students subject to this intellectual exhaustion . . . or more precisely to this sedentary life (sédentarité)."[5] The popular press, too, registered alarm over the state of French students, which one writer dubbed "L'éducation homicide."[6]

The Academy devoted four months of debate in 1887 to the physical and mental condition of students, a protracted discussion that included some of the leading physicians and pedagogical experts of the day. Although some took issue with Lagneau's prescription that in addition to gymnastics, military exercises be adopted based on the German model, his diagnosis of chronic exhaustion among secondary school students was largely confirmed.[7] Though the 1887 debate directed French public attention to the problem of fatigue, throughout the 1880s a series of government commissions issued reports documenting the spread of typhoid fever and tuberculosis, as well as "myopia, dental lesions, digestive difficulties, spinal deformations, weakness of constitution, anemia, phtisie," among students. In part, these revelations were a consequence of a system of public education that for the first time placed French schoolchildren under the scrutiny of public hygiene officials.[8] Lagneau and numerous other physicians argued that these afflictions were a result of the tyranny of overwork "imposed on our young people," and they pressed the Academy to take a strong stand against the excessive overwork of the secondary schools.[9] Even if "cerebral fatigue" did not produce "a true pathological state," Lagneau remarked, it created a state of "nervous overexcitation, a mental hyperactivity, which notably modified their character."[10]

Most experts took it for granted that French students were exhausted. Even the technical term for "exhaustion" *(surmenage)* to designate human overfatigue entered the French vocabulary at this point, whereas, as one commentator pointed out, it had largely been a veterinary term. Prior to the school-fatigue crisis, *surmenage* generally referred to the spoilage of meat that occurred when animals arrived at the point of slaughter exhausted.[11]

Highly competitive entrance examinations for the Lycées, instruction in Greek and Latin, extensive homework, and recent additions to the curriculum, like natural history—all were held culpable for the epidemic of intellectual exhaustion. One sympathetic critic of Lagneau's complaint described French students subjected to this severe overwork as debilitated by "a more or less permanent weakness of the intelligence, a loss of all initiative, of all force of will, of all moral energy, and withdrawn." As a result, he said, they "remained remarkably slow, heavy and stupid for the rest of their existence."[12] An entirely new diagnostic term (or "species of fatigue nosology") emerged from these debates—"la céphalagie scolaire" (scholar's brain). Indeed, the Academy concluded that nervous difficulties, hyperasthesia, neurasthenia, intellectual slowness, and "the profound alteration of cerebral faculties are often the result of an excessive or premature intellectual challenge."[13] Michel Peter, a renowned expert on the educational system during the Third Republic, referred to its "victimes scolaires" and demanded that the government enact the educational equivalent of a factory act to protect schoolchildren "condemned to forced labor" against exhaustion.[14]

In response to this crisis, the Academy issued a strong statement condemning the educational system for ignoring the problem and calling for reform in "conformity with the laws of hygiene and the demands created by the physical development of children and adolescents."[15] Yet not all French medical experts agreed with the anecdotal character of the Academy's proceedings. Alfred Binet publicly condemned the impressionistic evidence presented to the Academy, which he characterized as "a literary duel rather than a scientific discussion."[16] Binet called for a national commission to measure the fatigue of students and to replace theoretical speculation with the systematic, experimental study of fatigue in the schools.[17]

The education crisis of the 1880s opened the way to the experimental study of students and education in the decade to follow. The debate on "school fatigue" was not limited to France alone but quickly extended to the Scandinavian countries and Germany, where the problem of the "overburdening" of schoolchildren was a constant theme in the pedagogical literature of the 1890s.[18]

Mosso's *La Fatica*, which included a chapter demonstrating that excessive mental work resulted in decreasing muscular efficiency, spawned a generation of fatigue researchers concerned with its mental nosology and symptomology. German experiments using the ergograph on school-age children tried to determine the extent of muscle fatigue engendered by arduous mental activity, for example, memorizing Latin or poetry.[19] In 1896 a German educator, Friedrich Kemsies, tested the motor power of schoolchildren with Mosso's ergograph at different times during the school day, reporting a direct correlation between mental fatigue and the inability to perform physical work.[20] Ergographic studies also showed that "gymnastics caused the fatigue created by intellectual activity to disappear," acting as a "special kind of rest for the student exhausted by intellectual work."[21] By 1900 such studies of intellectual exhaustion became, in the words of the Swiss physician Théodore Vannod, "the order of the day."[22]

The most prominent of these early pioneers of mental fatigue was Hermann Griesbach, a professor at Mulhouse in Alsace, who hypothesized that mental fatigue could be measured by changes in the tactile sensitivity of the skin. Griesbach argued that "as long as we have no physiological standard, which could measure fatigue, the extent, or the acute or chronic character of fatigue remained completely in the dark."[23] Basing his investigations on the early nineteenth-century studies of tactile perception by the psychophysical psychologist E. H. Weber, Griesbach tested schoolchildren with a device he developed— the *aesthesiometer*—to measure the perceived distance between two-minute sensations on the surface of the skin ("spatial limen"). Griesbach revealed that as mental work increased, so did the inability to distinguish the distance between the two sensations.[24]

Griesbach was also interested in the comparative social dimensions of fatigue. Experimenting on pupils at a Mulhouse *Gymnasium* and *Oberrealschule* and on young employees in local offices and in artisanal and industrial workshops, he documented not only the "enormous" intellectual fatigue of those in the first group, but also showed that the diminished capacity for tactile discrimination (his measure of fatigue) was less affected by mechanical work than by intellectual work.[25] In other words, intellectual work was more fatiguing than heavy industrial or tedious office work.

In the 1890s Griesbach's method was widely imitated and its adherents, like Vannod, used his aesthesiometer (or a variation of it called the *algesiometer)* extensively to trace the "pathological state of morbidity occasioned by an excess of cerebral work" among schoolchildren.[26] Three months of arduous examinations produced not only chronic fatigue but various symptoms: "The appetite disappears, energy diminishes, headache persists, nights are agitated and troubled by insomnia, in short a morbid state occurs which permits us to conclude that there is the presence of a true intellectual fatigue."[27]

Both Griesbach's aesthesiometer and Mosso's ergograph enjoyed a vogue as measures of mental fatigue in the late 1890s.[28] But these physiological tests were able to register mental fatigue only indirectly, either through the failing power of the mind to register changes in the skin or through muscle fatigue *per se.* If fatigue were "inscribed," so to speak, in the perceptual apparatus of the skin, it was also manifested in several other ways. One experimenter tried to evaluate minute changes in time perception as an indicator of fatigue.[29]

Griesbach's most persistent critic was the Heidelberg psychologist, Emil Kraepelin, who in 1900 launched a vitriolic polemic against these methods. For Kraepelin, it "was completely impossible to get any kind of reliable conclusions about the level of fatigue from either the spatial limen or . . . as Kemsies had attempted, from the individual curves of the ergograph."[30] Yet Kraepelin did not entirely share the skepticism of his mentor, the psychologist Wilhelm Wundt, that the psychophysical school could not come to terms with the experiential side of human perception. Rather, Kraepelin believed that it was possible to devise more sophisticated and more specifically psychological techniques to measure mental fatigue with much greater precision, to evaluate "the labor power of the individual in very simple mental activities."[31]

Kraepelin agreed with the basic assumption of the psychophysical approach. The search for an objective standard of measurement of mental fatigue had to take precedence over subjective feelings of fatigue, which were unreliable from any point of view. Kraepelin introduced the distinction between fatigue *(Ermüdung)* and tiredness *(Müdigkeit).* Inasmuch as feelings of weariness might arise without evident signs of fatigue, fatigue might also be present but imperceptible to the fatigued person. The objective rhythms of bodily "performance"—as Mosso and Marey also claimed—were distinct from the individual's subjective perceptions.

But even if the subjective manifestations of fatigue were unreliable, they certainly signaled its onset, and as Kraepelin acknowledged, played a crucial role in the inhibition of performance. Nevertheless, the course of fatigue remained in each case uniquely individual: "There are people who work slowly and yet fatigue rapidly, and there are those who can maintain a high level of work performance for a long time, without any considerable decrease."[32] Fatigue represented "the fundamental characteristic of the individual personality, which can however be influenced within specific limits, but which, on the whole, determines the performative capacity of the individual."[33]

Kraepelin's Heidelberg laboratory concentrated on the impact of fatigue on work performance (Arbeitsleistung). In hundreds of experiments his students studied the cognitive skills, tasks, and overall performance required of schoolchildren, measuring individually the onset of fatigue and the fatigability of each child engaged in specific tasks. These "work curves" became the plots of each individual's fatigue-biography. The operative word in Kraepelin's research was error, which for him constituted the most evident sign of fatigue. Whether in adding columns of figures or in memorizing nonsense syllables, the ultimate speed and accuracy of performance became the measure of fatigue. Significantly, from Kraepelin's standpoint, there was little distinction between "performance" (Leistung) in the case of mental fatigue and performance in physical labor—each could be measured in terms of output. With Kraepelin's studies of fatigue, psychology assumed the status of an objective science of aptitudes, reactions, and behavior modeled on biology, but governed by, and inseparable from, the selection and evaluation of individuals for specific tasks. His approach was based on the implicit postulate that "the nature of man is to be a tool, that his vocation is to be set in his place and set to work."[34]

Indeed, the distinctions in fatigability revealed by his work curves resulted in Kraepelin's prescription for "dividing the students according to their capacity for work," a system of "tracking," which for Kraepelin paralleled the imperatives of a civilization predicated on production.[35] As he once mused, "A short while ago Erb correctly characterized the nervosity of our species as a kind of malady of development. It emerges from the fact that a certain part of contemporary humanity does not have a sufficient capacity for adaptability to tolerate the increase and expansion of our life work without injury." Kraepelin left no doubt that he favored a system in which the "unsuitable" would be left behind while the energies of the more capable could develop and be enriched "so that the path would be open to a new species more capable of performance."[36]

The scientific study of mental fatigue thus permitted the development of methods of distinguishing between the superior student capable of adequate "performance" and the inferior student, whose impaired performance was a consequence of the exhaustion prevalent among nineteenth-century youth. However, as Kraepelin soon recognized, the unanticipated virtue of these tests was that they produced a kind of "training," or "practice" *(Übung)*, which at least for a time, offset the effects of fatigue: "Fatigue creates a general decrease in work performance, even if this decrease can be compensated for a certain time by the increased training."[37]

Kraepelin proposed several solutions to the "overburdening question" in Germany during the 1890s. In addition to selecting the students according to their performance capacity (which, it should be noted, he distinguished from intelligence), he suggested that the schoolday include a series of rest pauses scientifically designed to facilitate productivity.[38] He also proposed that the lesson period be compartmentalized into shorter, more manageable segments. Above all, Kraepelin concluded that the schools could not absorb any further expansion of the curriculum. "Any further extension in breadth would necessarily come to ruin on the borders of the child's labor power."[39]

In France, too, the psychological study of school fatigue led to proposals for pedagogical reforms. In the introduction to his *La Fatigue intellectuelle,* Binet bemoaned that while the French Third Republic had eliminated the radical vices of "l'ancienne pédagogie," which he enumerated as "fashion," "preconceived ideas," "gratuitous affirmations," and the "confusion of rigorous proof with literary citation," it had not yet based its pedagogy on "observation and experience." France, he argued, compared unfavorably in this respect to Germany, America, Sweden, and Denmark, where the experimental study of the school was generally encouraged.[40] Binet criticized the Kraepelin method for its tendency to draw major conclusions from small samples, but affirmed its general diagnosis of intellectual exhaustion as the school's problem *par excellence.* He cautioned, however, that fatigue was not exhaustion and that "in order to be able to prove that school children are exhausted, it is necessary to demonstrate that fatigue at evening time does not dissipate, and next morning children resume their work as fatigued as they had been."[41] For Binet, the critical distinction between fatigue and exhaustion was between the normal and pathological, between the adequate "speed of reparation," which rest provided, and the lack of reparation in exhaustion. A precise distinction between fatigue and exhaustion, he concluded, could be achieved only through medical expertise. Such precision could determine the legitimate boundaries of fatigue, its economy of dissipation, and its pathologically dangerous state.[42] Despite these reservations the school debate clearly disclosed a salient aspect of mental fatigue: it was identified with the demands of modernity—an aspect even more apparent in the discourse on the most extreme form of mental fatigue, neurasthenia.

PATHOLOGICAL FATIGUE: NEURASTHENIA

In the pantheon of nineteenth-century fatigue disorders, one complaint stood above all others for its ubiquity and relentless attack on the core of psychic and physical energy. *Neurasthenia*, a term invented by the New York physician, George Miller Beard in the 1860s, covered "all the forms and types of nervous exhaustion coming from the brain and from the spinal cord." Beard attributed the causes of neurasthenia to "overpressure of the higher nerve centers," claiming that it was a pathology peculiar to the American continent and unique to the American lifestyle. He identified neurasthenia as "the Central Africa of medicine— an unexplored territory into which few men enter."[43] Actually, judging from the more than sixty different symptoms from tenderness of scalp and ticklishness to "anthrophobia" (fear of men) that Beard identified, we might conclude that neurasthenia was more like the Grand Central Station of medicine than Central Africa. For Beard, neurasthenia was clearly a disorder of a modern civilization that constantly required individuals to draw "on a limited store of nerve force." Consequently, from the standpoint of its excessive demands on "nerve force," Beard judged civilization to be a colossal mistake.[44]

Though his major work, *A Practical Treatise on Nervous Exhaustion* (1869), was not translated into French until 1895, during the 1880s "Beard's malady" became the diagnosis of the day.[45] For European physicians, Beard's works had the "merit of disentangling neurasthenia from the chaos of complaints generally known as *nervous*. As a classification, neurasthenia quickly superceded all the elusive disorders of the will hitherto grouped under such French rubrics as "névroisme," "irritation spinale," or "névropathie cérebrocardiaque."[46] But in Europe the literature of neurasthenia became more nuanced than in Beard's writings, and European physicians were less inclined to replicate his clinical practice, which often relied on the application of electrical charges to the body or head by a galvanic battery.

Though European physicians disagreed with Beard's contention that neurasthenia was uniquely American, they were attracted to his view that nervous suffering could be attributed to the excessive collisions and shocks of modernity. The jarring contrast between the old world and the new, they claimed, amplified the effects of modernity. Neurasthenia was not "only a *maladie du siècle,*" wrote one physician, "it is even more a malady of civilization augmented in intensity as a result of the progress of the past century; it advances to the extent that mankind becomes more sedentary, more intellectually active, it is a manifestation of intellectual exhaustion, which is the principal cause of the onset of this neurosis."[47]

Neurasthenics were identifiable by their impoverished energy and by the excessive intrusion of modern urban society on their physical and mental organization. The great French neurologist and psychiatrist Jean-Martin Charcot compared his neurasthenic patients, and those suffering from "railway spine" and other travel-related disorders, with the more extreme hysteria.[48] According to the German sociologist Georg Simmel, "The metropolis exacts from man as a discriminating creature a different amount of consciousness than rural life."[49] Similarly, Émile Durkheim's classic study, *Suicide* (1897) provided a vivid description of neurasthenia: "Every impression is a source of discomfort for the neuropath, every movement an exertion; his nerves are disturbed at the least contact, being as it were unprotected; the performance of physiological functions which are usually most automatic is a source of generally painful sensations for him."[50] Like Des Esseintes in Huysmans' *A Rebours* (see chapter 1) Durkheim warned the neurasthenic "may live with a minimum of suffering when he can live in retirement and create a special environment, only partially accessible to the outer tumult." But if the neurasthenic is "forced to enter the melée and unable to shelter his tender sensitivity from outer shocks," pain is the result.[51] Pierre Janet, whose psychological theories were almost entirely based on fatigue, claimed that "neuropaths" betrayed their actual feelings by declaring that "they were born tired and that their disease has never been anything but fatigue."[52]

The identification of neurasthenia with modernity was also reflected in the symptomology of the illness itself. Neurasthenia is a cacophony of complaints that replicate "real" illnesses without any observable organic lesion. The neurasthenic's weakened state and unstable emotions are at the mercy of chimerical thoughts and images, representations that are themselves refractions of a will incapable of resisting the stimuli of the modern world. This arbitrary and unbridled subjectivity occasioned by a diminished state accounts for such diverse phenomena as somnambulism and religious ecstasy, as well as for the notorious unreliability of neurasthenic patients in portraying their own disorder. The treatment of neurasthenia thus required a modern approach: an interpretive strategy that carefully disentangled the illusions from their real basis, and at the same time, affirmed the material power of the will to resist them. Both the physician's authority and the physical substratum of neurasthenia were bulwarks against the potentially negative effects of fatigue on the future of civilization. The growth rate of fatigue and nervous disorders seemed to be directly proportional to the intensity of energy necessary to contend with modern society.

Neurasthenia was considered essentially a weakness of the will, or nervous system, that pathologically inhibited action. Charles Féré, a student of Charcot, who published numerous works on the subject in the 1880s and 1890s, claimed that neurasthenic exhaustion *(neurasthénie d'épuisement)* was chronic fatigue characterized by "excessive cerebral work, by intellectual, especially moral exhaustion, and by the persistent preoccupation with the struggle for existence."[53] Neurasthenics were "vexed by a sudden flush of enervation, by tumultuous joy, or by a sudden flash of rage, by a weakness of fatigued legs, by a languid vitality, by a sluggish digestion, and by diminishing activity."[54] Léon Bouveret, author of one of the first major texts on neurasthenia published in France, *La Neurasthénie (épuisement nerveux)* (1890), considered it a "malady of the nervous system without any organic lesion," characterized by a "permanent weakness of nerve force . . . which has many different manifestations, from a gradual and slow weakening to a general and profound prostration and ennervation."[55] In short, neurasthenia was not only a common complaint, its ubiquity extended to the symptomology itself, leaving no recess of the body untouched by its enervating effects.

Despite Beard's contention that neurasthenia was rooted in the cultural shocks of modernity, as Robert Nye has shown, in France it was, at least at the outset, considered to be a hereditary disorder.[56] The powerful influence of Augustin Morel's theory of "degeneration" combined with Lamarck's theory of the inheritability of acquired characteristics competed with the Beardian view that neurasthenia was caused by the shocks of civilization. Charcot too argued for a hereditary predisposition to neurasthenia but determined its initial cause to be a kind of "auto-suggestion," which ultimately led to "functional difficulties of a permanent nature."[57] No less than one-fourth of Charcot's famous patients at La Salpêtrière were classified as neurasthenics.[58]

Charles Féré, who for a time was Charcot's assistant at Salpêtrière, was a leading proponent of the hereditary link between the "nervopathic family" *(La famille névropathique)* and its propensity to neurasthenia.[59] Entire social groups, especially those regarded as inferior by physicians, were thought to be prone to the illness. Bouveret, for example, believed that "two races were more predisposed to Neurasthenia, the Jews, and the slave race."[60] Philippe Tissié introduced the proverbial "wandering Jew" (who frequently changed countries, climates, and doctors) into French medical vocabulary in 1887 as a hereditary variety of a trope that was "at once racial, epidemiological, and social."[61] Apart from these hereditary neurasthenics, most experts agreed, there were an even greater number whose suffering was a consequence of the pressures of civilization. A 1903 French survey comparing neurasthenic patients with normal but mentally fatigued subjects concluded that "the fatigue of neurasthenics did not differ at all in its modality and its effects from all cerebral fatigue, because its causes and its modes of production are essentially the same."[62] Bouveret also remarked on the frequency of neurasthenia in the intellectual professions, citing the growing clinical literature on the subject.[63] Intellectually taxed individuals were especially susceptible to neurasthenia because their mental state evoked an extreme disquiet that often "takes on the aspect of melancholia."[64] A German writer even claimed that the pessimistic worldview so often found among intellectuals was a direct result of their unfortunate "reflection addiction," their chronic tendency to displace psychic energy in abstract "theoreticizing."[65]

The similarities between mental fatigue and neurasthenia also connected the syndrome to other disorders of the will. What struck so many physicians when confronted with neurasthenic complaints was that the loss or weakness of energy seemed to be at the root of all other fatigue disorders. "Symptoms: exhaustion in the sphere of affective faculties, sorrow, inquietude, depressed passions are the most powerful causes of nervous exhaustion, and most often act like intellectual exhaustion."[66] Charcot, too, remarked on the parallel between neurasthenia and "what we call intellectual exhaustion among those pupils, anticipating a great effort of the will."[67] And the young Sigmund Freud, who was influenced by the great *savant* whom he encountered at Salpêtrière during his Paris sojourn, contrasted neurasthenia, which manifested as "weakness of will," with the *"perversion* of will" found in hysterical patients.[68]

The most important textbook on neurasthenia in *fin-de-siècle* France was written by Dr. Achille-Adrien Proust, father of the greatest literary neurasthenic of the age. The elder Proust, a prominent Paris physician and epidemiologist, served the French state for many years in his capacity as director of a service at the Ministry of Public Health.[69] Proust won lasting acclaim for his lifelong work in "defense of Europe against cholera" for which he undertook in 1869 a strenuous journey on horseback from France to Persia in order to demonstrate that the disease was indeed a rapidly spreading epidemic that required effective immigration quotas to combat it.[70] In 1897 along with another physician, Gilbert Ballet, Proust co-authored *L'hygiène du neurasthénique,* which soon became the standard work on the subject throughout Europe. Proust buffs have often speculated that his own son, who often spent as much as six and a half days of the week in his bed (writing of course), was the "model" for many of the graphic descriptions of neurasthenia found in this work.[71] Indeed, the description of the neurasthenia engendered by fashionable life in "society" leaves little doubt that much of the work was built on biographical sediment.

Like Beard, Adrien Proust associated neurasthenia with the moral and intellectual pressures of modernity. But Proust and Ballet also criticized Beard for his assertion that neurasthenia was unique to the modern age. It was found in antiquity, they argued, though only in the modern age did it become a significant malady. Neurasthenia was more common than it was sixty years ago, they noted, and it is "equally spread amongst all civilized peoples in whom the struggle for existence keeps up an incessant and exaggerated exaltation of the functions of the nervous system."[72] They also emphatically rejected the degeneration theorists' claims that the preponderance of neurasthenia in the modern age "owes this unhappy privilege to a sort of all-round degeneration [which] has invaded the late-coming generations."[73] Although they did not deny that the "famille névropathique" might predispose its progeny to neurasthenia, they emphasized that neurasthenia was one of those diseases "least dependent on heredity" and regarded the social imperatives of modern life as the chief cause of neurasthenia. For Proust and Ballet, its "predominance in towns, among the middle and upper classes, in a word, in all circumstances where intellectual culture or commercial and industrial traffic are carried to their highest degree of intensity," provided irrefutable evidence "that over-pressure, and especially cerebral over-pressure, must figure in the front rank of causes of neurasthenia."[74] For this reason, they also argued that neurasthenia was more prominent among men than among women, and more prevalent among the intellectually demanding professions—law, medicine, and government—than among the physically demanding ones: "The extreme rarity of the neurosis among the labouring class, and its almost exclusive limitation to the cultivated classes, to the world of affairs and to the liberal professions, in a word to the social categories whose circumstances involve them in brain work."[75]

Unlike Beard, however, Proust and Ballet did not think that intellectual work *per se* accounted for neurasthenia. Rather they attributed the evil to the moral pressure that accompanied such work. Echoing Ioteyko's discovery that fatigue represented a "prophylaxis" against overwork, they also noted that fatigue and the "embarrassment of cerebral activity" that produced it, was the best protection against this over-pressure," keeping it "within just bounds."[76]

Proust and Ballet devoted most of their work to a discussion of the social causes and cures of neurasthenia. As we have mentioned, there is more than a note of personal ire in the vividly described demands of fashionable idleness to which the scions of certain successful *pater-familias* are prone:

> One can easily be convinced of this by picturing to oneself the existence led, especially in the Parisian world, by those who are called in the current slang "society" men and women. Those who go out much and especially women, have their whole day taken up by the duties that convention and the vain care of their reputation impose on them: visits, dinners, balls, evening parties make their life one of continual constraint, and of obligations without respite.[77]

Not all writers on neurasthenia agreed that the ailment rarely afflicted the laboring classes. "A worker without culture," noted Charcot, was prone to neurasthenia "when a series of painful emotions struck," and the category of "traumatic neurasthenia" was generally reserved for working-class patients emotionally affected by accidents or other traumatic events.[78] Proust, however, believed that these traumatic cases were not necessarily identical with the neurasthenia typically found among middle-class patients. A German writer attributed working-class neurasthenia to the philosophy of Marxism and the heavy ideological burden that it laid on the proletarian soul: "It [Marxism] replaces apathy and hopeless resignation in the face of destiny with a kind of fanatical fatalism, a more hopeful resignation."[79] The paradox of this ideological illusion intensifies the neurasthenic symptoms. A collection of case studies published by Dr. Firmin Terrien argued that peasants were also afflicted by these disorders, but, in contrast to urban neurasthenics, they relied on a wider variety of personages—for example, local sorcerers—to find a cure.[80]

The neurasthenic patient is frequently described in these texts, often with an obsessive concern for detail. Proust and Ballet comment that there are two types of neurasthenic patients, those who speak little and answer badly the questions put to them, and those who seem excited and speak too much.[81] Bouveret graphically portrayed the former type:

> The patient is pale and thin, without strength or courage, and always sad and dejected. He sees everything from the worst side. He rarely smiles. He goes along with his head down, avoiding the looks of others, his eyes languid and dull. He hardly dares look people in the face when he speaks to them, and the vagueness of his look is as it were a sign of powerlessness, an avowal of his moral strength. He always has the gait of a tired man; he is usually very sensitive to cold, and is clothed in the summer almost as in winter; his speech is slow, broken and trailing.[82]

Proust and Ballet also charted two orders of "stigmata" among neurasthenic patients, purely mental symptoms and physiological manifestations like persistent headache, rachialgia, neuromuscular asthenia, dyspepsia from gastrointestinal atony, insomnia, and "hyperasthenia," or exaggerated sensitivity of the skin to pressure.[83] Mental symptoms, numerous and varied, ranged from "a conscious weakening of the personality with a more or less pronounced loss of power in all the faculties."[84] Neurasthenics frequently suffered from "aboulia"—a diminution of the will—of the intellectual and moral faculties. Neurasthenics also slept badly, "and having become such as a result of exhaustion, dream little." But, "often in the wake of a comatose sleep one observes supervening in them an aggravation of all morbid troubles."[85]

This descriptive mania, which we find in Proust (a point of similarity between the otherwise estranged father and son) and Ballet and in the other experts in neurasthenia, is not without its *raison d'être*. Neurasthenia was elusive and thus required attention to its smallest manifestations. Furthermore, since neurasthenics were frequently preoccupied with the minutiae of their condition, producing ever-more elaborate self-diagnoses, their descriptions were notoriously deceptive. In the presence of symptoms almost completely wanting in objective character, the physician has no other means of information than the statements and even the lamentations of the invalid; now there are few persons who are capable of precise observation and exact appreciation of their own functional disorders."[86] Neurasthenia created a dilemma for the physician: a virtual maze of shifting and unreliable "objective" symptoms and even more unreliable "subjective" monologues in dire need of the trained eye and ear. In this protean symptomology, we can see the mirror image of what we have already shown to be the objective and regular character of physiological fatigue.

How were physicians to understand, let alone judge, the endless complaints of these exhausted souls? This question occupied some of the best medical minds of the age. If neurasthenia frequently resembled both hysteria and melancholia, its accompanying plethora of physical disorders were equally elusive. Freud, too, commented on this aspect when he noted that hysteria and neurasthenia were often combined, "either when people whose hysterical disposition is almost exhausted become neurasthenic, or when exasperating impressions provoke both neuroses simultaneously."[87] Most physicians agreed that while neurasthenia often displayed physiological symptoms, they could find no organic causes or lesions. Although the "perpetual sensation of fatigue" remained the chief complaint of neurasthenic patients, they also produced symptoms so close to "real" disorders that misdiagnosis was common.

Neurasthenia was not simply a malady, but frequently an unstable mimesis of other maladies. In fact, a reading of the textbooks on neurasthenia underscores the multifarious signs that the disease manifested, an incessant orchestration of physical analogies. The neurasthenic panoply of symptoms constituted a kind of corporeal text, which Walter Benjamin once referred to as "a living image open to all kinds of revision by the interpretive artist."[88] The physician's interpretive stance gave coherence to the diffuse meanings provided by patients. Like a critic confronted with a disjointed and elliptical modernist narrative, the authority of the author-physician stabilized the chaos of appearances. In this act of interpretation the physician (father) resembles the artist (son) in trying to circumvent the arbitrariness of the allegory-symptom by solidifying it in an interpretive schema. Indeed, the similitude and mutability of neurasthenia, which resulted in cycles of activity and fatigue, also allows us to speak of a disorder of shifting symptoms, or perhaps of shifting signifiers. To explain the notorious unreliability of their patients, physicians referred to a "cyclical or circular neurasthenia," in which hyperactivity alternated with total inactivity.[89] Here we can locate a second order of modernity in neurasthenia beyond its modern etiology, a modernism of the symptoms and the narrative of the illness itself.

The elder Proust discovered that even his most intelligent patients incoherently described their disorder. Charcot applied the revealing *bon mot* "L'homme du petit papier" to neurasthenic patients who frequently appeared with slips of paper or manuscripts endlessly listing their ailments.[90] The lack of attention manifested by some neurasthenics could be measured, Féré proposed, by a variation on the aesthesiometer, an audiometer, which compared the "states of distraction" of neurasthenics with those of normal, but lazy subjects.[91] The inability to remember their own symptoms was widely reputed to be another universal characteristic of neurasthenics. Otto Binswanger, the leading German expert, noted that neurasthenics were frequently inattentive and showed disturbances in memory, especially in more complicated thought patterns: "If you question the patient more closely, so you will find at the outset of this memory disturbance that the memory image of an earlier occurring sensation is not lost, nor are simple thought sequences."[92] One writer referred to "fatigue-anesthesia" as the state of being too tired to remember to feel tired.[93]

Perhaps the most extreme reaction to this difficulty of treating neurasthenia can be found in the work of the Swiss physician Paul Dubois' *L'education de soi-même* (1910), which achieved enormous popularity at the turn of the century. Dubois treated his patients—some of whom declared themselves to be totally exhausted and remained in bed for years—by refusing to take their complaints seriously. His preferred method of treatment was argument by counter-example. Dubois merely pointed out that the contradictions in their behavior demonstrated the illusiveness of their symptoms. If a man complained that he could not leave his house, Dubois countered that he walked for miles about his own grounds; if a woman complained that she could not give lessons to her children, he pointed out that she read novels all day long. Dubois believed that the conviction of impotence in these patients was caused by an exaggerated pessimism and that there was no more need to recognize such fatigue than there was to be concerned with the ailments of hypochondriacs.[94] For Dubois, suggestion was the cause *and* cure of neurasthenia. Not only Dubois, but other popular authors counseled, as did Proust and Ballet, that the will had to be "educated" to resist flight into neurasthenia and to withstand the excessive demands of civilization.[95]

This peculiar characteristic of neurasthenia—disease of the unreliable subject, with its confusing combination of contradictory physical and mental symptoms—required a special kind of treatment not usually applied to physical disorders. As a disorder of analogic symptoms, Proust and Ballet affirmed that its treatment called for an attentive and patient physician willing to listen to, and above all interpret, the patients complaints:

> By proceeding thus with methodical inquiries, prudently directed and often repeated, the physician will be able little by little to check the statements of the patient, separate the true from the false, and arrange the symptoms according to their clinical importance, disentangling those of leading importance from those that are secondary.[96]

MENTAL FATIGUE, NEURASTHENIA, AND CIVILIZATION

The function of the physician, therefore, was to discern the true nature of the illness and "draw up a balance sheet." As Proust and Ballet remarked, "There are few patients whose examination demands so much patience and tact as that of neurasthenia."[97]

Since the patient's social milieu played a key role in any cure, treatment of neurasthenia was more difficult than diagnosis. Prevention, too, had to be concerned with environment. Proust and Ballet observed that boarding schools were particularly dangerous for children prone to the disorder and that the vices that festered there—"exaggerated development of the sexual instincts, debauchery and lubricity"—were especially to be avoided. They cited a recent study of German boarding schools that revealed thirty percent of the pupils displayed the neurasthenic symptoms of "languor, sadness and paleness." If hygienic measures were not strictly followed, they cautioned, if a child is not habituated to *will* to resist such negative influences, there may arise a "moral paralysis called *aboulia.*"[98]

They also noted that the many available therapies—cold ablutions, douches, plungebaths, sea or river bathing—were beneficial, but they remained skeptical of the American physician Weir-Mitchell's "rest-cure," which demanded complete withdrawal from the demands of society and preached complete reliance on the care of others. Proust discounted this approach as well as Beard's "electrotherapy," "hydrobaths," and other technical cures as secondary to the more basic suggestion effect of the physician. The medical expert was placed in the forefront of the "cure." The moral action exerted on the neurasthenic by the physician and his surroundings constitutes one of the most powerful therapeutic agents that can be employed. Above all, the physician must remain aloof from a patient's claims, and avoid ridiculing (as did Dubois) or ignoring them:

> The physician who wishes to secure the confidence of a neurasthenic must then listen to him attentively, examine carefully all his organs, and above all refrain from any bantering, from any ironical reflections, however strange may be the complaints or confidences that he receives. When this first point is gained, that is to say when he has acquired the confidence of the patient, the physician will thenceforward be able to reassure him authoritatively as to his condition by declaring to him, not that he is not really ill or that he is only a '*malade imaginaire,*' but that he has no organic lesion, and that consequently his illness, though demanding serious and perhaps prolonged treatment, is perfectly curable.[99]

Although all neurasthenics were plagued by fatigue, there was little consensus on what precisely caused their loss of energy. For example,

Féré observed that at La Salpêtrière patients with hysterical disorders demonstrated an extreme physical fatigue during their periods of "automatism," a symptom usually encountered only after arduous labor in normal individuals.[100] He took this fact to be evidence that neurasthenia was not simply without organic origin, but, that in fact all the "derangements of the mind" were produced by some form of extreme nervous excitation, or "irritabilité" of which exhaustion was simply its obverse side. Nevertheless, neurasthenia's apparent defiance of the "laws" of physiology, with its unreliable and analogic symptoms, absence of lesions, and confounding behaviors, only intensified the search for a materialist explanation of the disorder.

MATERIALISM, THE WILL, AND THE WORK ETHIC

By the early 1900s several prominent French psychologists began to despair of the ambiguities associated with the multiple symptoms and causes of neurasthenia, and proposed a classification *asthénie,* or "asthenia," to distinguish the specific "maladies of energy" or "diseases of the will" from neurasthenia *tout court.* Especially necessary, they believed, was to distinguish the physical loss of energy and the inability to act from neurasthenia's secondary manifestations, for example, hypochondria and inattentiveness. Physicians could not explain how the lack of will inhibited action nor how mental exhaustion occurred. Knowledge of the causes of psychic fatigue seemed to lag behind the advances that physiology could claim in understanding physical and even mental fatigue. For this reason, many psychologists believed that it was possible to approach the study of neurasthenia from a materialist perspective that more closely approximated the science of bodily fatigue.

Albert Deschamps, in his *Les Maladies de l'énérgie* (1908), noted that "today, the domain of neurasthenia is so vast, so imprecise, that it is necessary to undertake a work of revision." Deschamps agreed with the general picture already drawn by the neurasthenia experts: it was a malady of fatigue, most often afflicting individuals in intellectual professions with "incessant preoccupations." When it appeared among the working classes, however, "asthenia was most serious": "Those laborers who work until past midnight after having lunched on two *sous* of fried potatoes and a slice of sausage, become exhausted with extreme rapidity."[101] In this sense *exhaustion,* as Binet and Henri had already argued, was defined as an "accumulation of fatigues which were only incompletely restored."[102]

But in this "vast forest," Deschamps continued, "it is necessary to mark out a path to separate that which is 'simple nervousness'" *(név-ropathie simple),* from hysteria, from degeneracy, and from "psychasthenia" *(psychasthénie),* and to show that neurasthenia is not singular— "that there is no *one* neurasthenia, but the *asthenias."* To consider each diminution of energy under the general rubric of neurasthenia was an "error of interpretation," he claimed. *Neurasthenia* was a "defective term" since it was only one type of asthenia, not "all asthenia."[103] Despite its appearance with various symptoms, its cause was always the same: a "diminution of the specific energy in an inverse ratio to the potential capital of energy." In short, it was not merely the moral and social effects of modernity that produced neurasthenia, but the physiological effect of a specific loss of energy.

Closely following Helmholtz, Deschamps claimed that the organism was a vast reservoir of energy, nourished by multiple sources. Deschamps argued that energy was carried along its pathways by "nervous waves" *(l'onde nerveuse)* analogous to electromagnetic waves, which created a "perpetual circulation of energy in the organism." Excessive expenditure of energy, lack of adequate supply or, "a defective organization in the energy reservoir," would be classified as asthenia.[104] A general disfunction of the system would be termed *dysergie.* Other difficulties could be grouped under the following subspecies:

hypo-*ergie* = diminution of function
hyper-*ergie* = augmentation of function
para-*ergie* = deviation of function
an-*ergie*   = suppression of function[105]

Deschamps was not alone in his attempt to place neurasthenia on a firm physiological foundation at the end of the nineteenth century. Mosso, too, believed that "for the production of thought, emotion, or feeling a transformation of energy is required," but admitted, "the palpable proof of this we cannot yet give."[106] Théodule Ribot, the influential editor of the *Revue Philosophique,* the central organ of philosophical materialism in *fin-de-siècle* France, also attempted to provide the "diseases of the will" with a materialist foundation.[107] His major work, *Les Maladies de la volonté* (1884), is much more than a textbook of physiological psychology. It is a passionately argued anti-Kantian treatise, directed at his opponents at the *Revue de Metaphysique et Moral,* the journal of the staunchly Kantian Sorbonne. For Ribot, consciousness was only one aspect of the "world of desires, passions, perceptions, images and ideas" that are the product of bodily sensation: the will and the emotions were all associated with some form of "nervous activity." In his view, the psychophysical basis of the will dissolved the argument for a free will beyond the grasp of materialist explanation: "In every voluntary act there are two entirely distinct elements: the state of consciousness, the 'I will,' which indicates a situation, but which itself has no efficacy; and a very complex psychophysical mechanism, in which alone resides the power to act or to restrain."[108]

The "pathology of the will" was of concern to Ribot for two reasons: first, such disturbances revealed the material foundations of human psychology while proving that those mental representations that Ribot associated with reactionary ideas—religion and political despotism— were consequences of the pathological effects of physiology. For Ribot, the healthy will was a bulwark against the unwanted and destructive thoughts and images plaguing fatigued individuals. If desire represented the primitive form of affective life, the will represented its antithesis—a higher form of material power that counteracted its negative effects. Desire, he noted, "does not differ from reflex movements of a very complex kind." The so-called spontaneous actions of little children and savages furnished Ribot with unlimited examples of desire operating as a reflex. But, he added, "history shows us that in the case of despots, placed by their own opinion and that of others above the law," desire can also reassert its claims in politics. In Ribot's evolutionary schema of psychological evolution, the will represents the highest stage of physiological *and* moral development, while "desire marks an ascending stage between the reflex and the voluntary conditions."[109] Consciousness is adapted to the activities and needs of a species whose basic nature is acquired, structured, and fixed by the laws of heredity. The will normally overrules the natural and spontaneous manifestations of desire, but when unrestrained, desire asserts its power unchallenged.

Ribot claimed that this negative power of desire was greatly "augmented" when the will is weak, and conversely, that the work of education or morality consisted largely in its suppression. The fear that the cumulative effect of mental fatigue would create a society incapable of moral restraint was already a strong current among French degeneration theorists, and, as we have seen, a common theme of late nineteenth-century medical thought. Marie Manacéine, a Russian physician and author of the popular treatise *Le Surmenage mental dans la civilisation moderne*, which appeared in France in 1890 with a preface by Charles Richet, argued that the "agents which weaken, slacken and trouble the associations" destroyed the "foundations of the progress of humanity."[110]

In Ribot's theory, the will is *both* a republican (politically) and a material force (the identity of republican morality and materialism is a crucial aspect of Ribot's philosophy); it suppresses crude desire but also establishes universal law in its place. Since he considered consciousness a "reflection" of energetic processes, Ribot confronted the "metaphysicians" with the argument that the will "is not the state of consciousness as such, but rather a corresponding physiological state which transforms itself into an act." The essential relation is not between a psychical event and an activity, but "between two physiological states, two groups of nervous elements, one sensory and the other motor."[111]

For Ribot, Richet, Wundt, and other "psychophysiological" psychologists of the nineteenth century, ideas were representations of physical forces. Ribot classified all ideas into three distinct groups according to their power, or capacity, to transform into action: (1) intense intellectual states (passions, fixed ideas) that rapidly pass into action; (2) rational activity, or reflection, in which primitive sentiments have become weak and a bond between idea and act has been established; and (3) abstract ideas in which reflection predominates over the tendency to movement or action.[112] These states follow the evolutionary schema alluded to earlier, in which reflex and desire is the earliest evolutionary stage; and the will and abstract ideas are the more highly developed conscious forms of mental life.

What then is the will? Ribot defines it as an "inhibitive power." It acts—as do any number of physiological phenomena—as a suppression of excitation, as a "mechanism of inhibition," as the inverse of a reflex. Like Deschamps' "nervous waves," Ribot also posited a "nervous influx," which varies from individual to individual, in which "one form of expenditure prevents another, the disposable capital not capable of being employed for two purposes at the same time."[113] Ribot admits that this physiological mechanism still remained unknown and obscure: "The laws of the distribution of nervous activity, of that species of circulation of what has been named the nervous fluid,—who knows them?"[114] What *is* known is that the inhibiting impulses are frequently accompanied by emotional states, like a fear of persons, laws, customs, or God and that these emotions are always "depressive states, tending to diminish action." The conscious will therefore is only a reflection of its physiological substructure: this physical basis distinguished the "constitution of individuals, stable or fluctuating, continuous or variable, energetic or feeble."[115]

For Ribot, the pathologies of the will (irresolution, morbidity, pessimism) are not states of consciousness, but consciousness of inner states of depletion. Those who are "poor in ideas" almost always tend toward the least action or the most feeble resistance. The pathological will appears in "two great classes, according to which it is either *impaired* or *extinguished.*"[116]

But Ribot also classified the pathologies of the will along a moral continuum descending from the normally unimpaired will to the utter extinction of the will. *Aboulia* occurs when the desire to act exists but is impotent; when the *"I will* does not transform itself into impelling volition." To illustrate this type of inhibition he cited a famous case treated by one of the founding fathers of French psychology, Etienne Esquirol:

> A magistrate, very distinguished for his learning and power of language, was, as a result of troubles, attacked with a fit of monomania. . . . He has recovered the entire use of his reason but he will not go out into the world again, although he recognizes that he is wrong; . . . "It is certain," he said to me one day, "that I have no will except not to will; for I have all my reason; I know what I ought to do; but strength fails me when I ought to act."[117]

Another example cited by Ribot is the notorious case of Thomas de Quincey, whose *Confessions of an Opium Eater* (1851) epitomized his impairment and withdrawal from his studies of political economy, experiencing a state of "intellectual torpor" during the "four years during which [he] was under the Circean spells of opium." These cases, along with subsequent studies in the medical literature of Ribot's day, had the same common traits: a healthy intellect, no physiological impediment, but the "transition to act is impossible."[118]

Ribot's analysis of aboulia touches on an aspect of neurasthenia not immediately evident to those early investigators of the disorder who simply equated its onset with the social pressures of modernity. Neurasthenia was a kind of inverted work ethic, an ethic of resistance to work or activity in all its forms.[119] The lack of will or energy manifested by neurasthenics is the incapacity to work productively. It might even be argued, as in the case described above, that neurasthenia was oriented toward "performance," since some kinds of "spontaneous" or "automatic" work—the younger Proust's incessant writing, for example—were obviously exempted. The diagnosis of asthenia à la Ribot or Deschamps reversed the moral etiology of the earlier experts. Whereas Proust and Ballet, Bouveret, or Binswanger regarded moral pressure as the source of inhibition, Ribot and Deschamps emphasized primarily the material power of the will to resist work. Though often overlooked, the nosology of neurasthenia profoundly changed: the materialists displaced the malady of overpressure with the malady of impaired energy and resistance to the performance principle.

The psychological materialists frequently prescribed work as a therapy as opposed to the moral or "suggestive" treatment of Proust and Ballet. They also rejected the famous "rest cure" prescribed by the Philadelphian S. Weir-Mitchell because it capitulated to the disorder's requirements and assertions of incapacity. Patients who withdrew from their demanding lives into the infantile passivity of "Dr. Diet and Dr. Quiet" did not, the materialists claimed, usually return to their former pursuits, especially after a long period. Theodor Dunin, author of a 1902 German treatise on neurasthenia, noted that the malady was characterized by an "Unlust zur Arbeit" (aversion to work) and that "inactivity was the greatest enemy of neurasthenia." Among his neurasthenic patients the largest contingent was composed of "idlers" and "dilettantes of all sorts." Dunin cautioned physicians not to advise neurasthenic patients to give up their work since those who become unemployed as a result of doctor's orders are the most difficult to treat, and neurasthenics who are inactive for months and years can "only be restored by a return to activity."[120]

Dr. Paul Möbius of Berlin also advised that physicians speak to their patients of work, but he acknowledged the difficulty of treating them by the regimen of work cures. The neurasthenic's tendency to "do nothing" was exacerbated by the "life-style" of the "higher classes" generally associated with such patients. This was even more common, he added, with wealthy women who "had no preoccupation and could find none."[121] Though they claimed that neurasthenia was equally, if not more common among men, Proust and Ballet, following Weir-Mitchell (who treated women almost exclusively), considered the "neurasthenia of women" an especially debilitating form of the disease: "The dominant feature of this neurasthenic state is profound discouragement, powerlessness to exert the will, in one word *aboulia,* joined to a degree of *muscular asthenia* that is hardly ever seen except in this form."[122]

Let us recall the point of Ribot's reflections on the will. The chief purpose of the will is to resist and deflect the subterranean stream of images, associations, and phantasms that manifest when the will is impaired. The anxiety, fear, or terror that culminates in a "paralysis of the will" are consequences of a physiological depletion, or lack of nervous energy. Ribot contended that "it is from these states that depressed feelings, or *aboulia* results."[123] According to Ribot, "morbid inertia" is usually accompanied by intense feelings of fatigue. The "labor" of the will is often intense but inadequate to the task.[124]

Like many other enlightened *savants* in *fin-de-siècle* France, Ribot viewed psychophysical materialism as an assault on mysticism and religious doctrine. Charcot, for example, frequently drew a parallel between the psychic states experienced by hysterics and those experienced by religious mystics.[125] Ribot devoted a good part of his treatise on the diseases of the will to exposing the physiological basis of religious "ecstasy," or "annihilation of the will," as he called it.

This otherworldly state has often been counterposed to the Western work ethic. Max Weber, contrasting the goal-rational asceticism of the modern work ethic with the otherworldly spiritualism of the East, commented on the lack of intentionality that usually accompanied the ecstatic or mystical state. Ribot too thought that this kind of religious mysticism could be explained as the "extinction of the will" where a form of mental activity still persisted.[126] "When this state is attained," Ribot noted, "the ecstatic presents certain physical characteristics: sometimes motionless and mute, sometimes expressing the vision that possesses him by words, songs, and attitudes." Ribot illustrated his point with an example from the autobiography of St. Theresa, who, when seized by the presence of God, became "wholly lost in Him." In these "pathological" states the annihilated will is incapable of goal-rationality. Escape from the work ethic constitutes the central pathology of religious ecstasy.

Extinction of the will also accounted for somnambulism or hypnotic trance, in which consciousness is temporarily abolished.[127] Mysticism too was clearly a product of an earlier, simpler stage in the evolution of the will, an automatic stage in which the "conscious will" plays little role. The law of the evolution of the will, Ribot concluded, is evident in these states: "Dissolution pursues a regressive course from the more voluntary and more complex toward the less voluntary and simpler, that is to say toward the automatic."[128]

States of regression—religious experience, hypnotic trance, emotional trauma, or even anxiety—evoked an earlier preconscious and prerational life of the will. In these primitive forms of human experience the kind of energy and nervous force that exists below the superstructure of ideas can be grasped without reference to motivating ideas. Indeed, the images or thoughts produced by the will were symptoms, rather than causes of the disorders accompanying the depletion of energy. Like neurasthenia the "diseases of the will" required that mental ephemera be distinguished from the material foundations of the disorder. Pessimism, lack of initiative, fear of action, and above all, the refusal to engage in any productive activity were the "false consciousness" of fatigue. The will for Ribot is never the "mistress" of action;

rather, the work of the neural forces was always in command. In Ribot's psychology, all explanation was linked to the doctrine that impaired energy was the material "substance" of a pathology that produced irrational representations and gave rise to the regressive forces of religion or idleness.

At the end of the nineteenth century, two comprehensive psychological theories provided a more general framework for the understanding of neurasthenia and the will. The first, and at the time most famous, was that of the student of Charcot, Pierre Janet. The second belonged to the prepsychoanalytic Freud, though even at the outset his theories on the sexual etiology of neurasthenia distinguished his views from other specialists in fatigue. In February 1893, Freud wrote to Wilhelm Fliess that "sexual exhaustion can by itself alone provoke neurasthenia. If it fails to achieve this by itself, it has such an effect on the disposition of the nervous system that physical illness, depressive affects and overwork (toxic influences) can no longer be tolerated without [leading to] neurasthenia. Without sexual exhaustion, however, all these factors are incapable of generating neurasthenia. They bring about normal fatigue, normal sorrow, normal physical weakness."[129] Several historians have demonstrated the broad influences of the psychophysical theories of Gustav Theodor Fechner on the early Freud through Fechner's influence on Freud's teacher Gustav Meynert.[130] The extent of nineteenth-century biological materialism's impact on Freud remains controversial and deserves more treatment than can be accomplished here. However, Frank Solloway has argued that Fechner's role in Freud's early theories of stimulation and sensation was more important than that of the biological reductionism of Ernst Brücke, the Viennese physiologist and student of Helmholtz.[131] Freud's "project [1895] for a scientific psychology" nevertheless attempted to give scientific credence to the view that all representations were functions of the body's "sensory qualities," which Freud divided into the competing spheres of pain and pleasure.[132]

Like Ribot, Pierre Janet also regarded the will and fatigue as the keys to all psychological disorders, though he rejected Ribot's biological materialism.[133] Influenced by William James, whose *Energies of Men* (1907) provided him with the "principle of the mobilization of forces," as well as by Weir-Mitchell, whose rest cure became part of his prescribed treatment, Janet focused on the "inhibition to action," which he attributed to *psychasthenia*, a term he also hoped would replace neurasthenia.[134] As a student of Charcot, Janet did not deny a physiological dimension to the etiology of neuroses, but he viewed the emotions, especially traumatic emotions, as the chief sources of mental disorder.[135]

Janet's theory of psychological "tension" created a hierarchy of the energies required for different types of acts, from the most simple to the most complex. The demands of modern life only exacerbated the energy required of each individual, though Janet recognized that it was not simply excessive work, but the accompanying emotional demands that caused the onset of exhaustion in his patients.[136] Unlike Ribot, Janet did not see the emotions as mere "representations" of fatigue, but argued instead that emotion was a "variety" of fatigue and that his patients frequently betrayed the true origins of their disorders by saying that "they were born tired" or that fatigue was a constant companion.

Janet viewed the psyche as a permanent struggle between the economy of energy and fatigue in which the fear of action represented an economic withholding of reserves: "The renunciation of this or that activity, religious practices, the fine arts, study, or amorous sentiment, which may be observed so clearly in certain *psychasthenics* is an act of economy that is the result of a real feeling of poverty."[137] As we shall see, the implications of the "pathology of the will" developed by the physicians who theorized about neurasthenia and subsequently by Ribot and Janet, were ambiguous. Fatigue and energy represented the basic categories of a psychic economy that could be interpreted either in terms of a natural conservatism of the body's forces, or as the psychophysical affirmation of an energeticist faith in progress. Ribot, for example, seems to accept both versions, though he more strongly emphasizes the evolutionary and progressive character of the will. The tendency to conserve energy produced a kind of "juste milieu" in normal individuals; its absence in excessively fatigued or neurasthenic individuals made them incapable of resisting the imperatives of desire or the irrational associations that plagued them. Their pathological "inertia" was far more characteristic, as Ribot seemed to believe, of primitive societies and atavistic sensibilities. On the other hand, one could claim, as did the degeneration theorists, or more alarmist writers like Marie Manacéine and Proust, that neurasthenia was a constant threat to the diminishing reserves of energy required for civilization. But it is precisely this ambiguity—the progress attained from increased economy of energy and the threats to civilization arising from modernity's relentless extraction of force—that is evident in all of these theories of fatigue and neurasthenia. This problem can be seen most clearly in the debate over how these theories confronted primitive cultures, and more specifically, how they judged the "idleness" of savages, which became a *cause célèbre* in ethnography.

THE LAW OF THE LEAST EFFORT: CIVILIZATION AND FATIGUE

Studies of physical and mental fatigue often emphasized how these helped ensure an optimal state of equilibrium in which the demands of the economy of energy were syncretic with the imperatives of civilization. Fatigue acted as the regulator of labor power, much as a governor prevents a machine from exceeding its efficient speed and rhythm. Both Marey and Mosso agreed that biological time and body rhythms seemed to regulate naturally the pace of work and reduce waste in human labor. The science of fatigue projected the principle of efficiency onto the body. The "law of the least effort" expressed the tendency of all organic life to find the shortest path to its goal. From this perspective, fatigue could even be seen, as Ioteyko proposed, as a kind of beneficial prophylaxis that restored the equilibrium of the body and allowed the human motor to operate at optimum efficiency. The adherents of a psychophysical approach to pedagogy and psychology shared the view that the insights of fatigue research could be adapted to the factory, the school, and many other areas of society.[138]

"All natural movements are rhythmic movements, as all forms of life are symmetrical," wrote Féré, who devoted an entire chapter of his *Travail et plaisir* (1904) to the "influence of rhythm on work."[139] For Féré, ergonomic studies of fatigue and rhythm confirmed the essential symmetry between the laws of the universe and the laws of nature: "The regular forms of crystallized matter, the course of the planets around the sun, the alternation of day and night, the division of cells in the course of the development of organized beings . . . these are all manifestations of the general law of rhythm, the law of matter, the law of form, the law of life. . . . [Rhythm is the] principle of order in time and in space."[140] The rhythm of the body was the root of all art; the repetition of figures, of movement, of forms was the source of aesthetic pleasure. Féré explained that "by the reproduction of the same forms, whether simultaneously in space, or whether successively in time, art corresponds to the great laws of life, of rhythm and symmetry. All art obeys these laws."[141]

Féré's reflections on the aesthetics of rhythm were derived from numerous ethnological studies that appeared in France and Germany in the mid-1890s.[142] In 1894, the noted anthropologist and psychologist

Guillaume Ferrero published a series of articles in Ribot's *Revue Philosophique,* applying insights from the study of fatigue to anthropology. According to Ferrero, in mental as well as in physical labor, the tendency to inertia was universal. "Mental inertia" expressed the intellect's need to achieve a state of equilibrium. A certain immobility was necessary to restore the energy that the psyche requires for its acts. Extending Ribot's comments on the annihilation of the will, Ferrero argued that hypnotism and somnambulism also confirmed that complete mental inertia could be demonstrated among enlightened, civilized populations. In that state, similar to sleep, "The body is immobile, the face is an impassable mask, the figure maintains an expression of tranquility and calm."[143] This condition of rest, Ferrero emphasized, was evidence of the "law of the least effort" in human psychology. Ferrero believed this insight confirmed the arguments of French psychologists like Féré or Richet (who placed a primacy on excitation in movement) that any economy is regulated by the law of the least effort.[144]

For Ferrero, however, the social consequences of this downward pull of inertia were negative. The major portion of humankind did not engage in social labor without constraint or force. Criminal populations—thieves, vagabonds, prostitutes—were motivated solely by the avoidance of productive labor. The "inclination to idleness" is characteristic of degeneration, he claimed, since every mental effort "was instinctively repugnant to mankind." Following Ribot's hierarchy of the will, only voluntary effort at the highest stages of civilization produced mental labor. Ferrero also noted that attentiveness was abnormal, transient, and exhausting. At the outer limit of fatigue there is simply "functional inactivity." The human species naturally attempted to preserve its mental and physical force, to conserve its psychic energy, and to act in conformity with the law of the least effort. "Least effort" was the law of civilization.

Ethnological research offered an embarrassment of evidence confirming the view that in primitive cultures idleness or "inertia" was the natural state of the human species. Modern society, in short, constituted a violation of human nature. For Ferrero, the question for psychology is why does humankind progress at all? Why does inertia not rule the species, eliminating all incentives to progress, productivity, and civilization? What causes the human spirit to abandon this state of idleness from time to time? What are the inducements that provoke him, "little by little, to produce that great organization of work which characterizes civilized societies?"[145]

Ferrero acknowledged that his image of the primitive as an inert creature raised certain dilemmas. Numerous ethnologists had observed that "among the savage peoples they danced until a complete exhaustion of forces set in, until they fell to earth, overcome by fatigue, as if they needed to discharge, in a very short period of time, by means of a furious activity, all of the nervous force accumulated during their long days of idleness." Whereas barbaric peoples saw work as "an evil second only to death," in dance they seemed to engage in a "veritable choreographic frenzy."[146] How could the "inertia" so frequently observed among those savage peoples be "natural," given their readiness to self-exertion and energy in dance, in ceremonial or religious activities, and in sport—activities that they took more seriously than any socially useful tasks?

Ferrero found the answer not in the utility nor purposefulness of work but in the natural rhythms of the body: "The productive work of civilized man is regular and methodical, [whereas] the sport [or rituals] of savages is irregular and intermittent."[147] The savage, in other words, "devotes himself to sport when the nervous force accumulated in the psychic centers demand a discharge"; the worker or businessman must remain at work during fixed hours and cannot deviate from the rhythms established by modern industry. A forger or worker at a loom submits to a series of acts that demand a great deal from both intelligence and will. "Productive work demands an enormous volitional effort," concluded Ferrero, while the activity of primitives is "almost automatic." In dance or sport, they recoup what is lost in productive labor. The "repugnance" that savages display to work derives from their "horror of mental and volitional effort without which methodical and productive work is impossible."[148]

Ferrero's conclusion that the primitive was a prisoner of idleness and inertia did not long remain unchallenged.[149] Two years after Ferrero's articles appeared, the German economist and musicologist Karl Bücher published an extraordinarily popular book, *Arbeit und Rhythmus,* which explored the influence of rhythm on the work habits of "natural peoples."[150] Bücher argued that the apparent contradiction between the fatigue and exhaustion actually enjoyed by primitives or workers on bicycling holidays and their antipathy to productive labor was not rooted in any inherent idleness or inertia, but in the rhythms that accompanied the physical effort. Moreover, he pointed out, Ferrero's contention that hunting or warfare involved little intellectual or volitional elements was contradicted by "all direct observation." It was not work *per se* that the primitive avoided, but rather "tense, regular work." In fact, the labor of primitives was "extraordinarily toilsome," accomplished with inadequate technical means, complicated labor processes, and nevertheless achieved with extreme artistry.[151]

Even more wrongheaded was Ferrero's contrasting of the savage's "automatic" activity to methodical, productive work. Productive labor, which requires an economic, conservative pattern of muscular movements and a constant mental effort to reduce fatigue, also necessitates the substitution of automatic movements for those of the intellect and will. These automatic movements "maintain a certain equilibrium and ensure that the beginning and end of a movement are always with the same spatial and temporal boundaries."[152] Analogous to what Kraepelin had called "training," and Ioteyko the prophylactic character of fatigue, rhythmic work regulated the expenditure of energy. Often this rhythmic activity produced its own natural sound, or tone—the beat of a blacksmith's hammer, for example—which was often accompanied by a song or chant.

Surveying the anthropological literature, Bücher found that ethnologists observed such rhythms accompanying fatiguing labor in a variety of cultures. His research documented numerous examples of songs and music linked to different types of work and also uncovered translations of lyrics that often attested to the drudgery involved, as in this Lett milling song:

> *Millstone, greystone*
> *Won't you in the ocean drown?*

Bücher's purpose was not to document antipathy to labor but to find the link between time and motion in the economy of rhythm. "Rhythm," he believed, "arises from the organic essence of mankind. All natural activities of the animal body appear to be dominated by the regulation of thrifty energy use."[153] As the external manifestation of energy conservation, rhythm appears as the "origin and starting point" of work. With Bücher's argument, the savage was no longer condemned to the pervasive imagery of idleness and the cultural ethnocentrism of the traditional European work ethic. Instead, a new and homologous image of the primitive's work and play emerges, characterized by the same propensity to energy conservation as the physiologies of civilized man and the requirements of efficiency.

Bücher thus completely rejected Ferrero's contention that primitives were bound by the laws of inertia. He sought to resolve the contradiction between their alleged dread of labor and the exhaustion experienced in their rituals by proposing that rhythm—the energy economy of the body—was at the origin of both culture and production, not external constraint. For Bücher the order of things is inverted by civilization. If civilization deprives labor of its rhythmic and natural character, the primitive displays the essence of the body's laws in his work. Work was accompanied by the chanting of the workers, giving it a "festive character [so work] could not appear to the individual as a burden."[154] The ethnology of fatigue no longer located the sources of resistance to work in idleness or in the "horror of mental and volitional effort," but in the lack of rhythmic labor.

Bücher concluded that civilization robs man of his natural propensity to labor by destroying the rhythmic element in work. It was not the uniformity of labor in the machine age that caused monotony and fatigue, but its *externally imposed,* unrhythmic character that was antithetical to the bodily demand for more rhythmic activity. "Precisely the uniformity of labor is the greatest benefit for people, as long as they can determine the tempo of their bodily movements and stop themselves. Rhythmic labor is not *per se* spiritless, but to a large extent spiritualized *(Vergeistigte)* work."[155]

Several years later Ferrero's conclusions were also challenged by Ribot. In fact, Ribot argued, the pathological fatigue and inertia that Ferrero viewed as the normal state of humankind was identical to the pathological *asthenia* investigated by Deschamps and Janet. These asthenics embodied the "law of the least effort" in their obsessive desire for sleep, for rest, and for avoidance of mental activity. Idleness was a kind of "old age anticipated," an atrophy of "the superior elements."[156]

For Ribot, the "inertia" that Ferrero castigated was not normal but pathological. Yet without some inertia there could be no economy, no industry. The economy of energy simply represented a higher degree of efficiency in which the law of the least effort was only one element among others. Whereas in asthenia the result was the avoidance of labor, in the economy of effort the result is the simplification of labor and its more efficient realization. The "horrible exhaustion" feared by psychasthenics was nothing less than a regression encountered in both modern and traditional societies as the profound anxiety of any form of innovation. Fatigue was a refusal or fear of modernity. Janet had noted just this "aversion to all novelty" in his asthenic patients, a pathological traditionalism, which he called *misonéisme* (opposition to anything new).[157]

Ribot tried to show that the "law of the least effort" was socially adaptive. Indeed, social life was unthinkable without the kind of "shorthand" that this law provided. Collective psychology—language, morals, daily life, political institutions, religious belief, science, and art—all demanded this capacity offered only by "inaction or the minimum of action."[158] In the history of religion the "law of the least effort" brings about tolerance since the fanaticism of belief that divides the two parties ultimately "surrenders to the analogy" that can be found among all versions of belief. Even linguistics, for example, demonstrated that there was a kind of "principle of laziness" *(principe de paresse)* at work in phonetics, gradually eliminating excessive sounds.[159]

This is the productive side of *misonéisme*, the disposition of the spirit to find the shortest path in any endeavor. The pathology of neurasthenia thus reveals a paradoxical "normality" of progress, of efficiency, and of productivity. *Misonéisme* is inherent in all analogy, not unlike the susceptibility of neurasthenics to produce analogic symptoms as a kind of visible shorthand of their pathological fatigue. The inertia that Ferrero described was both ontogenetically and phylogenetically a more primitive state of the will, a stage whose pathology persists in the resistance of the energy-afflicted to modernity. Resistance to modernity is an atavism. But even if *misonéisme* produced anxiety in the face of change, modernity was significantly dependent on it. The law of the least effort did not produce inertia but innovation—a "rupture with habit, a new adaptation, [and finally] the consolidation of the new habit." Because the human will is characterized by a "mediocre perseverence and vigor," because attention quickly fatigues, *misonéisme* emerges as the protective instinct of the species that regulates physical and mental activity.[160]

Ribot's identification of *misonéisme* as the self-preserving instinct that permits us to alleviate extreme fatigue and achieve the law of the least effort in human activity returns to the central problem posed by neurasthenia. If at the outset, neurasthenia was identified by Beard and his European followers as the inability of mind and body to resist the onrushing stimuli of modernity, Ribot and Bücher attempted to find a social purpose in our natural propensity to resist inordinate expenditures of energy. In *misonéisme* energy conservation is no longer threatened by fatigue but is accelerated by it. In Ribot's theory, in Janet's psychology, and in the debates over primitive inertia, the allegedly pathological resistance of savages and neurasthenics to modernity was balanced by the normalizing function of fatigue—the response that enables society to restore depleted energy. As "survival of the fittest" was the mechanism that explained the evolution of new and viable species, fatigue becomes the mechanism that explains the emergence of innovation and modernity. Fatigue makes possible the "law of the least effort" in society, and even, as Ribot explained, assures its hegemony in the realm of ideas. Without the prophylactic aid of fatigue, "the law of the least effort" could not function as the regulator of economy and efficiency in mind, body, and society.

Only in its most pathological forms does the will manifest a true idleness or inertia pernicious to social progress. To be sure, extreme fatigue was still opposed to progress, and in its most debilitating contemporary guises—neurasthenia and psychasthenia—it contributed to the rejection of work, rationality, and novelty. In religion, this tendency to inertia provided mankind with a "philosophy of repose" (as in Buddhism), elevating rest to a supreme principle. Anxiety about fatigue was misplaced in the modern era, the materialists concluded, since materialism had ushered in a higher stage of civilization that brought with it the social economy of energy and equilibrium of force. Normal fatigue did not threaten modernity but defined the outer perimeter of excessive labor and energy expenditure. In this sense modernity can be seen as the product of fatigue, insofar as the recuperative tendency of the body's energy system, as Ribot assures us, "presupposes a balance between debits and receipts, between useful activity and rest."[161] This fatigue is a necessary counterweight to the society that expends energy by regulating its movements according to the law of the least effort. Fatigue no longer threatens modernity and productivity; it ensures its triumph.

CHAPTER SEVEN

~~~~~~~~~~~~~~~~~

The European Science of Work

SOCIAL ENERGETICISM

B Y the end of the nineteenth century a small group of progressive reformers and scientists began to articulate the implications of physics and physiology for their vision of a society that would increase productivity while conserving the energies of the working classes. The social doctrine of energetics was elaborated by scientists and social philosophers as the solution to the problem of maximizing productivity while reducing the waste and decline that seemed inevitable in industrial capitalism. As nature exhibited a propensity toward conservation and a decline of force in entropy, so too society had to balance these opposing tendencies. Energy conservation became the Continental answer to the Darwinian or Spencerian vision of society propelled by the natural laws of conflict and struggle. Whereas English social thinkers and scientific popularizers like T. H. Huxley viewed thermodynamics as a metaphor for capitalist superiority, on the Continent these discoveries pointed toward an equilibrium of economic expansion and social justice.[1] Grounded in the principles of scientific materialism (and Lamarckian biology), social policy could stand above the interests of social classes and political imperatives.

Most prominent among those to elaborate the social implications of energetics was Ernest Solvay, a Belgian chemist, self-made industrialist, philanthropist, and social philosopher. The Belgian socialist Emile Vandervelde aptly called Solvay an "authentic Saint-Simonian and a liberal," though Solvay's liberalism was predicated on the energetics of matter, and his Saint-Simonianism was filtered through Comte, Helmholtz, and Claude Bernard.[2] Unlike the unfortunate Saint Simon, whose life was a series of failed conspiracies, Solvay enjoyed an extraordinary success.

Solvay's fortune came to him in part as a result of the sad fate of the eighteenth-century industrial chemist Nicolas Leblanc, who in 1775 developed an innovative method of converting common salt into sodium bicarbonate (washing soda) for the manufacture of soap. Leblanc committed suicide in penury when his mentor ran afoul of the committee of public safety. But Solvay, the son of a Belgian salt merchant trained as a chemist, recalled Leblanc's process a half century later, improving it by using ammonia wastes to convert salt far more efficiently and cheaply into soda ash.[3] Obtaining a patent for his "ammonia process" in 1861, Solvay along with his brother Alfred and a British businessman, Ludwig Mond, established a small factory near the industrial town of Charleroi. By 1867 Solvay had built soda plants in France, Germany, and England, and by 1914 his empire grew to twenty-three plants, accounting for Solvay's enormous personal wealth. One consequence of Solvay's discovery was the worldwide decline in the price of soap from $140 per ton in 1850 to $22 in 1902—his contribution to improving personal hygiene in the second half of the nineteenth century.[4]

Solvay called his doctrine *productivism,* the "social equivalent of energeticism." Claiming that modern science was dominated by the sovereign concept of energy, Solvay argued that no aspect of the universe remained untouched by it: "It is the movement, the order, the unity, the law."[5] All organisms strive toward a "maximum energetic efficiency," confirming the law of progress in the "organic regime" or biological structure of the species. Society was simply a higher level of "energetic phenomenon," in which all human activity could be reduced to the conversion or exchange of energy. These conversions could be measured in terms of their social efficiency, or what Solvay called a "coefficient of social utility." Solvay summed up his idea as both a human and physiological principle: "The best existence" is nothing other than the "physical-chemical law of maximum work."[6]

Though Solvay denied free will, he claimed that the growth of the state and the expansion of knowledge necessarily led to the improvement of the conditions through which energy was converted into social force. His *productivism* conceived of society as an enormous industrial enterprise dedicated to increasing overall productivity while encouraging social justice. The highest priority of society and nature was to eliminate "abusive and excessive energy consumption."[7] Paramount among this wasted energy was class struggle, which Solvay considered a "collossal illusion" because it diverted individuals from their most purposeful activity—"maximum efficiency."[8] "To be a productivist," Solvay claimed, was to recognize that "the true march of progress, and the well-being of human beings, can be assured by the development of the production of material things, by all possible means, in a quantity and quality appropriate to temperament, general state of health, the country inhabited, and the mode of work."[9]

Solvay conceived of his "positive politics" as a process of "free socialization," undertaken voluntarily and coordinated by the state. The sole purpose of the state was to encourage initiative and to further the development of a scientific and technological elite, or "governmental class." The responsibility of this elite was to determine scientifically the best routes to advance humankind, an idea Solvay proposed to the Belgian Senate (to which he was elected in 1892) in 1894. Solvay's "Plan Social" also included a monetary scheme, *social comptabilisme,* which would eliminate exploitation, interest, and profiteering by replacing currency with a system of social accounting books.[10] These measures could be funded, Solvay believed, by a tax on inherited wealth and by an excess profits' tax, which would be used for a "social fund."

Solvay was not a lonely prophet of his gospel of energetics. Possessing the means to realize his dreams, in 1895 he founded the Institut des Sciences Social at the University of Brussels, the first of its kind. Six years later he dissolved the institute and endowed the school of social sciences and politics at the University of Brussels for twenty-five years, establishing a new Institut de Sociologie in Brussels Parc Léopold. The new institute was near his already established institutes of physiology and of anatomy. The physical layout of the institute, as its first director, the sociologist Émile Waxweiler, put it, "is already a kind of program in the largest sense of the word."

> Ernest Solvay wished it thus, despite the astonishment of specialists in physiology and anatomy, at the time of the Institute's construction, on seeing the sociologists, who had more literary than scientific renown, installed so near them. He intended, thereby to give bodily expression to the synthesis of his preoccupations, and he associated the buildings in order to affirm the affiliation of his ideas.[11]

In Germany, the doctrine of energeticism was promoted by Wilhelm Ostwald, a renowned chemist and scientific writer, whose scientific corpus included more than forty-five books and five hundred papers. Ostwald shared Solvay's enthusiasm for energy, calling him "the founder of sociological energetics" and dedicating his programmatic *Die energetischen Grundlagen der Kulturwissenschaft* (1909) to Solvay.[12] By far the most prominent of the German advocates of the new energetics, Ostwald's cultural works, including *Die Energie* (1908) and *Der energetische Imperativ* (1912), outlined the implications of the doctrine of energy for the human sciences, in particular sociology, "the supreme science in the Comtean hierarchy." Ostwald's scientific ideas came under attack from a new generation of physicists, especially Ludwig Boltzmann, at the Lübeck Congress of Scientists and Physicians in 1895. After this defeat, Ostwald increasingly turned to his popularizations of science, his social ideas, and his activities on behalf of the German Monist League, whose presidency he assumed in 1910.[13]

Ostwald baptized his country estate in Saxony, "Energie." In a universe constructed out of pure labor power, energy was not merely the source of social wealth but the means by which "all products of energy and distance could be measured."[14] No area of human endeavor was untouched by the laws of energy. The entire effort of civilization, Ostwald maintained, was devoted to converting raw energy into available energy, to transforming the energy that is otherwise dissipated into socially available energy. Law, for example, reduced the energy wasted in violence, while competition improved the techniques of converting energy to economic use. All progress can thus be measured by the efficiency of energy conversion, by the elimination of waste, and by the coordination of energies to maximize efficiency.[15] In 1911, Ostwald summed up his work in his famous maxim: "Don't waste energy, valorize it (Vergeude keine Energie, verwerte sie."[16]

FATIGUE AND THE EUROPEAN SCIENCE OF WORK

In the late nineteenth century energeticism provided a general framework for translating the physiologists and physicians' obsession with fatigue into a coherent social doctrine. Yet these ideas also had a practical impact in the emergence of a distinctly European science of work in the two decades before the First World War. By the late 1890s, Marey and Fremont's pioneering motion studies and Mosso and Ioteyko's studies of fatigue spawned a new generation of physiologists and sociologists in France and Germany who focused on the problem of industrial work.[17] These efforts were not concerned with the worker as a member of a social class, nor with the economics of wage labor, but with the economy of the laboring body, whose rhythm and movements were subjected to detailed laboratory investigations. For these fatigue experts, work was an essentially and exclusively physiological phenomenon. The human organism was considered a productive machine, stripped of all social and cultural relations and reduced to "performance," which could be measured in terms of energy and output.

The European science of work was located at the intersection between political economy, social hygiene, and biology. In 1906 a French physician observed that although the science of labor was only in its infancy, it was rapidly taking its place among "the great questions such as tuberculosis, alcoholism, housing or unhealthy occupations, which until today dominated social medicine."[18] An international community of scientists, physiologists, and hygienists who were engaged in the scientific study of work, established outposts in France, Germany, Italy, and Russia.

Dedicated to the formation of a new scientific discipline and zealous in its pursuit of the laws of fatigue, worker's nutrition, and reform of the workplace, these experts were united by several scientific journals, especially Richet's *Revue Scientifique* and Ribot's *Revue Philosophique*, which provided a thorough reporting of developments and a rapid translation of key works. Most important were international congresses organized to facilitate contacts and publicize results. The International Congresses of Hygiene and Demography (Paris, 1900; Brussels, 1903; Berlin, 1907; Washington D.C., 1912) brought together physiologists, public health experts, and reformers concerned with poverty, epidemic disease, and industrial health and safety. At the 1903 Brussels Congress the section devoted to "fatigue in industrial work" centered on the need to extend future research from the laboratory to the workplace. Though some delegates believed it was necessary to "undertake a complete physiological and medical exploration of different categories of workers," others were skeptical that such investigations could solve the problem of exhaustion in actual conditions of industry. Zaccaria Treves of the University of Turin argued that fatigue could not be measured only by declining productivity and output and that under existing conditions the subordination of the worker to the rhythms of the machine required greater psychic intensity and a "discharge of nervous energy" even as the individual muscles were spared exertion. Declining productivity became manifest only in the latter stages of fatigue while long-term damage to the organism and the social costs remained largely invisible. Clearly, the regulation of fatigue had to occur long before its effects became apparent.

Josefa Ioteyko delivered a passionate appeal to the 1903 congress, calling for the "institution of physiological laboratories in all countries to conduct research on the physiological conditions of creation, expenditure, and regeneration of animated motors." She proposed the direct observation of workers in factories and workshops, calling for a mandatory annual physical examination for workers and demanding that governments collect data on the effects of fatigue on worker pathology. Workers' fatigue, she claimed, could be measured by ergographic tracings, but in some cases evidence of the "pathological effects of exhaustion" were evident only from the accumulation of statistics.[19]

Ioteyko's proposals reflected a broad consensus at the Brussels Congress that it was "absolutely necessary to establish, by scientific methods and under the real conditions of industrial organization, the extent of the exhaustion among the working population."[20] Despite some disagreement over the value for industry of fatigue research confined to the laboratory, the delegates considered the science of work a promising means of resolving the social issues of the epoch. Direct observation of the working body and of the energy required by the specific conditions of work could contribute to resolving the worker question by "assuring a perfect functional equilibrium which permitted the worker to use his physical and moral education to his best advantage." It was vital, the congress concluded, that governments be encouraged to promote "studies of occupational fatigue."[21] Such efforts could deliver to reformers "indisputable" arguments on behalf of a shorter workday, higher wages, and increased productivity.

In 1905 the French Office du Travail sponsored the first studies to use the science of work to effect reform legislation. The Montpellier physician Armand Imbert was commissioned to conduct a series of experiments comparing a wheelbarrow *(brouette)* and a two-wheel cart *(cabrouet),* used for transporting stones in a masonry.[22] To Imbert's surprise when the Chamber of Deputies enacted a law regulating work of this kind in 1908, it used his studies but somehow reversed its conclusions, banning the "stone" cart despite Imbert's evidence of its superiority to the wheelbarrow.[23] Despite this unfortunate outcome, Imbert welcomed the government's reliance on physiologists for drafting legislation. His experiences also convinced him of the benefits of "moving to a direct, experimental and microscopic study of occupational work itself," and of taking into account "the effects of the length and intensity of the working day with the aid of facts and rigorously scientific procedures."[24]

In the same year as Imbert began his research on the carts, an Italian physiologist, Professor Pieraccini of Florence, collected detailed statistics on the output of six experienced typesetters. Noting the average number of lines set and the number of errors, which he compared with the time of day, he discovered the growing effects of fatigue on the number of mistakes as the lunch hour and end of the workday approached.[25] Studies of other occupations followed: Jules Amar's work on the filer in 1910; Imbert's investigations of the cutting of vines by women in the wine industry of the Midi region and his investigation of French glass polishers; and Charles Fremont's detailed 1913 work on riveting.[26]

Limited in both scope and impact, these studies represented the first efforts of scientists to demonstrate the efficacy of their techniques for social policy. Their authors considered the support and collaboration of government indispensable for even the least ambitious of undertakings. In his pioneering *Thèse* on the efficiency of the human machine, *Le Rendement de la machine humaine* (1909), Amar wrote that it was necessary above all to collect a "considerable number of measurements reported over a long period of time."[27] Amar was born in Tunis in 1879, son of an Algerian Jewish father who was a naturalized Frenchman. He arrived in Paris to study theology at the École Rabinique de France but soon abandoned his religious vocation to study botany, biochemistry, and, eventually physiology with Auguste Chauveau. His studies were interrupted by three years of military service, after which he became an assistant *(préparateur)* to the physiologist Georges Weiss at the Faculty of Medicine in Paris.[28] In 1907 as part of a plan by the Ministry of War to conscript the natives of the French colonies, Amar was chosen to undertake a long-term study of the working capacities of North Africa's indigenous population.[29] Under the auspices of the Office du Travail, and with the assistance of the French Ministry of Public Instruction, Amar was dispatched a year later on an official mission to North Africa, where he conducted his investigations on prison inmates at Biskra in Algiers—as he put it, unconstrained by either "bad will" or "numerical insufficiencies."[30]

Amar employed various innovative devices, including his own *ergometer (ergometric monocycle)*, which combined a bicycle with a respirator and an ergograph. Subjecting the prisoners to a series of tests to determine the optimum efficiency, or "yield," of the human motor, he selected his subjects according to occupation, excluding habitual criminals. Amar was convinced that his native subjects were "not unique as motors" though "the nature of their diet is the principle factor of their difference from us." In contrast to the bulk of ethnographic research conducted at that time, Amar's was entirely "physio-logical and social."[31]

Amar's report delivered in June 1909 to Georges Clemenceau, pres-ident of the Council of Ministers, was highly critical of colonial adminis-trators whom he reproached for their ignorance of the basic principles of physiology.[32] Amar was especially contemptuous of the old-style colo-nial elites, who repeated shibboleths about the power of the will and believed "that man works with his nerves." "Diet is more determinant than the will or self-love in work," Amar countered.[33] Amar also found that rural workers were inferior to urban workers and that "the Moors and the Kabyles were superior to the Arabs in the *amount of daily labor* of which they are capable."[34] Nevertheless, he concluded, "the Arab makes an ardent and even a devoted worker or soldier, on the sole condition that his daily bread is guaranteed to him."[35]

Amar's conviction that the economies of human labor power were malleable and that the laboratory study of fatigue had to be superceded by the direct investigation of work also appears as the central theme of his 1909 *Thèse.* There he argued that "the yield of the human machine is *variable* according to the mechanical and energetic conditions of work. It depends on time, on speed, on the employment of this or that group of muscles which intensifies or reduces the expenditure of en-ergy. It rises or falls according to the working subject and the greater or lesser contraction of the active muscles. . . . Thus, we see how the capable worker is able to economize his forces. If there is a science of work, there is also an *art* of work."[36]

If fatigue represented the physical and mental limits of a body's capacity to work, Amar argued, then the greatest use of these studies lay solely in establishing "the limit at which the human forces are incapable of further exertion."[37] But precisely because it is a "pathologi-cal sensation," he added, fatigue is "not susceptible to exact measure-ment." For Amar, the science of work had to go beyond the study of fatigue: "It is therefore desirable to substitute for a sensation, fatigue, the direct estimation of possible work."[38] As a result of his Biskra experi-ments, Amar concluded that a science of work could in fact calculate the amount of energy output necessary to fulfill optimally and economi-cally a given task: "The estimate of the work done should be made according to the expenditure of energy," that is, on a scientific basis.[39]

Upon his return to Paris in 1910, Amar continued to conduct his experiments in Georges Weiss's laboratory where he remained until 1913. He concentrated on the most economical expenditure of force in specific industrial tasks, usually involving simple tools. A plane or a file was connected to a recording cylinder similar to Marey's tracing instruments, so that the speed or muscular activity of the work could be measured. At the same time the worker was connected to a respiratory machine that simultaneously collected and analyzed the chemical composition of the air expended. "The closest attention must be given to the movements of man and tool," Amar wrote, "for thereby great economies both in time and effort can be realized."[40] Amar devised a dynamographic measuring apparatus that could be adapted to all kinds of tools and that furnished "clear and faithful" recordings of the work done.[41]

Amar was convinced that all tasks using the same tool required a similar output of energy and that when a muscular action was exercised in the same fashion, it always resulted in the same tracing. By this technique, a worker's movements could be carefully observed and charted, "with a view towards eliminating all that is unnecessary."[42] Any deviation could be explained "principally because the worker is lacking in skill."[43] For example, Amar compared the results obtained by a "good workman, skillful and well trained," with those of a less-practiced apprentice. He found that the accomplished journeymen normally adopted an efficient economy of motion that starkly contrasted with those of an apprentice, whose "chief defects are irregular and spasmodic action leading to unduly rapid fatigue."[44] Under optimal conditions, a skilled worker, or one who follows the course prescribed by precise measurements, can minimize fatigue with no appreciable impact on performance. In 1910 Amar formulated this principle as "Amar's law": A muscle returns to its condition of rest in direct proportion to the rapidity of its performance of work."[45] This discovery, he claimed, had practical consequences:

> By correcting the trifling defects of position displayed by the worker, by applying conventional or rule of thumb habits, and by taking into account results of our dynamic and energetic measurements, we have been able to determine the normal position of the feet, the proper distance of the body from the vise, vertically and horizontally, and the positions of the hands with regard to the tool. When these conditions are fulfilled the worker's fatigue is diminished without injury to his daily output.[46]

Under experimental conditions, and with mathematical and scientific rigor, Amar studied the mechanics of the human body in microscopic detail. Following Imbert's lead—he credits him with being the first, after Marey, to use mechanical inscription to plot the "muscles and tools of workers"—Amar applied various techniques of measurement to different types of work, which he reproduced in his laboratory. In addition to the work of the hands, of the legs, and of different tools, Amar also investigated the transformation of chemical and caloric energy into work; the nature of respiration, diet, clothing, and hygiene. He conducted experiments in prisons and sometimes in small industrial workshops. He developed sophisticated techniques of measurement, inventing almost all of the basic techniques of modern ergonomic measurement. In December 1913 Amar was named head of a Laboratoire de Recherches sur le Travail Professionnel attached to the Conservatoire National des Arts et Métiers (see chapter 9). There he expanded the scope of his investigations, focusing on specific trades, in particular the filer. He pioneered the training of apprentices and the study of their aptitudes for different types of work, which were in part based on "morphological" comparisons of different body types.[47]

With Amar's systematic inquiry the search for the dynamic laws of fatigue was extended to the search for the dynamic regularities of work. Amar was motivated by a single principle: scientifically predetermined optimum speed and position achieve a maximum amount of work with a minimum amount of fatigue. Indefatigable in his pursuit of the laws of expenditure of energy, Amar used his theory to explore not only working with tools, but writing, playing musical instruments, engaging in athletics, sports, and the military.[48] He experimented on himself, but also on different national and ethnic groups, producing a comparative anthropology of labor power. In 1919 Amar reviewed with pride his first decade of work: "This method has for ten years given proof of its simplicity and reliability. During that time it has been applied to about a thousand persons—Parisian working-men, soldiers, and natives of North Africa. It is therefore of universal applicability, and for that reason eminently scientific."[49]

Amar was also convinced of the intimate connection between social efficiency and social harmony. An average of one third of the available energy of man is wasted," he argued, while a more "methodical" organization could easily increase industrial output. Increased output based on science could resolve the conflicts between capital labor on a new basis: "We are convinced that these interests [of the workers] agree with those of the employers. The science of energy estimates exactly the daily sacrifices of the workers while professional technics teaches the methods of increasing human yield." For Amar, the science of work promised a scientific and technological solution to the social question. Labor and capital could be united through the proper "education of movements."[50]

ARBEITSWISSENSCHAFT: THE SCIENCE OF WORK IN GERMANY

In Germany the term *Arbeitswissenschaft* was directly adopted from the French "science du travail" in the mid-1890s.[51] Two German schools emerged at the outset: the physiological school of Otto Fischer, Christian Wilhelm Braune, Nathan Zuntz, and Ludwig Max Rubner; and the psychophysical researches of Emil Kraepelin and Hugo Münsterberg.[52] Fischer's treatise on the human gait, *Der Gang des Menschen* (1895), applied chronophotography to the movements of a crack Prussian regiment at drill.[53] By far the most important work of this genre, Nathan Zuntz and Wilhelm Schumburg's *Studien zu einer Physiologie des Marches* (1901), surveyed all aspects of military drill: clothing, nutrition, the length and speed of the march, the causes and consequences of fatigue, the influence of rest, and above all, the "pathology of the march" and march-related illnesses (for example, heat-stroke).[54]

Often regarded as the classic work of the scientific physiology of labor, the Zuntz-Schumburg study summed up German applied physiological research at the turn of the century.[55] It also pointed to what its authors considered the most fertile field of application for the new physiology of the human motor: state institutions like the military and public education. Rubner, the leading German nutritionist and an advisor to the Prussian military board, was requested to develop food plans for Prussian prisons and to advise the government on other health-policy issues.[56] In 1903 he became vice chairman of the Imperial Health Board (Reichsgesundheitsrat) and later, a member of the Scientific Council for the Army. In 1913 Rubner founded the Kaiser-Wilhelm Institut für Arbeitsphysiologie, in Berlin, where he expanded his nutritional focus to a panoply of work-related physiological concerns (see chapter 9).[57]

In contrast to the explicitly military orientation (though often from opposed medical viewpoints) of the German physiologists, Emil Kraepelin employed the methods of experimental "psycho-physics" to study both physical and intellectual fatigue. A student of Wilhelm Wundt, the founder of quantitative experimental psychology in Germany, Kraepelin first applied Wundt's approach to the study of mental illness, which remained a major terrain of his activity and over which he exercised great influence as Ordinarius Professor of Psychiatry.[58] Kraepelin began his experimental studies of fatigue in his Heidelberg laboratory in 1892, publishing some of the earliest experimental studies of labor in Germany: *Über geistige Arbeit* (1894); *Zur Hygiene der Arbeit* (1896). Taking as his point of orientation the psychophysical approach first developed by Ernst Heinrich Weber and Gustav Theodor Fechner in the 1860s, Kraepelin argued that basic physiological processes could only partially account for differences in "performance capacity," especially in mental labor.[59] Above all, Kraepelin believed that "performance capacity" was primarily a psychological phenomenon and had to be studied from the perspective of psychology, rather than from a purely medical or biological standpoint:

> The issue of fluctuations in mental *(geistig)* performance first came to me in consequence of my efforts to measure the influence of external effects, especially toxins, on psychic life. Such experiments demonstrated that the duration of certain simple psychic processes was consistently not the same on different days under the same external circumstances. . . . Under these circumstances it proved necessary to first investigate what changes in mental performance were the consequence of inner causes, apart from changes in external conditions.[60]

Like Mosso, Kraepelin considered "the central goal of this young science the struggle against fatigue, since in fatigue lies the curse, the danger of labor."[61] But he rejected the ergograph or aesthesiometer, which he claimed did not measure mental fatigue but only some of its physical or tactile by-products. Instead, he proposed calculating the "changes which occur under the most diverse conditions of everyday life" through a series of simple tasks to be solved within a given time, usually five minutes.[62] One common procedure was to test the ability of individuals or groups to memorize nonsense syllables or numbers, demonstrating by counting errors that each individual evinced a different rate of fatigue independent of mood. The course of fatigue could then be directly observed "in the progressive fall of work performance." Under the effects of fatigue, Kraepelin claimed, thought as well as "bodily movements lose not only energy, but safety and subtlety."[63]

Kraepelin's focus on the "task" underscored the link between the German *Arbeitswissenschaft* and modern industrial processes, as opposed to the skilled or artisanal labor on which the French seemed to concentrate. For Kraepelin, industrial work was based on adjusting minute physiological and psychic processes to the rhythms of the machine. The "mechanization of work" increasingly demanded that bodies, or corporal mechanisms, be harmonized with the work process. Since "the muscle is the most highly perfectable dynamic machine," Kraepelin argued, this harmony could most easily be achieved by training a new generation of workers in psychophysical methods, which would result in a "constant improvement in the relationship between physiological work performance and physical energy performance."[64]

Kraepelin primarily emphasized the need to separate performance *(Leistung)* from such subjective aspects. His theory rested on the crucial distinction, which he introduced, between tiredness *(Müdigkeit)* and fatigue *(Ermüdung)*. For Kraepelin these were two "totally different states." For some people "the warning signal of tiredness persists despite fatigue," while for others "fatigue is not accompanied by such manifestations at all."[65] In hundreds of micro-studies and experiments, Kraepelin and his assistants attempted to eliminate the ephemeral qualities of tiredness, attitude, and satisfaction *(Arbeitsfreude)* from the objective course of fatigue. Kraepelin plotted fatigue on his "work curves," which traced the energy expenditure of the body or mind. Not only was each individual's rhythm and work tempo unique, but work performance could, he claimed, be increased through training or practice *(Übung)*.

German *Arbeitswissenschaft* universalized the concept of work, equating it with "performance" in the industrial workplace. Along with Hugo Münsterberg, who pioneered the study of "Psychotechnics" at his laboratories in Berlin and at Harvard University, Kraepelin compared the "rhythmically structured" work of past epochs with the rigid economy of force imposed on the body by machinery. Influenced by Bücher's ethnological work on the rhythms of primitive work habits, they saw an intimate connection between the development of modern technology and the "unnatural" character of modern work processes. Following Marey and Fremont, Kraepelin and Münsterberg focused on the corporal dynamics of industrial work, but unlike the French, they explicitly linked the progressive and inevitable *reduction* of the work of the body's movements to industrialization. Mechanized work, they observed, reduced physical effort to those muscles directly engaged in the actual process, eliminating all superfluous motion. Whereas the French physiologists tended to focus on single tasks such as vine cutting, hammering, or filing, Kraepelin was concerned with a uniquely physiological phenomenon, which he called "the principle of the smallest muscle."

Kraepelin considered this discovery his most important contribution to understanding the connection between the psychophysics of fatigue and modern industrial work. The steady displacement of force from the larger muscles to the smaller was the historical teleology of the new factory: as modern industrial technology required increasingly precise tasks, it inscribed smaller and smaller trajectories of motion on the body. As a result of this progressive dematerialization of labor power, energy could be more efficiently deployed (since brute strength was now obsolete) by concentrating on one or more movements. Exercises could be designed and applied strategically to affected muscles, permitting a slower decline and a lessening of fatigue in the work curve.[66]

Kraepelin's discovery of the displacement of work performance from muscles demanding greater expenditures of energy to those demanding less had, he believed, crucial social implications. He observed that with modern industry, the body's economy of energy steadily increased, and, through training—the repetition of the same activity—conservation of energy is also constantly augmented. Since the division of labor constantly improved performance while reducing energy consumption, the progressive differentiation of function in modern industry resulted in "increased savings of energy."[67] The "law of the least effort" was a social imperative.[68] It constituted a historical principle of energy conservation in the relationship between machinery and the human motor, a homology of technology and the body: "The loss of energy is always lessened while the economic performance of work is constantly increasing."[69]

Even more than Kraepelin, Münsterberg was convinced of the significance of laboratory studies for industry and of the "extraordinary saving of psycho-physical energy" that these studies would yield. In his laboratories in Berlin and at Harvard University, Münsterberg experimented on the "rhythmic movements of the hand, the foot, the arms and the head under different conditions of resistance [to determine whether] the greatest precision of rhythmic motion can be achieved for different muscle groups at the same speed." The results of "one-third of a million measured movements is the proof that every muscle group possesses its own optimum speed for the greatest possible precision," he asserted in his 1914 work, *Grundzüge der Psychotechnik.*[70] A "transformation of the psycho-physical capacity for performance," would not only improve industrial output, but also benefit workers by reducing accidents and by shortening the workday and workweek. By providing scientifically determined pauses and increasing free time, science could improve the conditions of work without reducing the overall productivity of labor.[71] Like the French, Münsterberg insisted on a strict neutrality for his industrial psychology as a science serving economic progress and social efficiency rather than pandering to "a reckless capitalism on the one side," or "a feeble sentimentality on the other."[72]

In contrast to Kraepelin, Münsterberg took into account precisely those subjective psychological "factors" that Kraepelin had exorcized from the study of work performance. In his *Psychologie und Wirtschaftsleben,* published in 1913 (an altered American edition, *Psychol-*

ogy and Industrial Efficiency appeared in 1912), Münsterberg concluded that "the psychology of individual difference is of decisive importance for industrial psychotechnics."[73] Such questions as the propensity for fatigue, exhaustion, and recuperation; the disposition to learn and to practice; the capacity to derive benefits from repetition; the tendency to forget; the personal rhythms of work; attentiveness and intelligence—all were critical factors determining the aptitude and success of a particular worker in a particular setting. What Kraepelin regarded as ephemeral, "character," or "mood," were critical, Münsterberg claimed. Münsterberg also observed that his experiments demonstrated that working in a common room produced better results than isolated activity. But he cautioned that effectiveness was enhanced when workers could see one another, but when the seating arrangement inhibited "the possibility of chatting."[74] As early as 1913 Münsterberg predicted that "the significance of such investigations could be so considerable, that the thought is not so far off, that in the future factories will employ psychological experts."[75]

Though both aspired to greater influence, Kraepelin and Münsterberg conducted their work in the cloistered atmosphere of German (or American) university laboratories with little outside impact. The highly experimental German *Arbeitswissenschaft,* and its remoteness from the world of production and work, soon produced a growing dissatisfaction with its lack of practical results. In contrast to the French, who by 1900 were beginning to close the gap between the laboratory and the workplace, Münsterberg complained of the lack of experimentation under real conditions and warned that the "schematic application" of the results of laboratory experiments to actual industrial settings could lead to many faulty conclusions. After 1913 he began to design more practical experiments to discover "the individual dispositions, associations, and reactions" of American workers.[76]

By 1910 critics both inside and outside the research community began to view Kraepelin's claims with increasing scepticism. His exclusive emphasis on work performance as a measure of fatigue was opaque and the results of his studies far less "homogeneous" than Kraepelin had initially assumed. Moreover, Kraepelin's dogmatic insistence that a line could be drawn between subjective weariness and objective fatigue was not, as subsequent experiments showed, borne out by his own laboratory data.[77] "Kraepelin and his students placed exaggerated weight on their repetitive and not always accessible experiments," commented one doubtful psychologist after carefully reviewing the literature.[78]

These criticisms coincided with a growing suspicion that there was no relationship between the abstract, laboratory psychology of work and concrete work—a charge that Max Weber leveled against Kraepelin in 1909 in his book-length review of the psychophysical school.[79] Weber concentrated on the limitations of Kraepelin's laboratory setting: His subjects worked steadily, expending maximal effort, and in contrast to workers outside the laboratory, had no economic motives. Their diets and alcohol consumption were rigidly regulated. No actual worker could be supervised to the same degree, as real workers were motivated by money and influenced by diet, sleep patterns, and sexuality. In a real industrial environment, moreover, frequent mechanical breakdowns occurred, and workers tended to work as little as possible.[80]

For sociologists like Weber, as well as for a growing number of experimental psychologists, Kraepelin's results—apart from his proof that rest pauses improved performance—were disappointing and the consequences of his studies for the workplace negligible. The difficulty of measuring fatigue, and "the numerous possible disturbances and errors of measurement," Weber concluded, "had completely destroyed the hope of 'exact' measurement of the extent of the effects of fatigue in the actual work of a single school class, not to mention for a thousand workers."[81] Even Kraepelin seemed to concede his lack of success when he replied that "the first truly exact investigations will not be experienced by either of us."[82]

Weber's trenchant criticism of Kraepelin (and Kraepelin's retreat from his most assertive conclusions) were damaging, though they did little to discourage the development of "objective" experimental psychology in Germany. Nor, it might be added, did Weber spare his ample wit and irony in reviewing the accomplishments of Ostwald and Solvay, whose "spillover" of a scientific "world picture" into a mundane "worldview" he denounced as "intemperate arrogance." Ostwald, however, fared better than Solvay, whose amateurish efforts at sociology Weber regarded as little more than a parody of the positivist precursors, Comte and Quetelet.[83] Weber's withering comments on the energeticists are more significant from another standpoint. They attest to Weber's own interest in developing an alternative approach to the problem of labor power: the sociological survey project of the Verein für Sozialpolitik (Association for Social Policy).

Weber and his brother Alfred were among the leading intellectual figures in the Verein, an academic association founded in 1872, to oppose both "social reactionary Manchesterism" and "social revolutionary Marxism." By 1890 the Verein had become a professional organization of academics, civil servants, and a few industrialists concerned with applying the achievements of social science to central issues in German social policy.[84]

THE GERMAN SOCIOLOGY OF WORK

Max Weber's role in the development of a sociological approach to the psychophysics of industrial labor did not merely derive from his disappointment with the Kraepelin school. During the year 1908 Weber spent much of his time absorbing the medical, psychological, and economic literature on the problem of work, resulting in his comprehensive essay, "The Psycho-Physics of Industrial Work." His interest in the subject was in part sparked by the work of Adolf Levenstein, a private scholar who produced a survey of German workers' fatigue, *Die Arbeiterfrage* (1912), the first of its kind in Europe. Levenstein, a unique figure in the history of empirical sociology, was a politically motivated outsider, about whom little is known apart from his survey, his collections of workers' poetry, and his remarkable volume of workers' letters on Nietzsche's *Zarathustra*.[85]

Inspired "by the wish to comprehend, as far as possible, the fundamental connections and consequences of technology on the life of the soul," Levenstein conducted one of the earliest large-scale attitude surveys.[86] Levenstein subtitled his work, "with particular attention to the social psychological side of the modern large industrial plant and its psycho-physical effects on the worker." He sent eight thousand questionnaires to workers in three categories: miners (in the Ruhr, the Saar region, and in Silesia); textile workers (in Berlin Forst); and metal workers (in Berlin, Solingen, and Oberstein). Although he at first welcomed the cooperation of the trade unions, he soon abandoned that course, observing that efforts to distribute his questionnaires in mass meetings produced uniform responses—the products of what he called "mass phantasy." Instead, Levenstein preferred to mail his questionnaires directly to workers, with a letter requesting that they (and their wives) fill them out (according to Levenstein he had a 63 percent return).[87] As the historian of social science Anthony Oberschall observed, Levenstein's *Arbeiterfrage* "illustrates once more the extent to which at the start of a new science an outsider and dilettante can be a true innovator and far more successful than the professionals."[88]

Levenstein's *Arbeiterfrage* contains an extraordinary wealth of information on social and cultural matters, such as books read, ideals, hopes, and expectations among the thousands of miners, textile workers, and machinists he queried. But his primary interest centered on fatigue. He accumulated significant data on the extent and types of fatigue experienced by workers. The subtlety and depth of many of the individual responses is striking: One metal worker, for example, who had spent much of his life on the land, aside from the last two years in factory work, noted: "After the weekend and two hours before the end of the shift I am already tired. But it is a different tiredness. Not tired like at home, when you work the whole day with a scythe or walk behind the plow. Here my limbs always tremble. And also in the country after work there was so much laughing. Here you only see sullen faces."[89] A weaver in a textile factory reported that even the greatest fatigue "cannot effect the performance of the weavers to the smallest degree. The mechanical weaver's loom does not tire."[90] A fifty-two-year-old miner wrote that "fatigue appears only when the work time is over."[91] If French and German physiologists and psychologists produced an "objective" approach to physical and mental fatigue, Levenstein's *Arbeiterfrage* provided the personal, "subjective" dimension. His survey offers a rich variety of material for studying the impact of fatigue on the worker's attitudes and emotions. Levenstein initially planned to write such a study under the title "Technology and the Life of the Soul, Contribution to a Psychopathology," but apparently little came of the project.[92]

Levenstein conducted his survey between August 1907 and April 1911. Before publishing his results, Levenstein had written to Max Weber and several other academic economists for advice. Weber tried to persuade Levenstein to permit university students to collaborate with him and advised him of the methodological limitations of his survey.[93] More important, his survey provoked the leaders of the liberal academic association, the Verein für Sozialpolitik, to undertake a social scientific survey of the "psycho-physics" of industrial labor.[94] In 1907, the Verein's annual convention proposed that it undertake an unprecedented survey of the impact of industrial work on workers' attitudes and circumstances. At the end of 1908, Alfred Weber and Karl Bücher wrote a prospectus on the project and the leading intellectual figures of the Verein, Alfred Weber, Heinrich Herkner, and Gustav Schmoller, were appointed to direct the survey and develop a research plan.[95]

Max Weber, the intellectual force behind the Verein's survey, mapped out the overall theoretical and conceptual plan in a series of articles written during 1908 and 1909.[96] In a lengthy prolegomena entitled "Methodological Introduction to the Survey of the Verein für Sozialpolitik on the Selection and Adaptation (Occupational Choice and

Occupational Fate) of the Workers in Large Industries," Weber outlined the purposes of the survey:

> The present survey aims to establish the following; on the one hand, what
> influences the large-scale industrial establishment exerts upon the individ-
> ual character, the occupational fate and the style of life of its working force,
> what physical and psychical qualities it helps develop in it, and how these
> (qualities) become manifest in the conduct of the daily life of the workers;
> on the other hand, how the development and the potential future develop-
> ment of large-scale industry is limited by those characteristics of the work-
> ers which are a result of their ethnic, social and cultural origin, of their
> traditions, and standards of life.[97]

Between 1908 the Verein completed fifteen separate investigations and
published the results.[98]

Weber's conception differed significantly from Levenstein's, whose
amateurish methods and conceptual unclarity he disparaged. The
Verein's survey, carried out between 1909 and 1911, also differed from
Levenstein's in its political aims. Levenstein's study was impressionistic
and oriented toward heightening public visibility of workers' problems,
whereas the Verein's survey was systematic and reserved in its judgments, explicitly refusing to recommend any legislation or reforms.[99]
Throughout, Weber insisted on the "value neutrality" of the Verein's
survey and its lack of either "directly practical" or "social political
goals."[100]

Despite these disavowals, the two projects reveal a decisive difference in orientation. Whereas Levenstein clearly wanted to chart the
destructive effects of fatigue and overwork on the psychic makeup of
the worker, the emphasis in the Verein's study was on "selection and
adaptation of workers for large industries" and on the impact of cultural, social, and occupational factors on worker output.[101] If Weber
criticized Kraepelin's method, he borrowed its overarching emphasis
on "performance" measured through productivity—a focus that remained the key element in his methodological approach and in the
Verein's surveys. In contrast to the laboratory fatigue study and Kraepelin's "work curves," the Verein's study emphasized the role of performance in actual work and the tendency of workers to "regulate"
their own productivity and to respond to economic, as well as physiological, factors in real-life situations.[102]

Weber argued that the starting point for any investigation of
large-scale industrial work should be the physiological and psychological factors influencing "performance-ability." However much hered-

ity, class background, nutrition, or culture might influence a worker, performance was the decisive factor. From an economic point of view the decisive issue was "profitability," and from that standpoint, "The worker is in principle nothing more than a (as much as possible) profitable means of production with its specific qualities and quirks."[103] Weber thus translated Kraepelin's psychophysical criteria to actual work settings. He recognized, however, that though a study might measure overall performance, worker productivity was extremely difficult to determine in the context of a profit-oriented firm. He also cautioned that even if many diverse factors were taken into account, output alone could not be an absolute measure of worker performance. Nevertheless, Weber did acknowledge the salutary effect on output of the enlightened policies of the Zeiss optical firm owned by Ernst Abbe, a paternalistic reformer who reduced hours while increasing productivity.[104]

The Verein's studies promoted a highly statistical style of analysis, focusing on measurable aspects of aptitude and on long-term "life chances," or "occupational choice and occupational fate."[105] Characterized by a careful description of the plant and the work process, the Verein's studies concentrated less on the individual attributes of the worker than on providing a social profile of the labor force: its changing age and gender composition, its social mobility, and its transformation over time. Also considered were the influences of age, social origin, family background, experience, fatigue, effort, work motivation, and leisure.[106] The new empirical social science was primarily concerned with determining the optimum yield of labor power conceived as a social phenomenon, or what Alfred Weber called a "total psychology of class."[107]

Kraepelin's individual work curve was superceded by the Verein's "performance curves" of men and women workers in a variety of industrial trades, especially textile workers. Unlike the French science of work or that of the German psychologists, the emphasis of the Verein's surveys was not on the quantitative "energy expenditure" or "performance" of the individual worker and the worker's body, but on the social body, or "corps," of workers conceived in terms of conditions of recruitment, structural components, optimal deployment, and strategic mobilization of forces. "The capitalist apparatus may be built up of a differentiated and interconnected system of work activities and from different hierarchies of social position," noted Alfred Weber, "but it has the tendency to suck every generation into this hierarchy, without any regard for its historical development, as something new, as a natural imperative of its own. Every generation is thus thrown back again on a general firmament of labor power, which it somehow purposefully realizes *(verwertet).* "[108] The combination of organic and energetic metaphor in Weber's description of the capitalist metamorphoses of labor power is instructive. In the vocabulary of the Verein's surveys, energeticism was translated into an empirical social science.

Though the "value-neutrality" of these surveys was assured by the politically neutral and nonpartisan approach of the investigators, they were constructed in a way that clearly linked improved work performance and productivity with reform.[109] This is not surprising since it is clear that despite disavowals of any *a-priori* political "interest" and despite Max Weber's claim that the surveys were intended only to test hypotheses and scientifically explore such issues as the relationship between occupational mobility and the life-cycle of the factory worker, in fact, the general purpose of these studies was to provide support for the liberal wing of the Verein für Sozialpolitik's social policy initiatives.[110]

At least three distinct purposes are evident in the Verein's surveys: The first aimed at decomposing an undifferentiated or Marxist notion of "class" through a more "status-oriented," or segmented, conception of the workforce—for example, by focusing on the decline of the skilled workforce; on the growth of a largely unskilled and female labor force; and on the emergence of the white-collar sector. The second aimed at increasing the profitability of firms by introducing new methods of calculating output and efficiency; and a third pointed to the beneficial consequences of reducing fatigue, overwork, and eliminating the social risks of aging or injury (these did include some specific recommendations, such as frequent changes of position to avoid monotony and financial support for older workers). In practice these goals were often at cross-purposes, however closely entwined they may have appeared to the authors of the studies.

Weber recognized that industrial enterprises were concerned with production costs, not with achieving greater joy in work. The actual economic gain that might be derived from greater control and surveillance of the expenditure of labor power, as opposed to the calculation of labor costs and profits, did not seem always justified from an economic standpoint. Though an initial rise in productivity might indeed result from such efforts, most managers believed a moderate increase in wages or piece-work systems served the same purpose at a lower cost. Improved technology was also a higher priority than motivated workers, and the degree to which the performance of the individual worker actually determined the quantity and quality of the product, or its profitability, remained economically unproven. Expensive administrative costs had to be weighed against the higher performance that might be achieved. In 1909 only a narrow group of industries employed any calculation of the yield of labor power, and these did so as part of the output of machinery, not of the workers themselves.[111]

The role of social science was to redraw the contours of industrial work in a new language that refused to employ either the vocabulary of class struggle or of market liberalism, removing the study of industrial work from the conflicting interests of workers, political parties, and ideologies. The Verein's efforts to create an empirical sociology provided detailed information about the "life fate" of the worker, about age, cultural background, and aptitude in order to disaggregate the abstract, politicized notion of "class." At the same time, these studies encouraged management to effect a higher rational "calculation," insofar as they argued that selecting workers based on aptitude and background, as well as introducing more effective systems of "motivation," could ultimately increase productivity and profitability. They also aimed at progressive reform to the degree that they were preoccupied with extending the well-being and longevity of the workers, and with reducing the element of risk faced by a modern industrial working class. In short, the underlying orientation of the Verein's surveys was to promote a politically neutral and rational calculus of energy expenditure and productivity.

The Verein's efforts assumed that increased social knowledge could contribute to eliminating the irrational aspects of the enormous social changes occurring on a broad scale in Germany and encourage state social policies toward the rational deployment of the nation's industrial energies. Although never explicitly proposed in these studies, the Verein's surveys consistently pointed to the proper selection of workers according to age, aptitude, and sex, and the development of standardized employment policies. As Marie Bernays found in her flagship study of the Rhineland textile firm, the Gladbacher Spinnerei und Weberei A.G., the period of highest performance for both sexes was between the ages of twenty-five and forty, though (obviously because of their child-rearing responsibilities) women lost their working capacity more rapidly than men in the higher age brackets—a fact that, Bernays claimed, explained the fewer number of women laborers in that age group. Bernays also found that social origins affected output: among male workers there was a direct correlation between the size of the city of origin and work performance. The village or rural outpost still provided more workers who worked intensively; middle-sized cities were less provident, while large cities provided the least efficient workers, which Bernays calculated as thirty percent lower productivity than middle-sized cities. Among women workers, too, the rural village yielded the best source of "labor power."[112] Bernays explained this phenomenon by pointing out that less efficient workers migrated from medium cities (where the firm she studied was located) to larger ones, representing "a selection of the less suited."[113] Alfred Weber concluded from these results that the flexibility of the "fate of the worker"—the mobility— over a lifetime was more significant for productivity than any static conception of either class or status.[114] Although none of the Verein's surveys proposed policies to recruit workers from the more productive provinces or to weed out older or female or other unproductive labor- ers through "artificial selection," it was clear that the thrust of these studies was toward rationalization and reform, in equal measure.

Ultimately, the surveys remained inconclusive on the actual rela- tionship between social background, culture, and shorter hours and rest on work performance. While some of the studies pointed directly to the beneficial effects of shorter working days and increased rest periods on productivity, others showed that higher performance and increased labor intensity (due to shorter hours) actually produced increased fa- tigue and exhaustion, as in the Stuttgart automobile firm (Daimler) studied by Fritz Schumacher.[115] In point of fact, these aspects seemed to be secondary to basic worker physiology or purely technological imperatives for determining the long-term productivity of the worker. In his 1909 essay, Max Weber still claimed that payment schemes were the best means of raising work performance. A year later, Alfred Weber accused his brother of ignoring the results of the Verein's studies, which had shown that worker performance owed less to methods of remuner- ation or cultural factors than to the technological requirements of in- dustry, the age of the worker, and the general level of skill.[116]

Most striking were the results of Bernays' unique reversal of the relationship between fatigue and productivity. Instead of trying to mea- sure the objective impact of fatigue on work performance as Kraepelin had done, Bernays investigated the impact of performance on fatigue. Ironically, she found that the more capable (*je tüchtiger*) the worker, the more fatigued and hostile he or she was to the work. Average- and lower-paid workers of both sexes, by contrast, experienced less fatigue ("not fatigued . . . only fatigued at night") than did highly paid produc- tive workers, demonstrating the principle that "higher wages coincided with lessened joy in work."[117] Those evincing "satisfaction" in work were also less efficient while "dissatisfied" workers were often "strivers" who attempted to move up from subordinate positions into supervisory roles.[118] Bernays concluded that "the 'unfatigued,' 'unexerting,' and 'satisfied' worker is in every instance less efficient, while the 'most fatigued,' 'overexerted,' and 'most dissatisfied' worker is the most useful." This pessimistic conclusion seemed to indicate that the raising of productivity could not in the long run eliminate fatigue.

The sociology of labor thus adopted the scientific categories of fatigue, performance, and output as the *sine qua non* of any science of work. In contrast to the biological and psychological reductionism of the French *science du travail* and the German *Arbeitswissenschaft*, the industrial sociologists redefined "optimum performance" in broad terms such as the "life fate" of the worker and attempted to calculate "performative economy" in a social framework, taking into account the profitability of the plant. In contrast to the physiological emphasis on the yield of the individual laborer in France, the Verein promoted empirical social science in assisting firms in calculating optimum performance in relation to the basic requirements of capital and industry. By 1911, when the Verein debated the implications of its surveys at its general assembly in Nürnberg, the prospects were auspicious for a more decisive impact on both the state and industry than those of the more strictly physiological and psychological *Arbeitswissenschaft*. In his concluding remarks, Max Weber noted with pride that as a result of the first surveys "a number of large factories that had previously not employed them, had installed systems of calculation *ad hoc*, because they gained the impression that those questions we ask for our own purposes, might also be of value for firm managers and their cost calculations, e.g., for the correct leadership of their plants."[119] With the success of the empirical sociology of labor, the beginnings of a thorough rationalization of what later would be called "the human factor" in industrial production became established in Germany. In the future, Weber confidently predicted, "The working human being will be as carefully calculated as to his 'profitability' as any raw material or coal in terms of its usefulness for the plant, and on this prediction a considerable part of the hope which we might have for the progress of our work rests."[120]

INDUSTRIAL PSYCHOTECHNICS

By the beginning of the first decade of the twentieth century, the European science of work was firmly established as a branch of "social hygiene." Communities of advocates were established in France, Belgium, Germany, and Italy with outposts in America, Russia, and Japan. Although we should not neglect important national differences in the style and background of the science of work, this internationalism was an underpinning of its advocate's claims to neutrality and objectivity.

Moreover, the frequent international gatherings of experts in fatigue provided a forum for articulating the social policy implications of often arcane laboratory experiments. At the same time, the internationalization of the science of work in part compensated for the relatively marginal status the science enjoyed in a national or local context. The prodigious French were in the forefront, largely under the aegis of physiologists (Marey, Imbert), hygienists (Amar), or engineers (Fremont), who were on the margins of their disciplines. This is hardly surprising in *fin-de-siècle* France, where medical sophistication often compensated for technical inferiority. In Germany the leading figures came from the ranks of occupational health specialists *(Gewerbehygieniker)* or from academic neurology or psychology like Kraepelin or Münsterberg.[121] These differences may account for the largely physiological orientation of the French as opposed to the more psychological and sociological *(geistes wissenschaftlich)* trends across the Rhine. Almost from the outset, the French were open advocates of political and social reform, and wanted to extend the science of work to large groups outside the laboratory, whereas the German psychologists were more reluctant to promote their efforts in industry.[122] German sociologists, on the other hand, were more liberal and oriented toward reform, though less aggressive in promoting state intervention than the French. The German proponents of *Arbeitswissenschaft* advocated an "energetic labor law" *(energetisches Arbeitsrecht),* which viewed the human being as a labor factor in the expenditure of energy and proposed rest pauses, training programs, and state intervention only as long as these served the "energy economy." For them, the subjective aspects of labor were "in no way decisive for the cultural importance of work," which was evaluated in terms of national energy supply.[123] In all countries, however, the science of work was based on the premise that greater productivity would lead to social happiness— and not, as in the American import Taylorism, or later Fordism, on the view that unhappiness had to be compensated through external, non-work-related material rewards.[124]

By 1910 the science of labor divided work into two distinct dimensions reflecting the origins of the science in the theory of energy conservation and in the imperatives of modern industry: the physiology and psychology of labor power—for example, the economies of the body in alimentation, respiration, and the movements of the muscles and the limbs; and the sociology of labor power—studies of industry and the conditions of work, the technological requirements of tools and techniques, and the physiognomy of the workplace. "Examination of the economic problem of labor," wrote Ioteyko, "resolves itself into (a) the subject: the power, the apprentice, the workman; and (*b*) the object: the resistance, the work to be accomplished in typical industries."[125] This division also had programmatic significance pointing to the two fundamental directions of the science of work after its institutionalization during and after the war: testing and vocational training of laborers, and industrial psychology. For the advocates of the new science, state regulation of labor power and the conservation of energy appeared at the outset to be synonymous. Maintaining a modern labor force and efficiently using the working capital of the nation were based on an energeticist model of the working body that valued the concepts of fatigue, energy, and performance. As long as the conservation of labor power appeared to be consistent with the laws of nature and the demands of society, tension need not exist between the productivist assumptions of the science of work and the articulation of humanitarian goals. Although rooted in physiology, fatigue became "un fait social" affecting the productivity of the individual worker and future generations. Economists who neglected the "special characteristics of the human motor," Ioteyko argued, also deprived the *patron* of a firm of the advantage gained from recognizing that a shorter working day might even increase productivity more rapidly than higher wages.[126]

After 1910, in both France and Germany, the science of work began to transcend the laboratories of the first generation of practitioners. It was increasingly oriented toward the resolution of the "social question" by defining the optimum use of the body's energy economy apart from the divergent interests of labor and capital. For advocates of the new science, the law of "least effort," and the notion of "optimum yield" could overcome the dichotomy between profitability and social progress. The elimination of fatigue, overwork, and excessive motion could be achieved through reducing accelerating work tempos and by introducing rest pauses and the shorter working day. Productivism and reform could be united. As Ioteyko remarked, "Although this study is only in its infancy, it has the benefit of all preceding researches of pure science which will give it their authoritative support in its noble desire that the working classes may benefit by the physiological and psychological discoveries of our century."[127] One perceptive commentator even called Münsterberg's "psychotechnics the last act of the belief in harmony."[128]

The science of work emerged in an era when the idea of scientific neutrality between capital and labor took on enormous significance. Its optimism was not based on merely a positivist concept of progress through science, though clearly this idea was prominent. The science of work also combined the idea of an industrial society regulated by the state with a productivist concept of industrial expansion unfettered by the conflict between capital and labor. Standing above the titanic struggle of classes, its social vision comprised a concept of work expunged of all political and social experience. As long as it was able to maintain a safe distance from society, its ideals could be safeguarded from the distractions of politics and conflict.

By 1910 it was becoming apparent that the European science of work could not rest content with abstract protestations of its social efficacy. It needed an arena in which it could deliver proof of its irrefutable truths. The laboratory achievements of the science of work were superseded by a growing need of its adherents to demonstrate its efficacy in resolving the crucial social issues of the day: the length of the working day; methods of remuneration; the rest day; industrial accidents; and educational issues and the training of recruits. However, in different national contexts the politics of reform created a new set of challenges to the science of work. Especially the arrival in Europe of the American Taylor system before the First World War revealed the "fundamentally eudaemonistic worldview" of the science of work, the conflict between the desire "to raise the level of happiness of the whole through the increase of productivity, but at the same time to increase the happiness of the producers, both in and outside of their occupation."[129]

CHAPTER EIGHT

~~~~~~~~~~

The Science of Work
and the Social Question

BETWEEN PRODUCTIVISM AND REFORM

Dᴜʀɪɴɢ the last decades of the nineteenth century European liberalism became closely aligned with the scientific doctrines of the age. Its twin pillars were medicine and biology: "social hygiene" sanctioned the view that society was best served by equilibrium and threatened by chaos. For Continental reformers, society was a delicately balanced organism whose interdependent functions required state intervention to protect it from harmful influences and strengthen its internal forces of resistance to decay and deterioration. Social statistics could attest to the cost of neglecting the social milieu, as well as to the potential benefits of removing its deleterious effects.

Social positivism was a two-edged sword. Biological determinism could also be deployed to redraw the maps of social hierarchy: statistical measures could determine intelligence, establish racial and gender superiority, and demonstrate the inheritability of criminality, poverty, and disease. Neo-Lamarckian theories of adaptive evolution were invoked along with doctrines of inherited "degenerative" traits to emphasize the social roots of pathology: low intelligence, insanity, criminality, and other nonadaptive behaviors. The threat of internal decline was enhanced by the discovery of cholera, smallpox, and other diseases that, it was thought, accompanied the arrival of immigrants from less temperate climates. Social hygienists like Adrien Proust maintained that Europe had to be "defended" against subversion from within and without. The politics of social modernity turned a "litany of social pathologies into a discourse of national decline."[1]

The "biologization" of politics crossed the boundaries of class and ideology in its emphasis on scientific and objective solutions to the social question. The same scientific doctrines also justified an expanding "cluster" of ameliorative social institutions and reforms. The energeticist axiom was that improvements in the health, longevity, and security of the worker conserved and augmented the productive forces available to the nation. Louis Querton's *L'augmentation du rendement de la machine humaine* (1905) was the catechism of the new social energetics, mustering biological, statistical, and sociological arguments to support the claim that progress depended on state intervention for the "construction, conservation, and enhancement of the efficiency of the human machine."[2] Social positivism was a powerful constellation of ideas invoked on behalf of what the philosopher Alfred Fouilée identified as "reparative justice," a system of compensations for the deficits of obsolete "laissez-faire" liberalism.[3]

Solidarism, a doctrine developed by Léon Bourgeois, one of the Third Republic's leading politicians, emphasized the mutual and collective moral and social obligations of all productive members of society. Reforms aimed at reducing exploitation, promoting productivity, and increasing social justice were the logical extension of ancient republican virtue to the modern age.[4] Solidarists like Bourgeois appealed to Frenchmen to recognize the imperatives of industrial society, elaborating their arguments in the language of natural law rather than social imperatives: "Natural social laws are only the manifestation, on a higher level, of the physical, biological and psychic laws, according to which living and thinking beings develop."[5]

In his speech at the 1900 Paris Exposition, the Socialist minister of commerce, Alexandre Millerand, pointed to the positive results of such supportive clusters of social defense in enhancing social solidarity. These reforms reduced "the weaknesses of individuals," enabling them to overcome the obstacles of their social milieu as well as offering them "evidence of human solidarity." If the laws of heredity affirmed our solidarity with our ancestors, he wondered, "why are we not so inclined towards our contemporaries?"[6] Solidarist economists, including Charles Gide, Charles Rist, Paul Cauwès, and Raoul Jay, gathered around the anti–laissez-faire journal, the *Revue d'Economie Politique* (founded in 1896), and the reform-oriented Musée Social (founded in 1894), which provided a theoretical and organizational basis for the reformers' programs.[7] The *Revue*'s line stressed the interdependence of all branches of the economy, the social costs of labor conflict, the negative social consequences of poor health, and the necessity of reducing insecurity and low living standards in order to raise the "personal productivity of the worker."[8] Gide called his social economics a "science of social peace and well-being."[9]

Solidarism was the ideological bulwark of republican reformers who believed that improving the workers' health and milieu would improve their productivity while preserving the "capital of the nation." Raoul Jay summed up the essential calculus of French social positivism in 1904:

> A nation that permits the destruction or reduction of the mental or physical forces of the manual workers makes one of the worst calculations. Those physical and moral forces are a part of the national capital like the sun or machines. The industrialist who, in order to reduce the costs of production, permits his machines to deteriorate . . . would be considered a fool. . . . If we do not think the same of an industrialist who imposes an excessive labor on his workers or pays them an insufficient wage, that is because we know that he will never have to repair the damage caused by his criminal negligence. The damage is charged to the nation.[10]

LABOR POWER: CAPITAL OF THE NATION

By 1900 the science of work became a powerful intellectual weapon in the arsenal of the middle-class reformers. In the atmosphere of intense controversy over the length of the working day and week, in the protracted debates over the risks of industrial work, and in the controversies over wages and work norms, the science of work began to play an important role in the efforts of liberal reformers to mediate social conflict. As Imbert claimed, applied to the study of the "worker question," the "methods and techniques of the laboratory are in many ways capable of furnishing perhaps indispensable data to establish the most equitable solution of the conflict between labor and capital."[11] At the 1903 Brussels Congress of Hygiene and Demography, the leaders of the European and American social hygiene movement gathered to debate how the science of work could move beyond the confines of the laboratory and provide policy makers and legislators with arguments "in favor of specific methods for the organization of labor."[12] Four years later when the delegates gathered in Berlin, scientists and reformers called for the increased cooperation of the state in applying the science of work to reduce fatigue, accidents, and the hours of work.

The inordinate attention paid to fatigue in the crucial decades between 1895 and 1914 reflected the anxieties of a society entering the industrial age. But it also mirrored the realities of a new, mechanized factory, born in the heyday of the second, or Continental, industrial revolution after 1895. Concern with fatigue, with time and motion, and with the quality of the work environment reflected deep changes in the factory. The workforce no longer had to be subjected to the moral economy of industrial discipline; its character molded to the demands of industrial behavior outside the workplace. Modern industry relied on cheap and inexperienced labor, replacing skill, self-discipline, and pride in craftsmanship with the economies of technologized production. The modern worker had only to be taught to internalize the regularity imposed by technology and the tempo of the actual work. "Chronic fatigue or exhaustion," noted one of the resolutions adopted at Berlin, "is observed in all the factories where the intensity of labor is regulated by the machine."[13]

In contrast to those who "quite wrongly" regarded the struggles over the length of the working day and over the conditions of work as economic issues, asserted Imbert, the science of work could analyze their social character and causes without appearing partisan. "Economy of force" was a law as applicable to an entire society as it was to a single muscle: "The human motor deteriorates, which causes injury to the worker . . . and the efficiency of the motor diminishes, which causes injury to the entrepreneur who employs it, apart from the social costs that must be borne later as a consequence of that deterioration." For these reasons, "Experimental science was called upon to play the role of finding a truly equitable solution to these conflicts through research."[14]

Many European socialists shared the mental universe of social positivism, even if they were sometimes skeptical of the motives of liberal reformers. A society constituted by "the solidarity of individuals," wrote a leading French syndicalist, "is injured by the exploitation of the individual because by his personal exhaustion, the general energy of production is correspondingly reduced."[15] A brochure, which circulated with the title *Travail et surmenage* ("Labor and Exhaustion") and a cover depicting an enervated worker, claimed that "chronic exhaustion was the rule in the general experience of modern work."[16] To be sure, at the outset few economists and even fewer industrial entrepreneurs shared this perception of labor power as a precious national resource. Neither French nor German firm owners could be expected to agree with Jay's characterization of the industrialist "whose workers are fatigued or inadequately paid [as] a parasite who subsists to the detriment of the entire community."[17] Working-class leaders also feared that a decline in national competitiveness might ultimately hurt those workers who would most benefit from the shortened work week.[18] The discourse on preserving labor power as the solution to the "social question" emerged gradually at the end of the nineteenth century in a space carved out between the *laissez-faire* doctrines of the old Manchester-style liberals and the more radical views of the syndicalists.

After 1900, however, the energetic calculus was applied to various social issues: the length of the working day, the weekly rest day, industrial accident and health insurance, the length of military training, the proper method of education, and the role of women in the labor force. In both France and Germany in the period up to the First World War, the science of work contributed to a new constellation of knowledge and politics devoted to conserving the energy of the social corpus.

THE PERSONAL PRODUCTIVITY OF THE WORKER

In legislating a reduction of the working day, continental Europe lagged behind England, where the ten-hour "normal working day" for women and children was established in 1847 (an exception was Switzerland, which legally mandated the eleven-hour day in 1870).[19] The persistence of the traditionalist view that productivity could be raised only by maintaining or expanding labor time, along with an anxiety of "belatedness"—a desire to accelerate, not slow down the pace of industrialization—no doubt accounts for the stubborn refusal of Continental entrepreneurs to entertain the idea of a shorter working day.[20] With the upswing in the economy and the rise of a new and more militant working class in Germany and France after 1895, the eight-hour day became the universal demand of the international labor movement, superceding the wage issue. Gary Cross, author of a study of "worktime" in Europe, has persuasively argued that the demand for a shorter working day was not merely a "reformist" or economistic demand of the workers' movement. It represented an international, intraclass movement uniting skilled and unskilled workers around a single issue, which symbolically incorporated the notion of "social rights." Social historians, following E. P. Thompson, have generally tended to see struggles over labor time as part of a protracted struggle in which workers resisted the hegemony of entrepreneurs by defending a "preindustrial" sense of time-consciousness and by refusing to bend to the discipline of the factory. This view of tradition-bound workers resisting fixed time standards and, to my mind, an exaggerated emphasis on spontaneous workers' militancy in local strikes and conflicts has often permitted social historians to minimize or ignore the extent to which late nineteenth-century workers were mobilized for the universal right to "free time" rather than around local issues.

European socialists and trade unionists viewed the eight-hour day as offering numerous permanent benefits: protection against excessive exploitation, a hedge against labor intensification, a more productive leisure (and political activity), and ultimately higher wages owing to a shrinking labor pool. The international May Day celebration on behalf of the eight-hour day in 1889 directed the international workers' movement toward realizing this goal through political legislation at the national level. As Cross argues, the eight hours' movement was "the result of a thirty-year political and ideological struggle."[21]

As a result of this activism, the relationship between worktime and productivity became a matter of no small concern. The issue encompassed not only the reduction of the working day, but also the tendency of industry to exercise new management techniques to control the worker's useful labor time. The introduction of the time-clock (Kontrolluhr) and the punch-card system (Arbeitszeitkarte) in German firms around 1900 exemplified this tighter restraint over worktime, as did increased shop-floor hierarchy and surveillance elsewhere.[22] Workers' struggles for control of political and social norms on the shop floor increasingly became conflicts over time, an issue that transcended economics.

Worktime in European countries varied across national boundaries and across industries. In Germany the twelve-hour day was generally adopted in the 1870s but never became law—a condition that permitted differing practices within and among different types of industry. Debates on worktime also introduced political distinctions between workers of different ages and sex. When a ten-hour day for German women workers was proposed in the *Reichstag* in 1904 to 1905, a "Cassandra cry" of despair was raised by German industrialists who defeated the measure.[23] The demand for a reduction of hours for women or adolescents often permitted reformers to couch their arguments in terms of gender, that is, as lesser workers of more "delicate" constitution.

In France the law enacted on 9 September 1848 fixed the maximum labor time for mechanized industry at twelve hours, a standard altered only a half century later, when in 1892 a law was passed limiting children ages thirteen to sixteen to ten hours, and women over sixteen to eleven hours (then reduced to ten hours by 1900).[24] In 1904 the law extended the ten-hour day to men in workplaces employing women. The 1892 law led to a protracted controversy, not only between liberal opponents of regulation and socialist advocates of the eight-hour day, but also among those who wanted to maintain a physiological distinction between women and children on the one hand, and adult male workers, who remained outside the scope of the law and whose rights were discounted by the absence of such "overriding" biological claims. The 1892 law also led to the much-discussed "crisis of apprenticeship"—a virtual epidemic of firings of adolescent and women employees to circumvent the newly legislated regulation of hours.[25]

In 1900 Alexandre Millerand proposed a gradual reduction of the workday to ten hours over a six-year period. The bill's passage was supported by arguments from economists like Maurice Bourguin, who emphasized its consequences for the health and vigor of workers and insisted that their greater productivity would compensate for the shorter hours, actually increasing output.[26] To bolster confidence in the benefits of these reforms, Continental critics frequently adopted the image of industrial Britain, not as the home of "werewolf capital" (Marx), but as an enlightened nation of well-paid and productive laborers. At the end of the nineteenth century England and, perhaps to a greater extent, the United States were perceived as having solved the labor question by maintaining a better paid and a harder-working labor force whose higher productivity and political timidity were laudable. The English worker (always male in this caricature) was frequently depicted as an astonishing specimen of superior industry and physical strength, attributable to his higher wages, shorter hours, and better diet.

As early as 1871, the Anglophile Hippolyte Taine offered this instructive comparison. The French worker, he observed, "worked perfectly well during the first hour, less well in the second, and much less well in the third, diminishing still further after that, until, in the final hour he could do nothing at all. His muscular force sagged, his attention relaxed." By contrast, in England, "the laborer worked as well during the last hour as during the first. And, his working day is ten hours, not the twelve of the French worker."[27] A Leeds industrialist, employing both English and northern German workers, commented that "one Englishman was worth two Germans."[28] This image of the English worker was especially strong among pro-reform Continental liberals who focused less on the virtues of free-trade than on the benefits of the shorter working day for increased productivity and power.[29] Such comparisons were not restricted to employers or reformers. Victor Delahaye, a machinist by trade, a member of the First International, veteran communard, and the head of his Paris trade union, made this comparison in an 1883 report to the International Colonial Exposition in Amsterdam:

> Each worker in the United States of America annually produces the commercial equivalent of ten thousand, one hundred ninety four francs (10,194 fr.) and never works more than nine hours a day and six days a week, receiving an average salary which is twice the average salary in Paris and four times the average salary in France; whereas each worker in France does not produce the equivalent of three thousand three hundred forty two francs (3,342 fr.) and works an average of 12 hours a day. In other words, working three hours less per day, each worker in the United States produces three times more than each worker in France.[30]

Throughout the 1880s and 1890s the United States and English "model" economies not unlike the Japanese today, served as the ubiquitous example of the benefits of docile labor unions, cooperation between labor and management, and an enlightened social and industrial policy. On the Continent, where labor remained a scarce commodity, the salutary benefits to higher productivity promised by wage increases already appeared on the reform agenda in the wake of the economic crisis of 1873. A report of the German iron-investigation commission (Eisenenquete-Kommission) concluded in 1879 that "every wage increase, when justified under the circumstances, raises the capacity for performance and morality," whereas labor power decreases with each drop in wages.[31]

During the late 1880s and early 1890s there was an intensive debate in both France and Germany on the virtues of the so-called "English Sunday," or *semaine anglaise,* the Saturday half-day and Sunday rest day virtually unknown in France and Germany. In addition to the advantages of a uniform work week, reformers frequently argued that reducing the amount of leisure during the workday also led to higher productivity: the "West European and American habit of relatively short workdays, weekly rests, and fewer longer holidays permits a higher productivity than the division of work and rest practiced in Russia or Italy."[32] The German philosopher Eduard von Hartmann complained that for capital to lie fallow for fifty-two days a year was a waste, and that instead of Sunday a general rest day on alternate days should be substituted (an idea that was later adopted for a short period in the Soviet Union).[33]

In his seminal work on wages and productivity, *Über das Verhältnis von Arbeitslohn und Arbeitszeit zur Arbeitsleistung* (1876), the German economist Lujo Brentano expressed his hope that the resistance of German entrepreneurs to shorter hours would disappear "as soon as they overcome their fear that the higher wages and shorter labor time, to which workers organizations and worker protection legislation lead, will so greatly raise the costs of production that domestic industry will lose its competitiveness on the world market."[34] Recalling the protests of English industrialists against the ten hours' bill forty years earlier, he pointed out that oft-repeated cries of imminent collapse from the industrialists, and echoed by political economists, had proved unjustified. Citing the experience of British industrialists as well as his own observations of English miners, he argued that "it is observed everywhere that the workers of the nations with shorter labor time produce more than the workers of those with more working hours."[35]

Brentano did not explain this superior performance by higher wages alone. The long-term effects of higher wages on the workers ultimately enhanced their physical and moral well-being, which contributed to their greater productivity: "If the improvements achieved are of long duration, they will be used for better nutrition, for more attentive self-care *(pflege)*, for improved and greater moral leisure, and greater education—in other words, they will lead to raising the physical and spiritual needs of the worker."[36] He added that the pressure of higher wages will also lead manufacturers to reduce costs by increasing their reliance on machinery or by improving the "economic organization of labor."[37]

Brentano's optimistic arguments did not go uncontested. His glorification of the English worker was sometimes sharply lampooned: "If the English or Scottish worker earned enough in three or four days to carry on his life with this wage for the rest of the week, he often has no desire at all to go to work during the remaining days. He creates not only a "blue Monday," but also a "blue Tuesday," and sometimes even a "blue Wednesday.""[38]

German economists like Wilhelm Hasbach, who advocated free market policies, tried to refute Brentano by showing that the upward spiral of work performance and higher wages ultimately followed the law of diminishing returns. Work performance increased at first but it slackened as the "distance from subsistence wages also increased." Hasbach argued that even if productivity grew, the costs of higher wages to capital would render German industry less competitive, depressing wages and undercutting whatever benefits might be had from the greater productivity of labor.[39] Others claimed that with shorter hours "the worker goes to the tavern more frequently, consumes more alcohol, and therefore arrives at work in less of a condition to perform work despite the shorter hours."[40] Some economists asserted that the productivity of workers, despite advances in machinery and technology, actually declined because the labor movement encroached on their joy in work. Was not the German worker so possessed by "the cult of the horny skinned Faustus" (socialism), asked Alfred Weber, that he had become indifferent to the fate of the economy?[41]

During the heyday of the second industrial revolution in the late 1890s, the debate on wages and industrial performance sharpened as the use of new payment schemes to boost productivity increased. Traditionally, wages were not used as a stimulus to productivity but were a "price" set by workers who sometimes exercised extensive control over the organization of work, worktime, and output. After 1895, however, larger firms introduced premium and piece-rate systems to undermine these old methods, linking wages to work performance.[42] Accordingly advocates of the new management techniques extolled the positive correlation between higher wages and higher productivity in England and America, sometimes to the point of absurdity.[43] French visitors to American exhibitions during the 1880s and 1890s reported that American workers produced three times as much while working one quarter less.[44] Not all of this literature disparaged the French worker: more patriotic economists emphasized that the greater productivity of American workers could never achieve the skill and invention of French crafts *(industrie d'art)*, in which they still excelled.[45]

In Germany the most widely read example of this genre was *The Cotton Trade in England and on the Continent*, a survey of the English and Indian cotton industries, written in the mid-1890s by Gerhard von Schulze-Gävernitz, a laissez-faire economist. In his vivid brief for higher wages and the increased use of technology, this enthusiastic admirer of British imperialism argued that "by the quicker speed, the lengthening of the machines etc., a larger production per day is achieved; a larger production which on the one hand, allows a curtailment of the hours of labour."[46] But even more important, claimed Schulze-Gävernitz, it was not simply improved technology, but improved physiology that accounted for the difference between English and German workers:

> The physical superiority of the English factory operative when compared with the Continental is recognised by German observers, just as the operatives of the great English industries boast of their own physical superiority.

He attributed this superiority to higher wages and increased consumption:

The progress which the English operative has made in health as well as capacity compared with his forefathers, depends chiefly on an improved standard of living. The enormous progress in the nourishment of the people which England has seen during this century is a most important element favorable to the capacity for competition of English industry as opposed to the competition of Continental industry.[47]

When the British Gainsborough Commission completed its investigation of Germany at the end of 1905, it found, to the delight of German reformers, that despite longer hours the German worker was inferior to the English worker in personal productivity. A detailed comparison of German and American shoeworkers conducted in the same year showed similar results.[48] John Rae, the British socialist economist, summed up the European experience in his widely translated *Eight Hours for Work* (1894): "It is possible for Continental manufacturers to improve their competing capacity as ours have done, by reducing their hours of labor beneath the limits generally prevailing at present, and it is very plain from the elaborate evidence laid before the Trade Depression Commission that their present long hours have never been any advantage to them or any disadvantage to us."[49]

These accounts called attention to the striking differences in diet among European workers. One commentator pointed out that the French railway workers who constructed the Paris-to-Rouen railroad in the 1840s were inferior to their English counterparts because their diet of vegetables and soup was inferior to the roast beef diet of the English workers.[50] Hector Denis, the socialist reformer and close associate of Solvay, argued in the Belgian Chamber of Deputies that it was possible "to calculate the amount of labor power expressed in calories of energy."[51] Émile Waxweiler, analyzing the diet of the American worker, also observed that the worker in the United States "found a mode of existence far higher [*plus élevée*] than that of his European competitor, and thus more favorable conditions for the expansion of his productive force."[52]

Increased consumption also played a role in expanding the productive power of the worker. Not only physical energy was improved by "le Standard of Life" (standard of living), a term introduced (in the Anglo-Saxon usage) in Europe during that period, but mental capacity was increased as well.[53] "A great deployment of nervous force demands a substantial and costly nutrition: it is among the workers capable of great nervous effort that we generally recognize the highest abilities and the best wages, from the weaver of fine fabrics, to the mechanic, to the artist and even to the intellectual worker, whose capacity for sustained attention is the form in which the predominance of nervous effort is most often manifested."[54]

These accounts also called attention to the variations in personal productivity of the worker across national boundaries. Labor power was a scarce resource, and the social costs of its misuse had to be weighed against the increase in output that might result from enhancing the worker's health and longevity. In a detailed overview of labor during the nineteenth century, the Solidarist Laurent Deschesne made the striking claim that it was not higher wages that caused greater productivity, but greater personal productivity during the century that produced rising wages. Attributing his unique thesis to an American, A. Walker, who wrote *The Wages Question* (1876), Deschesne claimed that European workers' wages rose relative to increases in their personal productivity, a fact that had escaped the attention of economists. Despite Marx's pessimistic predictions, salaries tended to increase with the greater value that accrued to the entrepreneur.[55] As Deschesne admitted, the degree to which personal productivity could be improved, or calculated with the other factors of productivity, remained as yet unexplored.[56]

Deschesne proposed that "the personal productivity of the worker be distinguished from such general economic factors as the productivity of machinery or the general productivity of the enterprise. Personal productivity, he argued, is "dependent on the force of his labor and on his personal aptitudes, which is translated in part into what we can call his "effective utility" ("l'effet utile"). The "average productivity of the worker" could be calculated along with the other factors of production.[57] This calculus of energy and productivity, in addition to extolling the virtues of technology and the benefits of higher wages, focused attention on the subjective aspects of labor power: the lessening of fatigue and the improvement of the social milieu in which the work was accomplished.

INDUSTRIAL EXPERIMENTS: HOURS AND OUTPUT

After 1890 a small number of Continental industrialists began to experiment with the shortened workweek in the interests of expanded productivity. These efforts were undertaken for economic reasons, but they were also intended to serve as models for other less enlightened industrialists. One particularly influential experiment was that of the Belgian industrialist and engineer L.-G. Fromont, director of the Société des Produits Chimiques von Engis.[58] Encouraged by the personal example of Solvay's reform policy in his own plants, Fromont tried to apply the doctrine of energeticism to his workers. He recounted his experience with two ten-hour shifts at a hot oven in his sulphuric acid plant where the workforce was chronically exhausted, deprived of sleep, inattentive, and often either absent or intoxicated. The workers were constantly complaining of work-related sickness and were hostile to the management, whom they blamed (justly it appears) for poisoning them.[59] When management announced in 1897 that it would introduce a three-shift system of eight hours each, with essentially the same wage-rate, the workers threatened to strike because of the expected decline in real wages. Nevertheless, once established, the workers actually produced more in seven and one-half hours than in the previous ten, at higher wages, and with fewer complaints of illness.[60]

Under the new regime, Fromont reported that productivity rose over one-third over a six-month period, labor relations improved considerably, along with lower production costs and a decline in alcoholism and discipline problems.[61] Fromont's success was widely publicized in Belgium by the Solvay Institute and by the Office du Travail on behalf of a legal reduction of the working day.[62] In 1906, the major German industrial firm of Bosch also introduced an eight-hour day with similar results.[63]

In Germany enlightened factory reform is most often associated with Ernst Abbe, director of the Carl-Zeiss optical works in Jena. In 1901 Abbe became the first German industrialist to introduce the eight-hour day, carefully monitoring the workers' productivity after they agreed to his plan to maintain productivity at nine-hour levels with no reduction in pay. Abbe was concerned with the personal sources of higher productivity in an enterprise that employed highly skilled workers in the manufacture of precision instruments.[64]

Abbe regarded his experiment as conclusive proof of his initial observation on the social and physiological characteristics of modern work. Mechanical work is characterized by a "uniformity" that produces a cumulative and progressive "fatigue, always of the same organs, the same muscles, the same nerve centers and of the same brain parts."[65] The longer the working day, the greater the need for longer periods of unproductive activity, which he likened to the "idling" *(Leergang)* of any machine. Abbe also observed that the workers reported extreme exhaustion at the beginning of the experiment, but soon did not either notice their fatigue or that their productivity had increased. Performance during shortened working time was more concentrated because the "passive fatigue" of the longer working day subsided. Abbe summed up his experience in the following axiom: "For each person, in each kind of work, and for the daily product of a given amount of work time, there is a maximum; the reduction of labor time must result in a rise of work performance as long as the gain in the daily expenditure of energy from the lengthened rest time and energy saving from the "idling" *(Leergang)* is greater than the energy expenditure demanded from the speeding up of the tempo of work."[66]

In France, several experiments with a reduced workday and workweek also revealed no significant decline in productivity. In 1901 a Lyon textile firm with more than one thousand workers reduced the workday to ten hours in 1901 with no significant change in output.[67] The French military experimented with a reduced working day in its workshop at Tarbes over a three-month period from September to July 1904. The nine-hour day was subsequently extended to all War Department industrial workshops. Working capacity increased in the initial period, but the War Department was skeptical that any further reductions could be undertaken without significantly raising production costs.[68]

FATIGUE AND PRODUCTIVITY

At the 1907 Berlin Congress of Hygiene and Demography, scientists and health experts from various countries considered proposals for a shorter working day. From the standpoint of social hygiene, the greater efficiency achieved by the elimination of overwork, by the shortening of the workday, by the introduction of rest pauses, and by the weekly rest day *(repos hebdomadaire)* would, the delegates claimed, enhance overall welfare and productivity. The German industrial hygienist Emmanuel Roth proclaimed that any reduction of the workday was desirable from a "hygienic standpoint" as long as the gain in recouping the day's energy *(Kräfteersatz)* from the longer rest period, or from the improved efficiency, remained greater than the energy consumed by "the intensification of the tempo of work."[69]

Several years earlier, with the support of the provincial Inspecteur du Travail, Mestre, Imbert tried to demonstrate the usefulness of the science of work in resolving a protracted conflict over wages between dockworkers and the shipping companies at the port of Sète. Comparing the lower wages earned by the wine workers he had investigated (four or five francs for a ten-hour day) at Midi with the relatively higher paid dock workers (eight francs for eight hours), he concluded that the difference was warranted by the substantially higher degree of fatigue: "That inequality of wage does not correspond to the quantity of exterior dynamic work produced, but to the intensity of fatigue, that is to say, to the value of the internal energy expended in reality." It was reasonable that "the external efficiency of the worker" was the industrialist's legitimate concern, whereas "the real expenditure of internal energy," rather than performance *per se*, represented the worker's self-interest. This essential difference, noted Imbert, accounted for the difficulty of resolving conflicts between labor and capital and constitutes "one of the principal reasons why any agreement between them is as difficult to obtain as it is to sustain." For this reason, he claimed, the science of work could play the role of an objective and "impartial arbitrator if it is "called upon to provide a solution based on experimentation, if it is truly equitable, and if it is not prejudiced by economic considerations."[70] Imbert's distinction between work performance and the physiological toll for the worker, which was ignored by the German *Arbeitswissenschaft,* underscored his sincere interest in securing reforms based on an objective physiological assessment of the work, and not simply on output and performance.

Other proponents of the science of work were more cautious of legislating the length of the working day. André Liesse warned that to establish *a priori* a uniform working day or week for all industries in all countries ignored the specific and varied physiological expenditures involved in different kinds of work.[71] Amar, too, characterized the demand for an eight-hour day as "unscientific," arguing that the amount of work should be consistent with the type of work and the age and sex of the worker.[72] The German industrial hygienist Theodor Sommerfeld agreed that the diversity of work methods and work situations required a differentiated and specific approach rather than a single national policy. Kraepelin also urged that productivity ultimately had to take precedence over the results of the fatigue study.[73] Nevertheless, the majority of experts in fatigue believed that a maximum limit on working hours for each trade could be, as the entrepreneurial experiments demonstrated, based not on "a utopian conception, but on precisely known facts."[74]

THE PHYSIOLOGICAL LIMIT AND THE TEN-HOUR DAY

In France the difficulties encountered by reformers' efforts to secure a ten-hour law (debated between 1892 and 1904) and the controversial impact of a law enacted in July 1906 requiring that all commercial establishments provide a rest day during the workweek *(repos heb-*

domadaire), produced extensive public debate on the value of labor power and the physiological implications of reform.[75]

In 1910 René Viviani, minister of labor in the government of Georges Clemenceau proposed a new law to reduce the working day gradually for *all* workers in previously unregulated enterprises to ten hours.[76] Not only the economic implications of the law were at issue, but in the debates that ensued, the physiological arguments of the science of labor (especially those of Imbert) were enlisted on behalf of the proposed law.

Supporters of the law claimed that the "physiological basis of the ten-hour law" was predicated on two decades of achievements in the science of fatigue: "Excessive physical work progressively affects the functions of the circulatory apparatus, and produces the gradual hypertrophy and dilation of the heart, a general arterio-sclerosis, it restricts the circulation of the veins and hastens renal anemia. . . . The corporal development of the worker is retarded in comparison to those of other social classes. In order to guarantee a suitable nutrition, to maintain an equilibrium of the expenditure and consumption of the organism, we have seen that a daily work of 10 hours is in *general* an absolute maximum, that it is arrived at by two four-hour periods of work separated by a rest of two hours (the eight-hour day)."[77] Numerous experiments demonstrated that "four hours work are the maximum after which rest must occur."[78] The absolute maximum amount of work accomplished in the course of a day without injury to the human organism was calculated at 100,000 kilogrammeters per day. An overfatigued worker would be unable to rest enough to repair this loss of energy and often had recourse to "alcoholic excitation" in order to stimulate artificially the body into a false diminution of the sense of fatigue.[79]

Dr. Maurice de Fleury, a prominent member of the Paris Academy of Medicine, claimed that for every worker there was a "rational hygiene," which can furnish a daily dose of labor "without exhausting his forces," which he calculated at eight hours. Fleury warned that in America French workers employed in large factories in Chicago found the intensity of work and lack of rest periods during the eight-hour day intolerable: "They all complained to the consulate of their country that there was not enough time to smoke a cigarette, or even the leisure to whistle a small refrain. . . . An American worker furnishes as much manpower in eight hours, as a Parisian worker could not even furnish in ten."[80]

The French socialist Edouard Vaillant, who was also a physician, introduced the phrase "physiological limit" during his speech on behalf of the eight-hour day in the Chamber of Deputies in November 1910.

According to Vaillant, the worker's "physiological limit" was the maximum length of work time and effort that could reasonably be expended; the maximal amount of energy efficiently transformed and utilized—as heat and movement—during work. This limit could be physiologically precisely fixed: "The revenue [nutrition] necessary and the forces an organism must expend [thermal and organic wastes]" produce an equilibrium of work and recuperation. Fatigue sets in when the worker is compelled to labor beyond the point of recovery. The "wastes accumulate and lead to a state of exhaustion which might terminate in illness or even death."[81]

In numerous speeches and articles, Vaillant emphasized the connection between physiology and politics: "In work, in muscular or mental fatigue, there are limits of time and intensity which cannot be transgressed without peril. Exhaustion must be avoided. The worker has the right to profitable nutrition, to a regenerative sleep, and to the relaxation of the evening. He deserves a day of rest and vacation."[82] Vaillant was also a leading supporter of a national laboratory for the study of the physiology of work, which was eventually established in 1913.[83]

More than any other politician on the French left, Vaillant made extensive use of the science of work in his arguments on behalf of the social question. Vaillant considered Imbert and Amar's experimental studies conclusive proof that a reduction of hours was not simply the self-interest of socialists but the affirmation that their demands rested on solid scientific ground.[84] For Vaillant, scientific research into modern work would provide the basis for "an allocation of individuals according to their aptitudes, in diverse trades." Each individual would adopt the trade for which he had the most aptitude or the greatest productive capacity. Productivity would thus increase and permit a general reduction of the working day.[85] The equilibrium between a shorter working day and the intensity of work, or "optimal performance," could be precisely determined. That "moment" could be fixed for each occupation, and physiology could establish the limits of improving performance without injury to the human motor.

The illusory argument that shorter hours reduced competitiveness was belied by the fact that those nations least developed economically also had the longest hours and the lowest wages.[86] Vaillant envisioned an extraordinary future for a France that adopted the solutions that "could be furnished by the physiological study of occupational work" and applied to each occupation:

> The nation which establishes that study, which first reaps its results, which organizes its work in such a fashion that it can extract to the greatest

possible extent all of the energies of the workers, and all of the labor power over which it disposes—augmenting its productivity under these conditions without altering the organism of the worker one iota—one might say that such a nation will be best armed to establish a normal and natural system of production. With its productivity pressed to the utmost possible limit under these conditions, it would simultaneously be the country best armed for international competition.[87]

In February 1911 in response to Vaillant's urging, a commission of experts was empaneled to recommend further experimental study of industrial work.[88] The following year a Chair in "industrial hygiene" was established at the Conservatoire National des Arts et Métiers. In May 1913, the minister of labor, Henri Chéron, delivered a report to President Raymond Poincaré that called for developing a program for the study of occupational work, occupational aptitudes, and the living conditions of workers and their families. Chéron's report was the first official statement that acknowledged the significance of the science of work for public policy:

> The rapid development of modern industry has inevitably given rise to the squandering of forces, and in certain cases, even to the endangering of the health of the workers, and has become an obstacle to the legitimate amelioration of their condition, and, by the excessive work required of women and children, has even exhausted the available sources of activity.
> The means to systematically regulate labor, must however, always be constructed according to empirical procedures developed according to scientific methods. It belongs to the order of things that after a period of tentative efforts and extreme conflicts, there should follow a period of precise, comprehensive and carefully coordinated study, during which we appeal vigorously to the results of facts and experimentation. . . . It is a matter of collating the results of a large number of experiments, of elaborating the relations between the nature of work, its specialization and organization, and the current state of technology, personal aptitude, energy expended and work efficiency.[89]

A commission of leading physiologists and engineers was appointed (including Amar, Chauveau, Imbert, Paul Langlois, Henry Le Chatelier, Richet, and Weiss) to coordinate the use of statistics, laboratory studies, and other data to propose ways of ameliorating the problems of industrial work. In December 1913, Amar was named to head a new laboratory set up for this purpose.[90] The following year he published his *Le Moteur humain et les bases scientifiques du travail professionnel*, a six hundred-page treatise, which encompassed the history, method, and contemporary state of physiological research on labor.

THE DEPLOYMENT OF SOCIAL ENERGY: MILITARY TRAINING AND PHYSICAL EDUCATION

Though it would be an exaggeration to speak of a coordinated social policy based on the science of work, in several areas we can document its direct impact on debates over physical education, the length of military training, and the causes of industrial accidents. These examples demonstrate that the idea of a rational deployment of social energy under the auspices of state intervention was evident in state social policy in Europe at least fifteen years before the war.

During the course of the nineteenth century the gymnastic exercises required of French youth fell increasingly under the purview of the military authorities. A law of 27 January 1880, for example, required gymnastic training in the lycées, in preparation for the three-year mandatory military service.[91] By the mid-1880s "a war of methods" raged among those who demanded the reform of French physical education. On the one side were the Anglophile proponents of sports, like Pierre de Coubertin, on the other, the protagonists of "rational gymnastics," including Philippe Tissié, Fernand Lagrange, Georges Demeny, Etienne-Jules Marey, and Angelo Mosso.[92] These scientists of fatigue condemned the exhaustion and disregard of the body's rhythm inherent in the athletic training of the times. But even among the physiologists, there was no agreement over which method of gymnastics was preferable.[93]

The movement to reform physical education was fuelled, as Robert Nye has shown, by a broad coalition of public hygienists, moral crusaders, and physiologists who were convinced that the nation was in a state of physical decline and that a resulting low rate of population growth threatened the existence of France.[94] Mosso's remark, that "the catastrophe of Sedan represents in history the triumph of German legs," was often quoted as underscoring the need to improve the physiological condition of the French population.[95] Exercise was no longer to be a disorganized and diffuse distribution of exhaustion or pain, but a rigorous, standardized set of activities based on the repeated and calculated deployment of physical energy.[96] As Lagrange succinctly put it, "That which is hygienic in exercise is not effort, it is work."[97]

In 1891 at the initiative of Vaillant, Marey's assistant, Georges Demeny, received a commission from the city of Paris to create a course of physical education, the first of its kind in France.[98] The material for the course became a model of the social effects that the energeticist movement might achieve with state support.[99] As Demeny emphasized:

> Original studies of physical education have multiplied, the field of research has expanded. The studies by Mosso, Tissié, and Lagrange on fatigue, books by Swedish or American researchers, diverse congresses, the publications of Binet and Féré and their collaborators on physiological psychology, of Ribot on the will, and many others, have hastened the synthesis of scientific knowledge necessary to establish education on a natural foundation.[100]

To decide the issue of sports versus gymnastics, in 1900 Marey was appointed to head a state-sanctioned "Commission d'Hygiène et de Physiologie" to survey "the precise methods and the effects of different sports on the organism, and to compare their value from a hygienic standpoint." Marey's observations demonstrate an extensive use of his graphic and chronophotographic method in evaluating physical exercises and sports: "Today, physiologists possess all the tools appropriate for all such studies."[101]

The Congress of Educational Physiology, which opened in Paris in August 1900 also sided with the hygienists, condemning the French educational system for ignoring the benefits of gymnastic education. A few months later the Ministry of War followed suit, calling for a substantial modification of gymnastic education in the schools and the military, based largely on the physiological principles enumerated by the congress.[102]

As a result, Demeny was commissioned to develop a new program of physical education adequate to the demands of a modern military, resulting in the *Nouveau règlement sur l'instruction de la gymnastique militaire,* a manual of training issued in August 1902. The *Réglement* reorganized the system of gymnastic exercises in the French military and in the public schools, though it was severely criticized by Tissié and unleashed a torrent of hostile reaction from the military.[103] Nevertheless, after 1902 the French public educational system and the military gradually adopted a modified version of the Swedish system, which became the officially prescribed gymnastic method.[104] By the middle of the decade, there was a truce of sorts between the sports advocates and the hygienists, who conceded that sports might require gymnastic exercises, much as music required practice in "scales or harmony."[105]

In 1905 Josefa Ioteyko published a lengthy appeal for a reduction of French military service from three years to nine months. Perhaps the most controversial of the publications of the Solvay Institute, her *Entraînement et fatigue au point de vue militaire* did not conceal its unabashedly pacifist sympathies. Ioteyko was influenced by Jean de Bloch, a Polish born economist and statesman who became an expert on Russian finance and industry—especially the Russian railroad. Bloch also devoted a significant part of his career to introducing physiological methods into military instruction. A dedicated pacifist noted for his multivolume *La Guerre future* (1899), Bloch claimed that war would become obsolete as a result of the destructive power of military technology.[106] Long before the First World War, Bloch predicted that European wars would be wars of attrition, fought in trenches over long periods of time, without decisive victories: the physical condition of the troops would decide the outcome.

Based on his personal observations of the Boer War, Bloch caused a public scandal when he asserted that current French methods of training and the length of military service had a negative effect "on the force of resistance of states."[107] According to Bloch, the length of military service was inversely related to morale: "The spirit of discipline declines with time spent under the flag."[108] At the time Bloch's article appeared, the length of military service was hotly contested in the French press, which debated the three-year term as well as the length of time needed to train cavalry officers. Ioteyko argued that given the changes in modern warfare, nine months of military service sufficed. She contended that in the "struggle between fatigue and training," which was at the core of all systems of physical exercise, training had to be maximized at the expense of fatigue. Troops suffering from "hypertraining" were militarily useless. Proper training had to be energetically sound, that is, accomplished by the least effort. Since all measures to improve society had to conform to the axiom "maximize efficiency," excessive military service constituted a serious "regression" of "social energy."[109]

A rational course of military training demanded the scientific calculation not only of the time required, but also of the frequency, the rhythm, and the intensity of the most favorable movements. "If science can demonstrate, by rigorous laws, that it is possible, useful, and even indispensable to reduce the length of military service the consequences must be imposed without fail on the legislation of each country. The result will be doubly advantageous: reduced time spent in the barracks will not deprive the nation of its best subjects, and there will be significant reduction in the costs of military expenditure."[110]

Not all military specialists—even among those who shared her energeticist credo—agreed that length of military service could be determined solely by the laws of physiology. Ioteyko's critics pointed out that combat conditions required experience, a factor not measurable by training time alone.[111] Ioteyko and Demeny, who had studied the "conditions of march" while Demeny was employed at the French military academy at Joinville-le Pont, belonged to a progressive school of military experts who gave priority to the individual soldier: his physical condition, the length of march, and aptitude with firearms. More conservative defenders of traditional military training reproached their conclusion as naïve. Military training, they countered, was not "exercise" but the subordination of the individual to the will of the group, especially under the exigencies of combat.[112] The length of military service could not be determined by laws of physiology but by the long-term effects of training on discipline.[113]

Ioteyko's campaign for reduced military training, Tissié, Demeny, and Marey's efforts on behalf of rational gymnastics, the controversy over the shorter working day, and the general orientation of the science of work toward reform in the early 1900s—all demonstrate the social hygienists' progressivism in advocating a scientific approach to the problems of labor and society. In 1905 Imbert attempted to convince the French socialist trade unions (C.G.T.) of the benefits of his research, inviting workers to appear at scientific congresses devoted to fatigue and industrial accidents.[114] Ioteyko was also closely associated with the socialist cause, with pacifism, and with feminism. In her laboratory studies at the Solvay Institute of Physiology, Ioteyko compared the working capacities of men and women, concluding that "amongst women their powers of endurance when at work give them the strength for momentary effort; but woman is much more able to produce a sustained moderate effort than to make a great momentary effort."[115] Her conclusions, which were clearly intended to disprove stereotypes of women as inferior workers, were echoed by the Belgian socialist Guillaume de Greef, who asserted that "the sex which is pretended to be the weaker, is really the stronger, it has the greater powers of resistance to the forces which destroy life."[116]

However, the greatest political impact of the European science of work was in yet another area of social contestation—the extraordinary conflict over industrial accidents and accident compensation. More than any other social issue industrial accidents consumed a generation of European reformers after the passage of the 1884 German reform law. It was in this domain that the science of work and the study of fatigue manifested its decisive influence. Until the arrival of the American Taylor system, shortly before the First World War, industrial accidents were the central arena for the new science to establish its claims to neutrality and objectivity.

FATIGUE, KNOWLEDGE, AND INDUSTRIAL ACCIDENTS

The problem of industrial accidents and the legislative debates and conundrums that grew around this issue became the crucible of the modern European social state, tilting the balance toward a new conception of "social rights." Situated at the intersection of working-class politics, the law, and social hygiene, the controversy over accidents altered the idea of social responsibility and risk in Europe.[117] The accident issue challenged the existing relationship between knowledge and political power by focusing public attention on the crucial questions: To what extent do the risks of industrial life transcend the private interests of capital and the formal limits of negligence law? Is the state obliged to intervene, not only in determining the conditions of work (hours, ages, sanitation), but in guaranteeing safety and security through compensation? If such guarantees are warranted by the special character of industrial risk, to what extent is modern industry culpable for the accident? Does modernity in fact pose greater risks to the worker than preindustrial society?

Adopted in July 1884, the German law introduced a comprehensive system of obligatory "public-legal" accident, sickness, disability, and old-age insurance.[118] The French parliament debated the problem of worker's insurance from the early 1880s, finally adopting a national private insurance system on 9 April 1898. Unlike the German system, the French system did not cover illness, even when directly related to the conditions of work. As many contemporary critics recognized, the accident insurance legislation of the 1880s and 1890s, especially in Germany and France, tacitly acknowledged that the free labor contract could only be maintained by recognizing the fundamental *inequality* of the two parties involved.

The German law was novel in its articulation of the principle that industrial accidents were caused less by the design of the plant, or by the negligence of a particular owner, than by the special conditions of industrial work.[119] The French doctrine of occupational risk *(risque professionnel)*, first introduced in the 1880s, held that it was neither labor nor management that was at fault in the case of an accident, but that industrial work was itself a social threat. *Risque professionnel* socialized risk by providing for indemnity without proof of fault. One of the law's chief supporters, the influential reformer Emile Cheysson, noted that "insurance is the compensation for the effects of risk by the organized mutuality according to statistical laws, that is to say, according to the economic laws that govern the course of things." In this conception of risk and responsibility "numeri regunt mundum."[120]

Both the French and the German laws were restricted to industrial work, especially the larger industrial plants—a limitation justified by the corollary doctrine of the "modernity of risk." If risk was social, it was argued, then ultimate responsibility rested with the work, particularly the egregious risks posed by industry. In the French concept of "fatalité de milieu ambiant" (fatality of the surrounding milieu), as in the German "besonderer Gefahr" (special risk), danger was associated with large industrial establishments in which workers were "servants of the machine."[121] Opponents of the law protested that the implied connection between modernity and risk was highly dubious, pointing to far worse hazards of nonmechanized work, especially mining and construction. The doctrine of *risque professionnel* was subjected to its most scathing critique by Léon Say, the liberal economist and leader of the opposition, when he declared that "the great *risque professionnel* of humanity is that each human being is mortal and might lose his physical or mental faculties."[122]

The alarming increase in the number of accident claims, registered in the data assembled in Germany after the first five years of the law, was also seized upon by opponents of the new system. The "correlation between the institution of that insurance and the increase in the number of accidents," led some statisticians to conclude that insurance itself increased the risks of modern industry.[123] Critics of the liability system often distinguished between the number of "real accidents" and the number of "legal accidents."[124] These soaring figures, they claimed, proved that the law had actually produced what it was supposed to prevent. Did the new law not encourage the reporting of even minor mishaps, and perhaps even provide an incentive to commit an accident? The accusation of worker fraud became the center of an international debate by the end of the first decade of reform legislation. The statistical deluge appeared to confirm the claim already put forth in the French parliamentary debates, that compulsory accident insurance was "an incentive for the inattention and the negligence of the worker."[125]

By 1891, after only six years, the total costs of administration and claims in Germany more than tripled. Employers' organizations complained bitterly of "heavy sacrifices," charging that "workers wanted to make capital out of the smallest accidents while drawing out the period of cure."[126] Leading medical specialists confirmed this assessment.[127] According to Professor Fritz Stier-Somlo of Berlin, a renowned expert on medical ethics, fraud was rampant in all areas of social insurance, but most pervasive in cases of industrial accidents. Medical fraud, he said, was already "a mass experience" *(Massenserscheinung)*, and "pension addiction" *(Rentensucht)* a plague *(Volkskrankheit)*.[128] In his widely read polemic, *Unerwünschte Folgen der deutschen Sozialpolitik* ("Unwanted Consequences of German Social Policy"), published in 1913, the conservative national economist Ludwig Bernhard claimed that only the British economy profited from the fact that "the [German] working class has in part succumbed to the pension addiction that has crippled its energy and joy in work."[129]

The rise in accident claims following passage of the French accident insurance law in 1898 soon threatened to duplicate the German experience. In the first five years of the law, the number of reported accidents involving partial disability more than quadrupled.[130] In 1900, when the statistics revealed an enormous increase of 36,000 accidents over the previous year, there was a public outcry from legislators, from the industrialists, or *patronat,* and from the medical profession against "the epidemic of simulations," against the "professionals of industrial accidents," and against the fraudulent abuse of the law by those who took advantage of naïve medical practitioners not trained in "la médecine soupçonneuse" (suspicious medicine).[131] The new law, warned Dr. Hubert Coustan, was "a clear menace to the laboring Frenchman, [who] upon entering the hospital or the bedchamber as a result of an occupational accident, dreams of only one thing—not of departing cured, but of departing with an income."[132]

Workers' organizations responded that the emphasis on fraud only diverted attention from the real truth revealed by the explosion of accident claims—the devastating risk of industrial work. The secretary of the Confédération Général du Travail, Victor Griffuelhes, charged that French judges were being "counselled" on their responsibility to find cases in favor of the companies.[133] In Germany, the social democratic press accused the Institut der Berufsgenossenschaftlichen Vertrauenärtze, the professional organization of company doctors, of having only one purpose: "to send the worker back to work as soon as possible."[134] The demand for "free choice of physicians" became a socialist slogan in Germany in the first decade of the century: "A doctor who sees only an evil in the whole insurance legislation; an unjust burden on industry, and a provocation for the worker to make outlandish claims, will always be a one-sided and inappropriate judge of the injured victim."[135]

A few statistical experts were more sympathetic to the workers' claims. Georg von Mayr, a leading specialist in occupational accidents, contended that the rapid pace of work brought on by shorter hours produced a more perilous situation for the worker.[136] Raoul Jay made a point similar to von Mayr's during a heated debate on the rising claims in the Chamber of Deputies in 1900: if fatalities increased at the same rate as injuries, he said, then "neither were employers more negligent, nor workers more imprudent." It was the nature of the work that was at fault.[137]

In this context, adherents of the science of work attempted to resolve the questions surrounding the accident crisis, offering proof that an accident was neither a willful act of the worker nor the negligence of a malevolent entrepreneur. It was a statistical fact linked to the body's relationship to industrial work, a social risk. The science of work regarded its task as providing the irrefutable proof that the industrial accident was the product of fatigue, a physiological response to the accelerated work tempo and the length of the working day. As Imbert noted, physiology alone could establish the truth of the accident controversy, and thus bring about the improvement of relations between labor and capital, which "otherwise would be dominated by distrust, suspicion, and hatred."[138]

In both Germany and France the mandatory reporting of accidents facilitated a careful study of the time, day, and nature of the occurrence. Analysis of the distribution of accidents during the day and the week might, industrial hygienists speculated, reveal a correlation between fatigue and the cause of most accidents: "The number of accidents would increase as the workers became more tired, and the distribution of accidents according to the hours of the day in which they occur could furnish a means to evaluate the degree of fatigue of the workers who are victims."[139] If proven, this hypothesis would finally lay to rest the charges of deception and chicanery, while substantiating the accident as a social consequence of the modern work experience. Moreover, its advocates believed that science could establish the necessity of state intervention in the workplace, greater surveillance, and, above all, the need for a shorter working day.

Imbert believed that a dispute as thorny as the one in Sète, arising from conflicting claims of worker organizations and insurance companies over the causes of the high rate of accident claims, could be resolved scientifically. It was a "profound error," he claimed, "to regard the workers and the insurance companies as natural enemies."[140] The insurer's complaints of worker abuse, the employers' anger over higher premiums, and the workers' charge that just claims were being unfairly rejected, were, he added, not simply "an inevitable consequence of the conflict of opposed interests."[141] Hostile perceptions reflected the lack of scientific methods with which to evaluate the causes of accidents. In fact, the number of accidents was so great that deception could account for only a tiny percentage of the claims. The sheer number of accidents could be explained only by the amount of work and the physiological state of the worker: "A great number of accidents result directly from the physical or cerebral fatigue of the worker at the moment when he is victimized and it is easy to provide multiple proofs of this assertion, which . . . would result in the possibility of greatly reducing the number of victims."[142]

In 1903 Imbert (with M. Mestre, the Inspecteur du Travail of the département of Hérault) investigated the 2065 accidents in that départment in 1903.[143] They found that the accidents generally climbed uniformly in the period before the midday break, repeating the same pattern more acutely during the afternoon, with the highest proportion of accidents falling in the last few hours of work. From this pattern they concluded that "given our mode of organization of work, the influence of occupational fatigue on the causes of accidents was undeniable."[144] Statistics revealed that the number of accidents increases to more than double during each half-day, and conversely, that the pattern demonstrated the considerable impact of midday rest on the fewer accidents occurring at the beginning of each afternoon. Fatigue, Imbert noted, "renders the worker less able to avoid an unexpected accident, because he is unable to respond with an effort as intense or with movements as rapid as in the normal state."[145]

In his observations of the dockworkers at Sète, Imbert witnessed some of the political implications of his research. The private insurance companies were wrong to believe that the disproportionate number of accidents in Sète as opposed to other ports, indicated the likelihood of organized fraud by the workers. Since the tonnage of merchandise handled at Sète was greater than at other dockyards, fatigue was also more marked.[146]

Imbert predicted that a half-hour rest during each half day in all industries would significantly reduce accidents. Society could not remain indifferent to the deterioration of the energy-producing organism, he declared: "It is dangerous, moreover, in view of the full development of that organism and its future efficiency to extract a usage that is too premature and too intensive; society cannot be indifferent to how many hours of work are consumed, and whether those hours are consecutive or punctuated by one or many rest periods."[147]

Statistics collected in other French départements, as well as in

Germany, Belgium, and Sweden, confirmed the fatigue-accident correlation: Industrial accidents occurred in a definite relationship to the hours of the day.[148] An 1897 German study showed that there were twice as many plant accidents in the three hours before midday break as in the previous three hours, and that the greatest number occurred between 3 and 6 P.M.: "With the slackening of corporal and mental attention as a result of fatigue, the measures taken to ensure personal work safety are also left unattended, and the body is helplessly surrendered to the dangers of the plant and the accident."[149] As the German industrial hygienist Emmanuel Roth concluded, what was previously thought to be negligence or carelessness on the part of the worker, "in the great majority of cases now appears as the consequence of the onset of fatigue."[150]

In Germany the number of accidents on Monday was even greater than those on Saturday, indicating that "besides work itself, the conduct of life is of decisive importance for safety and for the question of fatigue."[151] Delcately put, reports of disappearance of "Blaue Montag" (Saint Monday) were premature.[152] Even the socialists, who rarely commented on the issue, began to complain of drunkenness during working hours.[153] The French figures also showed a higher accident rate for Monday compared to the other weekdays. Imbert studied the accidents in Hérault by days of the week, finding that although there was a higher frequency of accidents on Mondays, the figures were deceptive since "a good number of workers rest on Monday from the fatigue of Sunday, which diminishes, to a degree that is impossible to determine with precision, the size of the working population on the first day of the week."[154] Though Imbert did not comment on this high rate of Monday absenteeism, by 1900 most industrial hygienists agreed that the struggle against the effects of alcoholism on the workplace was an essential aspect of labor's well-being.

These investigations also yielded an unexpected result. Against all predictions, as the hours of work decreased, the accident rate rose more sharply. Roth admitted in his report to the 1907 Congress of Hygiene and Demography, that the "progressive rise in general accidents stands in a certain causal relationship to the reduction of labor time, and to some extent represents its reverse side."[155] If the reduction of labor time produced greater fatigue and accidents as a result of the intensification of work, the exclusive emphasis of reformers like Imbert on a shorter working day or rest periods was misplaced.[156] Increased fatigue was a product of the work-time ratio and not the absolute length of the working day; the real issue was not simply wages and hours, but control over the speed and tempo of work, over time and motion.

Belatedly, the practitioners of the science of work recognized that they were witnessing a major transformation of industrial work. Because their procedures limited them to laboratory investigations of specific tasks, the changing nature of work and its organization had eluded them. The accident question brought into focus a new element—the speed-up of the machinery and work pace to compensate for time lost.[157] As Roth pointed out, it was crucial that "the intensity of work, the energy expended for a given work process within a specific period of time not exceed a certain amount. The energy expended does not depend simply on the absolute amount of work, but also on the distribution of the work over time. Thus, the true art of work is not to exceed the amount of energy which the organism can tolerate without permanent damage."[158] By 1908 it was becoming obvious "that the question of fatigue of the working class could not simply be limited to the question of the hours of work, or even to the organization of work and the design of the workplace . . . but to a large number of other questions of a social, economic, personal and psychological nature."[159]

SCIENCE BETWEEN THE CLASSES

During the first decade of the twentieth century the physiology of work became part of a broader effort to ensure a minimum of physical and social security and to conserve the capital of the nation. By 1907 it was clear to the international community of scientists and reformers at the Berlin Congress of Hygiene and Demography that only a more extensive program of supervision and inspection of industry could ensure the adoption of the physiologists' program.[160] In some occupations it might be possible to predict when the energy expenditure and fatigue were so deleterious that at the end of the workday there was insufficient rest to recoup the day's losses. At the onset of the following day, the worker would find himself "partially impregnated with poisonous substances."[161] In any case, first-hand knowledge of working conditions, the hygienists concurred, was indispensable to calculating the extent and consequences of fatigue.

The politics of social modernity, the hygienists claimed, would result in considerable savings to society. Most accidents could be prevented simply by introducing half-hour rest periods during the morning and afternoons. Fewer accident claims would lead to cheaper premiums and to a considerable recovery of "social energy."[162] The resulting accommodation of the trade unions with the insurance companies would also contribute to reduced accident claims, lessen the cost to the firm, and perhaps raise wages (offset by lower insurance costs). Imbert sadly confessed, however, that even these "more or less restrained" proposals, featured in Millerand's 1903 report on the worker question, were met with considerable resistance.

On 8 October 1904, the *Revue Industrielle* (the industrialists' journal) carried a biting rebuttal by Philippe Delahaye that criticized Imbert and Mestre for *a priori* applying the explanation of excessive fatigue to the statistics. Delahaye denounced their suggestions for reform as "irreconcilable with the conditions of industrial work."[163] Imbert and Mestre responded in kind, noting that "one does not have to affirm or deny *a priori* the existence of exhaustion; on that subject we can form an opinion by observation alone."[164] Similarly, the trade unions resisted efforts to reduce labor time as long as they were not matched by wage payments for time lost.

It is understandable, therefore, that the initial response to the science of work by both labor and capital was mistrust and suspicion. In Germany, the Verein's surveys indicated that workers were skeptical of questions about "joy in work" or efficiency, remarking that alas, "money was the only bond that tied them to their activity."[165] Imbert's investigations of the dockers and Gautier's research on the wine and spirit workers of the South were greeted by the "hostile indifference" of the local labor organizations. Imbert persisted, however, in his desire to see workers take part in scientific congresses, eventually securing the cooperation of several C.G.T. members at the First International Medical Congress of Occupational Accidents in 1905.[166]

Even more significant in promoting the advances of physiology and industrial hygiene among the French trade unions was the formation of the Association Ouvrière pour l'Hygiène des Travailleurs et des Ateliers de Paris in 1904 (Workers Association for the Hygiene of Laborers and Trades of Paris). The Association was conceived as the trade union response to the Alliance d'Hygiène Social founded that year, an organization composed of Solidarists like Bourgeois and Gide, and philanthropists including several Rothschilds, devoted to "moralizing and socializing the popular classes." Unlike the Alliance, the worker's Association attempted to bring "a practical sense" to the problem, promoting among other reforms, the removal of unhealthy materials and conditions from the workplace. The Association advocated periodic and regular inspections, something which it said was, despite existing legislation, "practically nonexistent."[167] Employers were less than enthusiastic about opening their doors to physiologists or hygienists, as the Association demanded at its first congress in October 1904.

In his 1903 report Millerand complained that "the majority of leaders of industry are less than struck by the primordial necessity of measures involving hygiene . . . and have made the work of the inspectors particularly difficult. Emmanuel Roth pointed to the numerous Berlin sanatoria where cases of overwork and neurasthenia were endemic among working-class patients.[168]

Typical of employers' attitudes was the survey conducted by the chief engineer of the Berlin municipal waterworks, W. Eisner, which provoked a storm of indignation from his fellow delegates when Eisner presented it at the 1907 Berlin Congress. Eisner defended the unjustly "rejected" entrepreneur by demonstrating that fatigue—"the condition in which the worker, despite a well-intentioned attempt on the part of his mental functions and his body's limbs, is no longer master"— could not be shown to exist. Eisner distributed a questionnaire to the managers *(Betriebsleiter)* of fifty Berlin enterprises, which asked whether fatigue was noticeable before the normal close of the shift, either in the workers' general appearance or from a decline in performance. The vast majority of respondents indicated: "If we have an after-work party for our people we observe that they dance cheerfully until the early morning hours. People who are too fatigued do not do this."[169] Eisner concluded that fatigue "is so rarely found that it is useless for our purposes" and that "the relationship between 'fatigue' and work could hardly be demonstrated by experiments and generalizations." While Eisner's method (polling managers) and conclusions were overwhelmingly rejected by the physiologists and social scientists present, his claims were an indicator of the German managers' attitude toward the concerns of the science of work.

By the end of 1910 the science of labor became convinced that neither capital nor labor could accurately perceive that the efficient expenditure of energy transcended ideology. If capital had to be taught to see labor in terms of its enmity to productivity and profit, labor had to be taught that the work of the body could conform to the laws of energy rather than the imperatives of politics. The working body performed according to its own physiological laws, which were objective and irrefutable. The laws of fatigue were an implacable feature of industrial society, constituting—along with occupational risk and the economics of productivity—a modernity of the working body, a calculus of efficiency.

A neutral science of the workplace, its advocates claimed, could stand above the contending forces and provide guidance for new legislation, but it could not enforce its claims on society. If the efficient deployment of the energies of society could not be realized in an atmosphere of class antagonism, it fell to the state to provide the necessary surveillance to ensure "a permanent medical supervision of all manufacturing enterprises."[170] The science of work considered the state its natural ally, armed with the power to control the excesses of labor and management in the interest of society. The work of the body had to conform to the laws of energy, not the imperatives of politics. For Imbert, only an authoritarian solution could overcome the "fears and inhibitions" of the entrepreneurs, who had to be taught that individual concerns should "not be confused with the general interest." It was not only entrepreneurs that proved uncooperative. Imbert also criticized the socialists and the trade unions, which, he believed, also had to be educated to see that the scientific training of the working class, the cultivation of proper habits through physiological medicine, and the use of a "science of aptitudes," would result in the most efficient deployment of labor, reduce accidents, and "augment the ease, the speed, the accuracy, and the uniformity of act."[171]

Imbert's position expresses the widespread consensus among the representatives of the science of work from a variety of countries, and articulated at the 1907 Congress of Hygiene and Demography, that its most important goal was a "constant supervision" of the workplace. Such radical prescriptions naturally provoked some skepticism about the ultimate utility of the science of work. As Hector Depasse, a member of the Belgian Conseil Supérieur du Travail and a liberal politician, remarked as early as 1895: "It is possible, to a certain extent, to supervise the employment of time, but how do you supervise the employment of energy?"[172]

Before the First World War the hope that the state would perform the task of ensuring a thorough rationalization of the workplace in the interests of both capital and labor was utopian. Though the state in both Germany and France greatly extended its supervision of industry it was not until after the war that anything approaching effective control became institutionalized. Moreover, as we shall see in the following chapter, from the standpoint of industry the American import, Taylorism, offered a far more profitable form of rationalization, a solution that was haltingly adopted before the war. Nevertheless, the accident issue was a laboratory for the state social policies that accompanied the postwar socialization of the industrial sphere and the rationalization movement in both Germany and France. The crisis of simulation and fraud, and the political debates that it engendered, foreshadowed a new role for science in the resolution of social conflict. In the debates on the productivity of the worker, the length of the working day, the military and education question, the working body was subjected to a new energetic calculation that placed fatigue at the center of a politics of energy conservation.

CHAPTER NINE

The Americanization of Labor Power and the Great War 1913–1919

THE CHALLENGE OF TAYLORISM

IN the years leading up to the First World War, the European science of work faced its most profound crisis with the coming of Taylorism to Europe. Apart from the advent of machine-driven production, no other development in the history of industrial work had an impact equivalent to Frederick Winslow Taylor's ideas of industrial organization. No other method of organizing work was (and continues to be) as controversial as Taylorism.[1] The impact of Taylorism in Europe has been considered from many points of view: the enormous popularity of his writings, the system's implementation by industrialists, the reactions of workers, and its cultural impact on ideas of efficiency and organization. The confrontation between Taylor's ideas and the European science of work—Taylorism's most important competitor—has received far less attention.[2] This confrontation was all the more intense because of the apparent similarities between the two approaches. Each claimed validity as a science of work, each claimed to analyze the worker's task and movements in minute detail, each claimed to improve productivity and efficiency, and each claimed to be able to transcend the constraints of class interest and ideology in the interest of a rational and scientific economic organization. Both approaches were profoundly modernist in orientation, promising to emancipate industry and technology from the inhibitions and prejudices of tradition and class conflict.

Frederick Taylor claimed that his system applied engineering skill to organize the labor process and the firm according to rigorously scientific principles. Broadly conceived, "scientific management" (as it was later called) rationalizes the component parts and the general functioning of the enterprise in a series of stages in order to increase productivity and eliminate the waste of labor power and materials. These stages included (1) the division of all shop-floor tasks into their fundamental parts; (2) the analysis and design of each task to achieve maximum efficiency and ease of imitation; (3) the redesign of tools and machines as standardized models; (4) the linking of wages to output; and (5) rational coordination and administration of production. Taylor was convinced that his system would eliminate the source of workers' discontent and industrial conflict, that individualism, sobriety, and competition would replace collectivism, dissolute behavior, and *soldiering* (artificial constraints on production).[3]

More narrowly conceived, the Taylor system was concerned with determining the most efficient method of accomplishing each task in the labor process and with the division of all tasks on the shop floor into replicable units. The Taylor system was also a method of linking wages to productivity through "time and motion" studies, keyed to the speed and output of the individual worker.

In the broader sense, however, the Taylor system shifted effective control from the shop floor to management, enlisting the expertise of the engineer in redesigning the work process from beginning to end. For its admirers, which included capitalists like Walther Rathenau and Marxists like Lenin and Antonio Gramsci, Taylorism was the first truly scientific method of organizing modern industrial work based on efficient procedures. Taylorism provided industrialists and managers with a means of breaking the resistance of workers' organizations to technological progress, of rapidly training unskilled laborers for modern industrial plants, taking control of the factory out of the hands of paternalistic and conservative entrepreneurs, and putting it in the hands of competent, trained professionals—engineers. Even Taylorism's harshest critics have not concealed their recognition of its significant social impact. Marxists and non-Marxists alike have insisted that the Taylor system ended the skilled laborer's monopoly on expertise, dissolved the traditional foreman's authority over the shop floor, weakened the power of unions to control wages, and gave management a powerful method of exercising control over the entire production process.[4]

More recently, historians of Taylorism, especially in the United

States and Britain, have cast doubt on the extent of the system's initial acceptance by industry, pointing to the relatively few firms that actually employed Taylorist methods before the First World War.[5] But, even if Taylorism alone was not responsible for the great transformation of work at the turn of the century, many changes advocated by Taylor were already adopted before his ideas were widely disseminated. Especially in the most technologically advanced sectors of industry—in the electrical, chemical, and automobile industries—the school-trained engineer usurped the monopoly of skills once enjoyed by the master craftsman. Plant managers wrested authority from the traditional shop foreman, destroying the "community of experience and skill that was the foundation of artisanal worker solidarity."[6] The significance of Taylorism cannot simply be judged by the number of firms that explicitly adopted his methods. Taylor's real importance lay in the creation of the first management-oriented industrial ideology and in his ability to synthesize and promote in a coherent framework the broad changes that were already taking place piecemeal in various industries.

In this sense, the debate *about* the Taylor system was in many ways far more important than the *extent* of its application. This debate, particularly in Europe, was over the scope and significance of a transformation that had been occurring in "the new factory" for more than two decades. In France, for example, Taylorism promised to liberate industry from the constraints of a society still encumbered with traditional social attitudes, militant workers, an economically traditionalist industrial *patronat,* and the hegemony of craft production. In Germany it offered industrialists the chance to maintain the national lead in competition with England and eventually to overtake the productive preeminence of America. Despite the initial reluctance of many entrepreneurs and shop foremen to accept the new system—which, it should be emphasized, also limited *their* power on the shop floor—the enthusiastic adoption of Taylor's ideas by a few major auto firms like Renault and Peugeot in France, and by electrical firms like AEG, Bosch, Borsig, and Siemens in Germany, demonstrated the readiness of a sector of modern industrial management to regulate the "time and motion" of the worker and to supervise the details of production to a degree heretofore unthinkable.[7]

The claims of the Taylor system to provide a scientific solution to the "worker question" appealed to industrialists and engineers in Europe primarily because its modernizing and rationalizing thrust linked greater productivity with social peace. Taylor himself emphasized that the novelty of his system was a "revolution in mental attitudes," which attempted to divert the concerns of both capital and labor away from the division of the surplus and toward common efforts to "increase the surplus."[8] By making technical competence indispensable for those in charge of work and labor relations in a modern plant, the Taylor system was a golden opportunity for European engineers to advance their profession.[9]

Worker reaction was, at the outset, less hospitable than that of management. The wave of strikes and protests that often greeted the introduction of the "time-motion" study, the efficiency expert, piecework wage systems, and planning departments were fuelled by the workers' perception that Taylorism was synonymous with "organized overwork." It was apparent to those most directly affected by the new system that Taylor's one-sided approach focused only on output and profits, and that the rewards of a premium or bonus wage was little compensation for the increased work tempo and surveillance that accompanied the new method. During the famous Billancourt Renault strike of March 1913, directed against the time-motion experts, the French syndicalist leader Alphonse Merrheim condemned the system as "the most ferocious, the most barbaric" system of work devised by capitalists. Taylorism, he said, "eliminated, annihilated and banished personality, intelligence, even the very desires of the workers, from the workshops and factories."[10] Émile Pouget, a C.G.T. leader, denounced the system as "the organization of exhaustion." In the same year the conservative journal of the German *Verein deutscher Ingenieure* remarked that the workers of the recently Taylorized Borsig plant in Berlin-Tegel, despite their usual submission to different methods of work, received the new system "very badly."[11]

Nevertheless, as we have shown in chapter 3, by the 1880s Marxism had largely become a productivist ideology as evident in Marx's positive attitude toward the expansion of labor power in *Capital*. For European Marxists, Taylorism created a profound dilemma because it combined a welcome method of rationalizing and modernizing production while increasing and accelerating exploitation. For Marxists, in addition to its promise of a more efficient use of labor time, the Taylor system could be adapted to a vision of society that required the ever-increasing productivity of labor to realize its aims. The negative effects of Taylorism could be attributed, as Otto Bauer later argued in his book, *Rationalisierung und Fehlrationalisierung* (1931) ("Rationalization and Mis-rationalization") to its abuse at the hands of overly greedy capitalists. As a result many trade unionists—and eventually Merrheim himself—were won over to the Taylor system as an inevitable consequence of industrial progress. Socialists could argue that it was simply "a means of reducing the worker's gestures to the strictly necessary and without surplus fatigue, the application of the method of 'jiu-jitsu' to industry."[12]

By 1913, the rapid diffusion and wide popularity of Taylor's ideas among European engineers, factory managers, and trade union and socialist leaders presented a major challenge.[13] To sophisticated European practitioners of the science of labor, the American Taylorists and their British disciples appeared hopelessly regressive, "grotesquely childish," and still worse, uneducated in the scientific developments necessary to a nonpartisan understanding of the problem of work.[14] They rejected the notion that the extraordinary interest of European engineers and firms in American "scientific management" could be attributed to the practical superiority of the Taylor system over the experimental determination of ergographic laws. But they could hardly fail to acknowledge the success of the new system. The initial impact of Taylorism demonstrated that the power of its ideological appeal rested on its anticipated economic and political advantages and on its reliance on engineers and white-collar plant managers as opposed to academic, laboratory-trained physiologists, psychologists, and physicians with limited knowledge of actual work environments.

In contrast to the science of work, the Taylor system drew upon the plant-based engineer, whose expertise came from the shopfloor and from the observation of workers rather than from the laboratory. Taylor's preference for the professional engineer remained a thorn in the side of the "ergonomists," who regarded the engineers as *arriviste* professionals whose own claim to superior competence over the domain of work was suspect.[15] The Taylor system also challenged a more fundamental premise of the science of work: that job satisfaction and greater productivity were linked—and that both could lead directly to social justice and greater happiness. Instead, Taylorism assumed that by offering higher wages as premiums for productivity, the "natural" unhappiness of workers could simply be compensated by nonwork-related material rewards.

Despite these differences it is difficult to ignore the obvious points of convergence between the two approaches, a perception that accounts for the tendency of otherwise sharp observers to group them together.[16] A gloss of the similarities between the Taylor system and the science of work makes these differences appear all the more significant by contrast. Both schools took as their starting point the decomposition of each task into a series of abstract, mathematically precise relations, calculable in terms of fatigue, time, motion, units of work, and so forth. Both shared a preoccupation with economizing motion and achieving greater work performance through adapting the body to technology. Both the Taylorists and the advocates of the science of work were in the *avant-garde* of work-related professionals who wanted to modernize European industry and society by increasing productivity: both envisioned a reinvigorated society based on expanding the productive forces available to mankind. Fatigue experts and Taylorist engineers were also latecomers to the industrial workplace, and both rested their claims on the authority of science. Profoundly antitraditionalist, both viewed the "atavistic" attitudes of workers and the old-fashioned paternalism of factory owners or shop foremen with equal suspicion. And above all, each shared the utopian hope that it was possible to resolve industrial conflict scientifically and rationally in the interests of economic progress. For both it was the body of the worker that constituted the point of contestation in the industrial sphere, not the politics of workers' organizations or ideological issues. The rationalization of production was predicated on the rationalization of the body.

Their paths diverged over the centrality of fatigue and the objectivity and competence of scientists in labor relations. Whereas Taylorists stood on the side of management, the science of work was more ambivalent in its allegiances, claiming to transcend the struggle between capital and labor. Limited to the terrain of the firm, the goal of the Taylor system was to raise the level of productivity of the individual worker and enterprise, whereas the science of labor endeavored to preserve and maintain the worker's energy in the interest of society. If Taylor emphasized higher wages and the rewards of leisure and consumption, the science of work emphasized the long-term benefits of conserving the energy of the worker and of the nation. Above all, Taylorism was concerned with the needs of management to reduce costs and maximize profits through an intensification of productivity, rather than with an appeal to the state to enact laws limiting exploitation or requiring shorter hours. Certainly, the overt similarities between the two approaches to work only reinforced the perception of labor scientists who initially regarded the Taylor system as a serious, powerful competitor.

The success of Taylorism provided the science of work with a significant opportunity to demonstrate its claim to scientific superiority, but it also underscored its fundamental weakness—its manifestly academic character. For this reason, the problem of how to deal with the American interloper produced three conflicting responses: First, there were those who were alarmed by the Taylor system's overt management orientation and its blatant disregard for the worker's health and well-being; second, there were others who regarded it both as a powerful yet potentially useful competitor; and finally there were those who viewed the Taylor system as a potential ally, which could provide an access for academic experts to the industrial workplace unparalleled in the years before the World War.[17]

Taylorism initially threatened to eclipse the central concerns of the science of labor: the rational determination of optimal efficiency for both labor and capital, the scientific regulation of energy expenditure, and the rational determination of the laws of fatigue. As Taylor's defenders pointed out, however, the science of work did not address issues such as the organization of the firm or the standardization of machinery and tools. Its focus was principally on the measurement of the workers' productive capacity and on the maximization of output—the areas of greatest overlap between the two approaches.

TAYLORISM IN FRANCE 1913–1914

The controversy surrounding Taylor and his system in France on the eve of the First World War was perhaps more significant than its actual impact on industry, which remained, until after the war, limited to a few large industrial firms, chiefly those in the automobile industry. In many respects the Taylorism debate was symptomatic of the limited French industrial experience and of the aura of success that surrounded the American industrial experience. Taylor registered his surprise at the more than two hundred articles that appeared on the subject in the wake of the Renault strike of 1913.[18] In this debate, the science of work played a major role, especially after Taylor's main French supporter, the prominent chemist-engineer Henry Le Chatelier, inspector general of mines, and professor at the Sorbonne, published a series of articles favorable to Taylor in the engineering journal, *La Technique Moderne*, during 1913. Politically conservative and anti-Dréyfusard, Le Chatelier was the P. T. Barnum of French Taylorism, an indefatigable lobbyist for the marriage of science and industry.[19] As early as 1904 he advocated the modernization of French engineering and the direct involvement of scientists and engineers in the production process.[20] His translations of Taylor's works began with *The Art of Cutting Metals (L'art de tailler les métaux)* in 1907, followed by *The Principles of Scientific Management (Principes d'organisation scientifique)* (1911), and *Shop Management (La Direction des ateliers)* (1913). In addition, Le Chatelier's numerous articles and translations appeared in his journal, the *Revue de Métallurgie*, which became a forum for Taylor's ideas.[21]

For Le Chatelier, Taylor's most important idea was "to increase the yield of work, without increasing the fatigue of the worker, and to thus bring about a considerable increase in wages."[22] By discovering the most efficient method of executing a given task through precise observation and measurements, Taylor also discovered the *optimal* conditions of work: "He doubled and tripled the efficiency of machine-tools, and he increased in the same proportion, the daily output of the workers."[23] Le Chatelier also recognized the applicability of Taylor's ideas for the small shop common in France, stressing that even in crafts *(l'atelier)* each task required a complexity of organization "that surpassed the capacity of the worker." As Taylor had shown in his famous example of a pig-iron monger named Schmidt, "The alteration of rest and work, the speed of each of the movements, the weight lifted in each effort, considerably modified the fatigue for the same amount of work."[24]

Le Chatelier recognized that Taylor's approach was opposed to that of the trade unions: "Their dominant preoccupation, is, on the contrary, to limit the production of each worker to assure a sufficient amount of work for the greatest number of them, and to reduce unemployment." Le Chatelier conceded that the new system could be "very onerous," requiring that men be placed above the workers who know their manual work better than the workers themselves. Even if French industrialists had initially been reluctant about admitting engineers to their plants, he confidently predicted that just as they had once reconciled themselves to chemical or mechanical research laboratories, "in the near future" they would regard his ideas just as indispensable. He concluded that Taylorism was the solution to the "grave crisis of industry" since the "incessant struggles between capital and labor were an impediment to further progress and a threat to the progress already achieved."[25]

In 1913 after workers at the Renault plant in Billancourt were defeated in a strike against the presence of the time-motion man on the shop floor, Taylorism became "the object of passionate discussion in the French technical press."[26] Le Chatelier defended management, pointing out that the workers did not reject the presence of the time-motion man and the stopwatch *per se,* only their lack of participation in the decision to employ such methods. He attempted to counter what he called the "singular deformation of the ideas of the great American engineer" by its defenders and detractors.[27] Le Chatelier also wrote to Taylor reassuring him about the outcome of the strike:

> The strike at Renault is over, all the workers have returned without I believe, having obtained anything. Despite the defective conditions and excess rapidity under which your system had been applied, the workers obtained an important salary increase, and they did not want to lose their

premiums. Do not concern yourself with that strike, it has not at all inhibited the application of your system in France. On the contrary it has contributed to its general recognition, and sales of your most recent book have rapidly accelerated.[28]

The "vulgarisers" of Taylor, Le Chatelier argued, had reduced his system to "chronometrics" *(le chronométrage)* and bonuses, while in fact, these represented only minor parts of a system that included the standardization of tools, the reorganization of management, and the professionalization of plant design. The system's critics, he argued, failed to recognize that the "purpose of timekeeping was to determine the best conditions of work corresponding to the minimum of fatigue, and that Taylor himself recognized that "the human factor" was today "the dominant factor, the primary aspect of all industrial organization."

Taylor, Le Chatelier insisted, did not "want to exhaust the worker by an excess of work," but instead offered a choice: "Either reduce fatigue by greater production and by a salary equal to or more than the level of production, or reduce both fatigue and production." He added that Taylor's system was simply the application of the methods of the military to the factory: No one would be shocked at the study of the fatigue of a soldier carrying a heavy pack at different speeds, and under different conditions of march. Why then, condemn the timekeeping in the factory that is considered normal in the barracks? The essence of Taylorism was not a disregard of the physiological or psychological aspects of work, but the convergence of all aspects of industrial science.[29]

Le Chatelier's brief for Taylor emphasized the elements of his system that most interested the ergonomists—his neglect of fatigue. His reading of Taylor, it might be argued, anticipated the ergonomists' criticisms, for example, that the Taylor system ignored the physiological effects of work, especially the problem of fatigue. The debate that ensued as a result of the Renault strike provided the occasion not only for a sharp exchange between pro- and anti-Taylorist scientists and engineers, but for division within the ranks of the ergonomists themselves. Differences over how to respond to the challenge of Taylorism put Jules Amar and Charles Fremont at odds with the most decisive opponents of the system, Imbert and Jean-Marie Lahy.[30]

Early in 1913, Le Chatelier wrote to Amar in order to win his support for the new system, suggesting that he would correspond with Taylor about Amar's work and indicating that he would be favorably disposed to Amar's appointment to the new research laboratory for the study of work in the Conservatoire National des Arts et Métiers.[31] Le

Chatelier, it should be recalled, sat on the influential commission appointed by the Ministry of Labor to create the new laboratory. He also wrote to Taylor to inform him about plans to build the laboratory and to describe Amar's work in glowing terms. Le Chatelier underscored his praise for Amar's scientific approach to fatigue: "He measures human fatigue by means of a respiratory coefficient, e.g., the relation between carbon monoxide and oxygen in the gas respired."[32] In August Amar published an article in the journal *La Technique Moderne,* which mentioned Taylorism *en passant* but succinctly reprised the achievements of the science of work from its inception in thermodynamics to Amar's own "law of rest." In his conclusion, Amar compared the science of work with Taylorism:

> Directed toward research of the mechanical and physiological factors of human work, it [the physiological method] assures maximum yield for industry without ceasing to offer the guarantees required for the normal functioning of the organism. In that sense, it *complements* [*(amende),* my italics] Taylor's system, since he did not particularly perceive—as was already mentioned—nor understand the problem of physiology. It [the Taylor system] is, in a word, the science of energy of a disciplined, efficient, and absolutely healthy worker.[33]

Le Chatelier, in return, contributed an introduction to Amar's *Le Moteur humain* (1914), which praised it for providing both workers and firm owners with the means of "reducing the wear of the human machine." Le Chatelier noted that the Taylor system had produced "sharp protests" by the workers, who in their ignorance had misunderstood Taylor's purposes and condemned it for its "organization of exhaustion." For this reason, the problem of fatigue took on greater importance, though it presented serious difficulties for those trained in engineering. "It is not certain," noted Le Chatelier, "whether engineers, who are strangers to physiological questions, are absolutely qualified to pursue the necessary studies, and it would be more rational to leave this to the concern of the physiologists."[34]

Subsequently, Amar adopted the compromise outlined by Le Chatelier. The Taylor system could be advantageously employed in tandem with the science of work, which he called a "universal" science. The result of that collaboration would be "that the social efficiency of at least half the population can be improved." But, he added, the "Taylor system is inadequate from the physiological point of view."[35] According to Amar:

The American *savant* did not have the means of appreciating the degree of fatigue, or a way of knowing the speed, the rhythm, or the effort, which in any maximum amount of work, results in the least expenditure of energy. We insist on physiological conditions because the true interest of industry lies not in the realization of maximum output or the employment of human energy at the greatest speed irrespective of the health of the worker; physiology, on the contrary desires to conserve and husband human energy in the interest of both the employee and the employer.[36]

"It is possible to organize human work," Amar wrote, "in strict observance of the principles of Chauveau without fundamentally contradicting the method of Frédéric *(sic)* Taylor." Scientific methods could measure speed, determine the proper movements, and standardize the choice of tools; in short, create an "ideal efficiency" in a real workshop or factory. The relationship between Amar and Le Chatelier amounted to a quid pro quo, with Le Chatelier's offering Amar his political influence and his access to Taylor, and Amar's offering a relatively positive assessment of the future collaboration between Taylorist engineers and ergonomic scientists. At the close of his *Le Moteur humain* Amar joyously depicted the future collaboration of scientific management and the science of work as one of great promise:

> The Taylor system, complemented by a recognition of the physiological conditions of work, applied by men of good will, endowed with wisdom and know-how, provides one of the most scientific solutions to the social problems of labor power, to relations between employers and employees. It is a matter of adopting it in principle and studying its method of application, since science works according to the beautiful metaphor penned by a great writer, "to cast up over the storms of the present, the peaceful rainbow of the future."[37]

Taylor's French supporters placed him in a direct line of descent from Leonardo da Vinci, the eighteenth-century engineers Vauban and Bélidor, Diderot through Chauveau, Marey, and Mosso—though, of course, for the French, Taylor was *n'est pas proprement un savant.* For Fremont, who conducted the first chronophotographic experiments in Marey's laboratory almost two decades earlier, the Taylor system was the logical result of the convergence of pure science and the practical problems of industrial production since the seventeenth century.[38] Taylor's insight into the labor process was the fruit of twenty-five years of experience in industry, allowing him to substitute rigorous methods for empirical practices.[39] For Amar and Fremont, Taylor's system and the science of work were complementary, representing a true unity of theory and practice.

The debate on the Taylor system among the French advocates of the science of work in the years before the First World War underscored its most serious limitation: lack of concrete experience in the workplace. Amar's extensive laboratory research into the physiology of work—fatigue, metabolism, respiration—in various tasks was a far cry from industrial application. Nor, as we have seen, did Imbert's efforts to negotiate solutions to labor conflicts with the aid of data on fatigue and statistics on accidents elicit much support from the trade unions and the socialists whom he tried to convert to his principles. Jean-Marie Lahy, a younger physiologist and critic of the Taylor system (see the following section), remained convinced that science could still provide "the means by which the patron and the worker could reciprocally bring about a solution to the problem of work."[40] But he, too, recognized that a psychologist or physiologist's use of research or instruments from the laboratory to solve issues of the organization of work was "highly speculative," since "the factory or the workshop was not a branch of the laboratory." He did, however, believe that the laboratory might provide assistance in solving industrial problems.[41] The industrialists, he charged, had attempted to "denude the men of science of their initiatives" and that only the "man of the laboratory," freed of partisan interests, could establish the rules of a scientifically organized workplace.[42]

JEAN-MARIE LAHY: THE SCIENCE OF WORK AGAINST THE TAYLOR SYSTEM

The most assiduous opponent of the Taylor system in France was Jean-Marie Lahy, a physiologist and psychologist whose early work was similar to Marey's youthful *oeuvre* in its concern with blood pressure and with the application of the graphic method in physiology and psychology.[43] A militantly anticlerical freemason, Lahy also published a series of pamphlets scientifically refuting Christian dogma and ethical doctrines.[44] Lahy was also sympathetic to socialism, sharing Imbert's vision of the science of work as a neutral instrument that could also be used on behalf of the working class. Unlike Amar, however, Lahy combined laboratory research with studies of workers on the job, producing some of the first detailed analyses of aptitude in such occupations as typists, tramway conductors, and linotype compositors.[45] In the study of typists, for example, Lahy argued that a distinct ensemble of "psychophysiological" signs were objective indicators of a particular skill, for example, swift hand-eye coordination, or good memory.[46] Though he initially conducted the linotypist study in 1908 with the cooperation of the Paris union, Lahy challenged the stereotypes of union leaders toward the recent influx of women workers in that trade. He demonstrated that the widespread view that female labor was less productive and deserved less remuneration was based on prejudice—a result not at all to the liking of his union sponsors. Lahy strongly opposed any reduction of wages for the women compositors and took issue with the trade union leaders who regarded female labor as inferior or as a threat to the integrity of the family.[47]

Lahy's counterattack against the Taylor system, which began in 1913 with the appearance of several articles in the *Revue Socialiste* and in the more popular *La Grand Revue,* demonstrated its inadequate base in science.[48] These articles were later expanded into his comprehensive *Le Système Taylor et la physiologie du travail professionnel* (1916). Unlike Amar and Fremont, Lahy was not impressed by Taylor's claim to scientificity, nor by his assurances that his system did not result in greater fatigue for the worker. Taylor's publications, he noted, were not as comprehensive and complete as Marey's, which had appeared years earlier, and his method of "chronometry" was rudimentary, "reducing the measurement to the time passed and the speed of the work."[49] For Lahy, Taylor's challenge was serious only because he subscribed to what Lahy called a "sort of scientific fetishism without positive value."[50]

Lahy characterized the Taylor system as "superproduction for everyman," a conception of work that rested on a "triple error," at once psychological, sociological, and industrial. Taylor did not perceive the worker as anything but a "perfectly adaptable machine," whose "only goal was to increase output."[51] Taylor's "ignorance" *(méconnaissance)* of basic human physiology, of the writings and research of the fatigue experts, and his single-minded desire "to obtain a maximum of efficiency," jeopardized the health and safety of workers. Lahy denounced the Taylor system's view of the worker "as one of the pieces of the grand checkerboard that forms the factory." Completely partisan in its emphasis on output, ruthless in its "carrot-and-stick" methods of wage premiums, and devoted to giving management complete control over production, the Taylor system ensured the absolute hegemony of management in the new factory: "Neither the participation of the worker nor the control of the public power can modify it given the current state of legislation."[52]

For Lahy, the Taylor system was built on false premises. Taylor wrongly assumed that a worker could be arbitrarily "assimilated to a machine," and like a machine subjected to infinite wear and tear. Nor could the worker be "abstracted" from the political and social circumstances of his or her life. "Each time work has been organized on a new basis," he claimed, "the reformers see as their first priority the perfectioning of a technique, and do not consider the worker except as an element in the production process, as the complement of a tool."[53]

According to Lahy, much of the confusion that accompanied the introduction of the Taylor system into France was caused by Taylor himself: his works were not well-conceived, his ideas were contradictory, and his formulations imprecise.[54] Lahy was decidedly unsympathetic to the more favorable reactions of his associates, Amar and Fremont, and blamed Taylor's European success on the shock that his system produced among physiologists, psychologists, and engineers who for years had pursued "their efforts to found a scientific organization for occupational work in isolation." When confronted with the apparently successful, and practical, American system, they readily capitulated, considering their results too meager, since "their rigorously pursued research apparently had not received any practical consecration." Such practical success had eluded the science of work for good reason. Its ideal of a scientific organization of work was entirely opposed to Taylor's, which was conceived as part of the efforts of American industrialists to create only "a greater efficiency of the worker who had to become an integral part of the economic mechanism." The science of work was not merely an adjunct to Taylorism, but represented an alternative to the "regime imposed on intensified production by industry." Judging Taylorism from its own "psychophysiological" standpoint, Lahy concluded that he could only condemn Taylor's prejudices "in the interests of the worker, the *patron,* and of the race."[55]

Lahy also disputed Le Chatelier's claims to Taylor's originality. Rather, much of what he proposed were already widely practiced in industry—that is, standardization of tools and work processes—or were the fruit of earlier discoveries of which Taylor was apparently ignorant—for example, the time-motion study developed and refined by Marey and Imbert. Taylor's studies were also "far from the precision of those pursued by us [in France], reducing the measurement of motion to the speed of work in occupations that leave little to the initiative of the worker."[56] Even Taylor's disciple, Frank B. Gilbreth, who had claimed the industrial chronophotograph as his own invention, eventually conceded that Marey had developed a method of "inscribing movements" much earlier, but not, Gilbreth inaccurately added, "through the medium of photography."[57]

Lahy contended that Taylor's system was applicable solely to the most elementary motions and was unsuited to work in a large factory, which required vigilance, intellectual effort, or precise movements—all of which demanded mental energy and produced mental fatigue. It was also inadequate for dealing with the serious psychological problem of industrial work, which Münsterberg had identified as "monotony."[58] Lahy charged that the Taylor system so grossly neglected the true nature of the modern factory that it threatened to destroy it. The factory, he claimed, was a delicate organism composed of different functions. In total disregard of this fragile balance of interests, the Taylor system puts everything "in the service of perfection." It places the engineer above the foreman *(contramaître)*; at any given moment the time-motion expert *(chronométreur)* controls the movements and conditions of each worker and each task; the traditional hierarchy is dissolved by the segmented and specialized authority of experts; and the entire operation of the firm is subordinated to the will of the administration. The internal operation of the modern factory can neither be studied nor perfected, except by taking into account how all of its diverse interests are put into play.[59] These collective interests were ignored by the Taylor system, which subordinates the internal functions of the firm to business imperatives—that is, to an exaggerated rate of production, to products of inferior quality, to an excessive pace of work, and to an oppressive control over the individual's ideas and actions.

Even Taylor's most cherished claim, increased profitability, was illusory. His system led only to "commercial hyperactivity," ultimately denying to those workers who submitted to it the wage advantages that had been promised them. Taylorism's America-centered approach required a seemingly inexhaustible pool of unskilled workers not readily available in a tighter European labor market, where a method of selecting workers based on psychophysical aptitude testing would be more beneficial. The workers' initiative was destroyed by the chronometric system, which stifled the evolution of technique by freezing the worker in a set of movements imposed from above: "For the worker who would have an active part in the progress of his technique, there is no chance except until he arrives—having accomplished his task—at the point of the complete exhaustion of his forces."[60]

Lahy viewed the Renault and Peugeot strikes as conclusive evidence that the workers would eventually reject the Taylor system: "The worker, exhausted in the name of a science whose object he cannot understand, cast off by those who are bedazzled by the powerful attraction of the new methods of work, will ultimately rebel against the new organization."[61] In classical socialist fashion Lahy catalogued the global catastrophe brought on by Taylorism's economic myopia: "Blind to the real aims of the factory, it can never perceive the collective ends toward which it is organized; from this blindness all of the latent conflicts between nations arise: hostilities deriving from price wars; flooding of markets of neighboring countries; and its main consequence, the killing of national initiative."[62]

Lahy was clearly the most thorough and penetrating critic of Taylor to emerge from the European science of work. Nevertheless, we can discern an ambivalence even in his most compelling arguments. Lahy shared Taylor's ultimate goal, if not his methods or social ideas: the reduction of wasted motion, the standardization of tasks, the improvement of the firm's efficiency. After the war, when the eight-hour day became standard in French industry and when the Taylor system was modified according to at least a few precepts of the science of work in various industries, Lahy softened his attack. In the preface to the second edition of his book (1921), he conceded that the Taylor system "partially addressed the problems of the organization of work," and if it were modified and accompanied by methods of "preserving the individual and the race," it might indeed represent progress and "be beneficial for workers of all categories."[63] In interwar France, Lahy became the founder and leading proponent of the movement for the "humanization" of work, contributing to its institutionalization and professionalization. His biography attests to the ultimate rapprochement between the science of work and Taylorism during and after the war.

GERMAN TAYLORISM AND
THE SCIENCE OF WORK

In Germany too, the Taylor system arrived a few years before the First World War as an "Americanizing" and rationalizing ideology, a new ingredient in Germany's effort to maintain its industrial supremacy. The Taylor system, as Judith Merkle has pointed out, also possessed enough of a paternalist, "corporatist" sensibility to eventually win acceptance among conservative and Prussianizing elites in Wilhelmian Germany.[64] As in France, the Taylor system appealed to the more competitive sectors of German industry, but in Germany it was adopted in those industrial firms oriented toward the United States as the most serious threat to German industrial hegemony.[65] In Germany too, the engineering profession paved the way for adopting the new system. Unlike France, however, the Taylor system reinforced the drift toward the reorganization of German industry—especially the replacement of the shop floor bosses *(Werkmeister)* by professional managers, the introduction of time studies, and new performance-based wage systems.[66]

These aspects were singled out by German engineers who praised the new "weapon of the USA" for its standardization of equipment and functions, for deposing the old craft-oriented system, and for the technical reorganization of management.[67] As early as 1903, the German Association of Engineers, the Verein deutscher Ingenieure (VDI) urged the imitation of American methods of factory organization, especially Taylor's differential wage scales.[68] Before the war, Taylor's ideas were adopted in the Borsig, Bosch, Siemens-Halske electrical firms, and in the Daimler-Benz automobile enterprise. In 1914 to 1915, Gilbreth led a joint German-American team assigned to advise the Berlin lighting firm, Glas-Glühlicht Gesellschaft Auer in its plans to rationalize its operations.[69] By 1914 several firms had already experimented with different aspects of the system, and the publication of Taylor's works on management, generated strong interest in his ideas among a few factory managers and professional engineers.[70]

Significantly, the German reception of Taylor's ideas was smoothed by their presentation in the language of energetics. The German translation (1912) of his study, *Principles of Scientific Management,* was introduced as "a system of budgeting human labor power," and its translator, the Berlin engineer Rudolf Roesler, remarked on the many parallels between Taylor's ideas and those of Wilhelm Ostwald, whose *Der energetische Imperativ* appeared the previous year (see chapter 7). "It appears to me remarkably interesting," wrote Roesler in his introduction, "that two men who did not know about each other, or until recently knew nothing of each other, arrived at the same results in so fundamentally different ways."[71]

The legacy of the German *Arbeitswissenschaft*—the industrial psychology of Kraepelin, Münsterberg, and the empirical sociology of the Verein für Sozialpolitik—also led to a more psychological criticism of Taylorism when it was first introduced into Germany. Among the earliest critics of Taylorism in Germany was Münsterberg, who in 1910 returned to Berlin after almost two decades at Harvard University.[72] Far less harsh in his judgments than Lahy, Münsterberg adopted a middle ground, reproaching the Taylor system for its lack of interest in the results of experimental psychology and physiology, but affirming its superiority in organizing industrial work based on the tempos and rhythms of the modern factory.[73] During his American sojourn, Münsterberg acquired a familiarity with the Taylor system and with the growing needs of management for a psychological approach to the problems that Taylorism inevitably produced. In 1913 he published his *Psychologie und Wirtschaftsleben,* which in a modified American edition entitled, *Psychology and Industrial Efficiency,* was an American bestseller.[74] In these works he evinced a clear admiration for the system, which he credited with "heightening of the individual's joy in work, and of the personal satisfaction in one's total life development."[75]

Münsterberg shared the Taylor system's emphasis on the rationalization of the workplace and on the adjustment of workers to new technologies and production processes. He praised it for replacing "everything which usually is left to tradition, to caprice, and to an economy which looks out only for the most immediate saving" with a scientific approach based on experimentation.[76] He also agreed with Taylor's assertion that a science of work should serve the interests of the entire society and not the immediate interests of labor and capital, adding only that the purpose of industrial psychology should be to enhance the efficiency of both.[77] In his compendium of applied psychology, *Grundzüge der Psychotechnik* (1914), Münsterberg spoke of "new possibilities which have been opened up to adapt the mechanism of economic demand to the rhythms of muscular activity," though he cautioned that "this adaptation must consciously be placed in the foreground and systematically tested according to the best capacities of psychophysical variation."[78] Through a combination of scientific management and psychotechnics, he claimed, "This new formation *(neugestaltung)* permits a better organization of the necessary bodily movements, and through it, the lessening of fatigue. The secondary movements are better utilized, fewer inhibitions become necessary—the revolution consists fundamentally in a better interplay between psychic forces made possible through the new regime."[79]

Münsterberg reproached the Taylor system for its overreliance on self-interest and coercion, for its single-minded focus to the task, and for its disinterest in the subjective experience of the worker—all of which, of course, he claimed industrial psychotechnics would remedy. Münsterberg was in the forefront of those who advocated aptitude testing, personnel departments, and even counseling in the modern firm. After his return to Germany, he wrote to over one thousand American manufacturers, inquiring into the mental traits they considered essential in their employees. In subsequent years several American firms hired him to determine, for example, the "mental traits that predisposed trolley drivers to accidents."[80] Industrial psychology was only one aspect of Münsterberg's vision of inundating the economy with the methods of applied psychology: he was a pioneer in advertising, marketing research, and the commercial uses of film. For Münsterberg, "the highest goal of economic experimental psychology was to banish from the world the excessive amount of spiritual dissatisfaction with work, spiritual dissolution, depression, and discouragement."[81] In the sphere of industrial relations, his much underestimated contribution to social modernity can be viewed as ushering in a "psychologically sophisticated Taylorism," presaging a new stage in labor relations, which arrived after the First World War: the advent of industrial psychology concerned with adaptation, motivation, and satisfaction.[82]

A more skeptical evaluation of the Taylor system appeared in the Verein für Sozialpolitik's house organ, the *Archiv für Sozialwissenschaft und Sozialpolitik,* which devoted much of its 1914 issue to a discussion of the "psychological and psychophysical" aspects of the "new industrial religion."[83] The economist, Wilhelm Kochmann, criticized Taylor's faith in wage premiums as a motivating factor, pointing out that the system offered no guarantees that "the worker's energies would not be excessively overused," or that in the long run the worker's performance would result in a loss of physical force that "could no longer be compensated for."[84] In addition, he observed that Taylor ignored the social policy implications of his system, and that in countries like Germany, where there were social insurance laws, the costs of ill health or the number of accidents caused by overwork might "easily overtake the gains made by the rationalized work."[85] Finally, he cautioned that the economic effects of a rapid speedup of production could not be calculated and that the overproduction of goods might clash with the natural psychological limits of the potential consumers themselves, "who are not yet mature enough to make use of the enormously increased mass of goods."[86]

The young Viennese sociologist, Emil Lederer, also writing in the *Archiv,* was more negative than Kochmann in his evaluation of the social impact of the new system, which he said was already extensively introduced into German firms (though not yet called the Taylor system). In Lederer's view, the essence of the Taylor system was simply the speed, or tempo, of production, which widened the distance between worker and management, created meaningless work, and relied on compulsion.[87] Despite Taylor's stated goal of social harmony between labor and management, it created intense class solidarity across whole firms and industries, rapidly replacing the old craft solidarities.

In the service of mass production, Taylorism produced a militant mass worker.

Lederer predicted that Taylorism would draw ever-larger numbers of workers into struggle against capitalism. It would come as a rude shock to trade union and socialist leaders who believed that its "poisonous fangs" could be excised.[88] Deeply imbued with the Marxian idea of progress and an unsullied faith in technology, they were under the illusion that Taylorism was only a technique of rationalized production, and, simultaneously a mechanism to create the conditions under which socialism could eventually be realized.[89] Lederer was unsparing in his caricature of the trade union leaders whose positive attitudes toward increases in productivity were unshaken even in the face of violent protests by the workers.[90] Social democratic passivity in the face of the Taylor system was evidence of the Marxist faith in the automatic process of history in which every technological development is justified as a future benefit to the working class, while present dislocations are temporary "stages" on the road to increased productivity and expanded socialization.[91]

Arriving on the eve of the First World War, the Taylor system kindled discontent within the community work-oriented physiologists, psychologists, and hygienists that had promoted a distinctly European science of work. Taylorism was more practical and successful than its European counterpart, but it generated divisions among "ergonomic" scientists that refracted a deep, but unacknowledged ambiguity in their own industrial ideology. The claims of the science of work to social neutrality and nonpartisanship frequently conflicted with its goal of increased productivity and the elimination of wasted social energy. In this regard, both Taylorism and the science of work employed an energeticist calculus.

In Germany as in France, the response of the European science of work to the Taylor system was ambivalent. Although its productivism was compatible with the American import, Taylor's apparent disregard for social reform and for the preservation of the working body was alarming. While some European scientists viewed the new system as a logical step in the division of labor in modern industry—notwithstanding the need for a compensatory physiological and psychological approach rooted in physics and biology—others reacted more negatively, condemning Taylorism for abandoning the most cherished principle of science, its social neutrality. If the Taylor system seemed to confirm the indispensability of research in fatigue and psychological expertise in industry, for others it epitomized the crucial distinction between maximal and *optimal* productivity of the worker. Taylorism's promise of profits first and foremost, could be had only by ignoring the physiology and psychology of the industrial worker and by imperiling the nation's energy reserves.

Despite this ambivalence, signs of a rapprochement between the two approaches began to appear in the months leading up to August 1914. At the National Congress of the Verein deutscher Ingenieure in September 1913, German engineers attempted to reconcile German *Arbeitswissenschaft* with Taylorism. In his speech Georg Schlesinger, the German counterpart of Le Chatelier, proposed a combination of "psychotechnics" with the principles of American scientific management.[92] Schlesinger remarked on the almost total ignorance of Kraepelin's ideas among German engineers but noted that this did not preclude their eventual adoption as part of the firm's reorganization. He also pointed out that the Verein für Sozialpolitik's empirical investigations had only a limited "historical value" because they "confirmed existing facts without offering any proposals for the solution to future problems."[93] Schlesinger readily admitted that engineering generally regarded the workers only incidentally and warned that in the future plant engineers "must take into account the results of psychology and national economy far more strongly than before, if the theme 'man' [Mensch] is to be exhaustively confronted."[94]

German scientific management could be more flexible than the American Taylor system: Schlesinger added that for Germans, "the actual science of management *(Betriebswissenschaft)* ends when the division of tasks, time study, standardization, systematization, study and adjustment of the worker to the new mode of work has occurred. With that, the material aspect *(sachlicher Teil)* is accomplished, which no reasonable and cost-conscious worker could in the long run oppose." But, he noted that other considerations like the wages and personnel "can be accomplished in a number of ways . . . if it is just and corresponds to the average worker."[95] In short, the two systems were ultimately compatible. It was not good intentions or a new intellectual synthesis but the World War—especially the reorganization of national industry for war production—that accomplished the amalgamation of the science of work to the Taylor system envisioned by Le Chatelier, Amar, Schlesinger, and Münsterberg.[96]

PSYCHOTECHNICS AND THE GREAT WAR

The First World War launched the first global mobilization of the productive capacities of industrial society for mass destruction. It was also a war of exhaustion—of the troops, of the home economies, and ultimately of the traditional European nation states. The war seemed to be the apotheosis of the dance of energy and entropy: the enormous power and force of the universe unleashed in the service of death and destruction—a fitting end to the century of thermodynamics. From the beginning, the combined forces of the Allies militarily canceled the forces of the Central Powers. In the stalemate that followed, the modernity of combat was expressed by a ratio of firepower and fatigue, mortality and exhaustion. Little wonder that the casualties who succumbed to the daily barrage of artillery that passed over the trenches like waves washing over a beach, were called "wastage."[97] Even the term most frequently applied to the war, *attrition*, (German: *Zermürbungskrieg;* French: *guerre de usure*) suggests a fatigue, and in many ways the problem of fatigue was central to both military and nonmilitary aspects of the war. The war greatly accelerated the integration of the science of fatigue and "psychophysics" into the vocabulary of modern economic life.

After 1914, the restructuring of the European economies for war production significantly eased the introduction of scientific management into industry, especially in the state-controlled munitions factories. The French "Union Sacrée" and the German and Austrian "Burgfrieden," the great national pacts to suspend class war and institutionalize collaboration between trade unions and management—particularly in skilled trades—created an administrative hierarchy of worker-management agencies from shop-floor committees overseeing production to socialist participation in wartime ministries. During the early years of the war, social conflict was suspended while the length of the working day and the tempo of work increased in the face of the demands of military and civilian production. The social policy achievements vaunted by most European states often lost ground after 1914 or were abrogated in the interest of victory: restrictions on overwork, protection of women and children in the workplace, and the nominal health requirements of industry were sacrificed to the war economy.[98] By 1915 a new generation of unskilled workers, composed largely of women and adolescents, were at work in industrial and munitions plants with little or no training, at lower wages, and with longer hours.

Similarly, the reorganization of agriculture and the monumental problems of feeding a nation at war required new adjustments to severely restricted and rationed diets, including the introduction of new synthetic foodstuffs. The fitness of troops became an important medical concern, as did the deployment of their skills in the war effort. The war required the maintenance of their physical and psychological capacities to sustain morale and combat. It also demanded repair of their bodies and psyches and the reintegration of the displaced, the crippled, and the disabled into wartime and postwar economies. All of these requirements were predicated on a new configuration of knowledge and practical techniques, of physiology and psychology and industrial processes.

The war did not so much invent new techniques of combating fatigue, diagnosing and treating psychological illness, improving efficiency or boosting output, as much as it provided a field for testing established techniques and for engaging the expertise of individuals in the service of their respective nations. The war employed these techniques to a greater degree than before 1914 and—perhaps more important than the immediate gains that science afforded the military effort—provided the advocates of the science of work with a vast laboratory to demonstrate the utility of their knowledge. As Fritz Giese, a leading German expert in industrial psychotechnics in Weimar Germany, remarked of his own wartime service, "The war not only presented psychology with new knowledge, it also created new subject matter, which, without that sad occasion would surely have remained estranged from it."[99]

The war lent legitimacy and urgency, among other things, to the aptitude test, to the neurasthenia diagnosis, to the industrial ergonomist, and to industrial reeducation of the injured and maimed. Experts on fatigue, on production, on industrial hygiene, and on nutritional physiology were enlisted in the staggering redeployment of national energy. In Germany, advocates of the Taylor system played an important role in the initial reorganization of the war economy. The industrialist Walther Rathenau, head of the *Allgemeine Elektricitäts-Gesellschaft* (AEG), and appointed to manage the Office of War Raw Materials *(Kriegsrohstoffabteilung)*, endeavored to keep the military industries supplied with materials. The fusion of private and public corporations resulting from this need was envisioned by Rathenau and by his confidant, the Prussian aristocrat and mechanical engineer Wichard von Moellendorf, as the forerunner of postwar state socialism.[100] A strong supporter of "rationalization" as German scientific management was often called, Moellendorf advocated the peacetime adoption of a "planned economy" *(Planwirtschaft)* along the lines presented in his

Deutsche Gemeinwirtschaft, published in 1916. The rapid integration of new workers, especially women and adolescent males into the munitions industry, increased the need for scientific management as a method to deploy swiftly untrained workers in industries previously dominated by skilled males. The rapid standardization of machine and tools, the emphasis on task versus skill, and "differential" wage systems were all introduced, often with the approval of trade unions.[101] As well as encouraging technological solutions to manpower problems, wartime propaganda applauded the disappearance of the traditional distinctions between skilled and unskilled workers.[102] Even before the war, the French army had demonstrated an interest in Taylor's system and had consulted with Taylor over its military uses. Far more important were the domestic uses of the Taylor system, which were consistently promoted by the French government during the war. The important role played by the pro-Taylorist socialist leader Albert Thomas in effecting a compromise between management and labor during his tenure as armaments minister (December 1916–17) was decisive, both in securing broad trade-union acceptance for scientific management and in furthering worker participation in the design and implementation of the Taylor system.[103] For Thomas and his secretary, the C.G.T. leader Léon Jouhaux, Taylorism promised to maintain high productivity at high wages with fewer working hours. Thomas justified the left's support for Taylorism by emphasizing that improvements in productivity did not exclude the participation of the workers.[104]

In 1916, Etienne Clémentel, Minister of Commerce, created a technical staff to promote the "experimental study of occupational movements and the most economic mode of functioning," in industrial work.[105] Wartime emphasis on planning and coordination, the growing power of the state over industry, and the rising status of engineering as a profession created a new consensus in almost all the countries at war that a united front of science, technology, management, workers' organizations, and the state could further the war effort through greater efficiency. As one French propaganda slogan put it, "Lost time in work, steals a part of the nation's capital."[106] Despite growing strikes against the detested time-motion study, the premium wage system, and long hours of work in both France and Germany after 1917, the war brought Taylorism and Taylor's ideas into the mainstream.[107]

The widespread adoption of Taylorism during the war gradually modified the initial skepticism with which the European science of work once regarded the American system. The war provided a sufficient range of activities affecting the physiological and psychological aspects of both combat and industry to facilitate a compromise between the two approaches. Although wartime constraints often reduced the capabilities of severely strapped scientific institutions, the groundwork for a peacetime expansion was set.

THE KAISER WILHELM INSTITUTE FOR LABOR PHYSIOLOGY AT WAR

In Germany despite severe cutbacks in manpower during the war, the Kaiser Wilhelm Institut für Arbeitsphysiologie (Kaiser Wilhelm Institute for Labor Physiology) was enlisted by the Ministry of War to undertake a series of studies on the nutritional requirements of troops and domestic forces, as well as on the optimal use of labor power in the munitions industry.[108] Founded in 1913, the institute was the brainchild of the physiologist Max Rubner who proposed the idea in 1912 to make practicable the discoveries of physiological energetics for the productive power of German industry and for the military. The task of the new institute, as Rubner envisioned it, would be to "fulfill the universally acknowledged need for detailed knowledge of occupational and industrial hygiene."[109]

In his draft proposal, Rubner conceived of "work" in the broadest sense, encompassing all aspects of human performance: "Under the concept of work performance *(Arbeitsleistung)* we usually mean the performance of the human being as a work machine *(Arbeitsmaschine)."* He added, however, that the original ideal of the body as an energy-producing machine was giving way to alternative forms of work "whose realization depends more on the intellect than it does on mechanical energy. This is evident in those thousands of kinds of work— the supervision of machines, the concentration on detail brought on by the division of labor—for which attention, skill, conscientiousness, and duration of performance is most decisive."[110]

Rubner's contribution to energetics, as we have noted in chapter 5, was to demonstrate conclusively in his classic 1894 experiment the validity of the law of energy conservation for biology.[111] In his subsequent works and in his plan for the new institute, Rubner extended his theory to the requirements of everyday life. In his more popular works, Rubner argued that the laws of thermodynamics were a sound basis for a total science of social hygiene, encompassing not only nutrition but all aspects of social endeavor.[112] His program for the institute included the detailed study of the nature and conditions of industrial work, the requirements of diet and hygiene, and the promotion of his most coveted social ideal: "rational nutrition."[113]

In 1915, Gerhard Albrecht, one of Rubner's assistants at the KWI, identified the wartime role of the institute as "determining the limits of labor power, the influences on the work performance of the mind and body, and determining how future work can be structured, not only to be profitable, but also to be economical without damage to the worker."[114] The institute also claimed to provide a more scientific method of determining the relationship between work and fatigue than the Taylor system, which, in Rubner's view, rested on "less than secure foundations."[115] If the Taylor system placed economic considerations in the forefront, the institute focused instead on the biophysiological aspect, the elimination of fatigue: "As a result of numerous observations of demanding work and fatigue, two significant means of preventing or even eliminating fatigue emerge, namely the precisely regulated distribution of pauses, and the exclusion from the task of fresh, not previously engaged muscle groups (so-called accessory performance)."[116]

Before 1914 the institute was oriented toward placing the Taylor system on a more scientific footing. During the war, however, the KWI was "in the first instance" concerned with questions of wartime nutrition and diet, producing both laboratory research and information for the public on foodstuffs.[117] Given Rubner's interest in the energetics of diet, the institute's chief objective was to find adequate means of replacing or supplementing scarce foods with suitable substitutes, including the much-disliked food "surrogates," and various means of "stretching" existing foods. With the decline of imports and domestic food cultivation, the availability and quality of foodstuffs shrank rapidly after 1916, and in addition to bread shortages, there were serious shortages of animal feed as well. By war's end, more than 11,000 food substitutes were available.[118] In conjunction with the Imperial War Nutrition Office, the institute researched the uses of rye-corn, various bread surrogates, and food adulterants *(Streckmittel)* for civilians and the military, and the ways to improve animal feed with chemical additives.[119] These experiments were sometimes conducted on human subjects provided by the Ministry of War, as well as on animals, with the goal of "principal clarification" and "realization in practice" as rapidly as possible.[120] The KWI's studies of fatigue and efficiency were also employed on the home front: for example, dockworkers in Kiel found a notice advising them that "the fatigue of a permanently stressed muscle, can be considerably relieved by the brief, disciplined work performance of a different, relatively less fatigued muscle."[121]

The war also extended the range of the KWI's efforts to develop aptitude tests for evaluating the distance perception of military drivers, pilots, and railway personnel.[122] By 1917, the institute concluded with pride and self-satisfaction: "The measurement of reaction time, and the refinement of sense perception *(Sinneseindrücke)* for military purposes have led to practically useful results and instruments."[123] One of the institute's co-workers, Professor Piper, who died in combat, developed a "reliable examination for the selection of artillery personnel who require a good sense of distance calculation."[124]

Looking toward the future, the institute set its sights, not only on the readaptation of men affected by "psychic disturbance" to civil life, but above all on the "preparation of a new generation with as high as possible performance capacity in the postwar era." Rubner predicted that psychological testing and psychophysical aptitude tests could result in the "selection of the most fit forces *(Kräfte)* for industrial work."[125] This new psychological orientation led, incidentally, to a protracted jurisdictional squabble between Rubner and Otto Lipmann, who along with William Stern founded a privately funded Institut für Angewandte Psychologie (Institute for Applied Psychology) in Berlin in 1906. Beginning in 1917, Rubner resolutely blocked Lipmann's application to the German government requesting that his Institute be given the status of a Kaiser Wilhelm Institut, parallel to that of Rubner's and similar to those already established in physics, chemistry, and other natural sciences.[126] Rubner's territorial academic instincts need not excessively concern us here, but his positive evaluation of the applied experimental psychology pioneered by Münsterberg is of interest:

> The opportunities for the application of psychological methods in everyday life are certainly more numerous than one had previously imagined. During the war, the measurement of the rapidity of tasks modeled on actual techniques performed by an automobile driver has proven very practicable in excluding unsuitable persons from this occupation. Psychological processes can never be entirely separated from their physiological basis and foundation, and a sharp differentiation between them is therefore impossible.[127]

The institute also called for greater cooperation in implementing its ideas in the workplace, suggesting working in tandem with the technical personnel charged with the design of "work machines," noting in particular the success of Georg Schlesinger's Charlottenberg Institute of Psychotechnics in redesigning industrial technology in order to accommodate the reintegration of war-injured men into production.[128]

In 1914 France called up its ergonomic reserves.[129] Literally so in the case of Lahy, who interrupted the final preparations of his manuscript on Taylorism for twenty months of military service in 1914 to 1915.[130] First at the Argonne, later at the Somme, Lahy used his spare time to set up experiments to study the reactions of the machine gunner with many of the psychophysiological methods that he had used in his laboratory in Paris. Although conditions at the front were difficult, he was able to contrive makeshift laboratories in an abandoned house at the Somme, in a cave at the Argonne, and in an unused airplane hangar. With the cooperation of his superior officers, Lahy devised a series of experiments to determine the aptitudes of the machine gunners using techniques, which, he admitted, were as precise as the "defective conditions of experimentation" allowed.[131] Testing reaction time to visual and auditory stimuli, Lahy found a "unique rapidity" of reaction time among those machine gunners highly rated by their officers as compared with those gunners whose abilities were rated lower by their superiors. He then created an "index of fatigability," based on the changes in speed of reaction over time.

Lahy also tested for "sang-froid," an essential prerequisite for what he called "lucidity of spirit" among the gunners. To trace variations in respiration and pulse rate while he fired a carbine close to the gunners, Lahy employed both the *sphygmograph* and *pneumograph* (a respiratory tracing device invented by Paul Bert). The physiological tracings clearly showed the most "cold-blooded" of these gunners returned to normal bodily function far more rapidly than those of any other group. This test measured *functional plasticity,* a term Lahy used to denote lack of nervousness in combat.[132] He wrote that in war as in industry, it was necessary to create a division of labor among the combatants, for which psychophysical testing was an efficient and objective method preferable to trial and error.

Tests similar to those of Lahy were developed by Professor Jean Camus of the faculty of medicine at the University of Paris and were used to determine the aptitudes of candidates for aviation training. Psychophysics was recognized early in the war as a way to select potentially successful pilots and to speed up the training. Laboratory tests were also devised to locate those men whose "absolute mastery of the self, and rapidity of decision permitted them to rapidly exercise the correct maneuver," without emotion.[133] In Germany, too, as early as 1915 aptitude tests developed by the "psychotechnicians" Walther

Moede and Curt Piorkowski were used by the military to select aviators, truck drivers, and radio operators, among others.[134]

Amar too applied his talents to rehabilitating and reeducating "the glorious victims of this most horrible of wars," especially amputees. In order to "stimulate industrial, commercial and agricultural labor," he wrote, "no effort must be lost or squandered. At stake is the condition of national wealth."[135] The amputees and severely disabled could, he argued, still be "capable of complete utilisation, representing a sometimes integral value" for industry.[136] The effort to reintegrate war-cripples into the workforce was necessary because of the "degradation" suffered by individuals having to rely on public assistance—not to mention the great cost spared the nation by removing dependents from the welfare rolls.

In order to expedite the reentry of these victims into the workforce, Amar created a comprehensive program of physical rehabilitation of veterans:

> Determine the general state of the organism (heart, pulse, touch); the articulations of the muscles and freedom of movement; evaluate the disposable physical force, and take into account moral forces; condition the work according to the basic and fundamental data; and put to use the best aspects of prosthetic technique.[137]

Most extraordinary was Amar's "working arm" *(bras de travail Amar)*, an industrial prosthesis to which an individual could attach several tools, including a complex mechanical hand that could be used to type or play the violin! Amar also devised prosthetic tools for specific jobs, for example, the "one arm punch" for punching railway tickets (adopted by the Paris-Lyons and Mediterranean Railroad).[138] His efforts were rewarded with support from the munitions industry, which began retraining war amputees in 1916, and by the Ministry of Labor which sponsored vocational rehabilitation and professional reeducation. In his preface to Amar's text on industrial reeducation, Le Chatelier singled out his work with amputees for its adaptability to the Taylor system, which, he claimed, could also be used to retrain and employ war-crippled laborers to great advantage.[139]

Apart from the manpower needs served by the psychophysics of aptitude testing and the rehabilitation of the wounded, the psychological effects of combat, especially pathological fatigue and neurasthenia, also emerged as a universal concern. The war brought in its wake a voluminous literature on the psychology of the emotions, in particular the nervous disorders of the combatants.[140] One problem confronted by almost all medical specialists was the psychology of "fear," or "anxiety." Fear, remarked Ioteyko, was of "capital and decisive importance, exercising its sovereign influence over almost all military campaigns, [sometimes even] changing the direction of battles and often deciding the outcome."[141] The psychology of courage somehow took second place.[142]

The neurasthenia experts employed by the military on all sides insisted on the enormous role that psychic predisposition played in the onset of all war-related disorders. Battle fatigue played a central part as the key physiological aspect of the individual's ability to resist the onset of pathological fear or panic: "Fatigue weakens the power of mental activity and causes nervous irritability. A fatigued man is powerless to marshall his reflexes."[143] Soldiers with a marked propensity to fear, terror, or, in its worst form *cataplexy* (fear-generated paralysis), were marked by a lack of psychic or physiological equilibrium. For the experts terms like *constitution emotive, constitution anxieuse,* and *hyperaesthesia*—all described a predisposition to extreme and paralyzing exhaustion in combat.[144]

Equally serious was the propensity of combatants to retain the marks of a traumatic experience, or of an injury, long after the event had passed or the injury had healed. The traumatic shock, or "accident neurasthenia," first encountered in the medical diagnosis of industrial accidents after 1884 was more common in war. Renamed "sinistrose de guerre," such cases preoccupied medical specialists who were treating the pathogenic effects of fatigue on men exposed to the daily barrage in the trenches.[145]

Less combat fatigue than exhaustion with war itself, by 1916 the neurasthenia of the trenches was commonplace. Increasingly a concern of military leaders, it was evident in a profound "inertia and indifference to the pleasures and distractions of life"; the diminishing of all physical and intellectual effort; and the "search for solitude" experienced by soldiers after relentless months at the front.[146] A German study of more than 2500 soldiers hospitalized for neurasthenia revealed that it affected "all classes of the population," though still most frequently found among those who came from "intellectual milieus" and among workers from large cities.[147] Neurasthenia was also more prevalent among officers than among the men of the ranks. One can clearly discern a class division between the hysteria or hypochondria frequently encountered among the troops and the more elite neurasthenia, characterized by "a general dissolution of energy, by an incapacity for decision, by a feeling of complete physical exhaustion, and by such extreme despair that "herculean officers often cried like babes.""[148]

Neurasthenia remained primarily an affliction of performance. Often presaged by a state of *cafard,* or sadness and distraction, the war-weary neurasthenic was usually also the victim of a poor diet, of lack of sleep and creature comforts, and inevitably of physical fatigue— all of which conspired to bring on the disorder, as did the normal emotions of war. Its "double aetiology" was schematically characterized as a ratio of inverse proportion between predisposition and what psychologists called "occasion," or external event.[149] If the pressures of peacetime accelerated the onset of neurasthenia, the war magnified the disorder a hundredfold.

According to Maurice Dide, whose multivolume *Les Emotions et la guerre* (1918) became a standard work of war psychology, the "hysteria of war" was caused by an "affective incoherence," which did not permit the individual to "dissipate" adequately the shocks of the war. Dide warned medical personnel who dealt with war neuroses to circumvent the "arguments furnished by their patients" and act "directly on their "subliminal being." For those combatants (both hysterics and neurasthenics) afflicted with "asthenia" (see chapter 6), he proposed that "a special milieu be constructed," preferably with specially built isolation chambers, so that the victims could rest undisturbed. Others, however, were more skeptical of the emphasis that the psychophysical school (as well as the novel psychoanalytic approach) placed on the latent causes, arguing that it was the war itself that provoked these illnesses, not the predisposition of the soldier.[150]

The war led to the discovery of various new fatigue syndromes. For example, the "syndrome émotionnel" was opposed to "le syndrome commotionnel"; the latter was a nervous reaction to a violent or intense shock, usually caused by an explosion. The "commotional" disturbances, or shell shock, noted Gilbert Ballet (co-author of *L'hygiène du neurasthénique*) "retained the mimetic expression of fear," as opposed to fear itself. In contrast to shell shock, the emotional syndrome had an internal, "autogenous" cause brought about by "a series of intense, affective impressions," in short, by the terrors of combat. The former was accompanied by fatigue, hyperactivity, fear, and disorientation; the latter by affective indifference, amnesia, mental inertia, and an inability to act.[151]

As the war dragged on, so did the dispute between the advocates of a disciplinary solution to the problems of psychological fatigue and those for whom neurasthenia or anxiety required appropriate medical or psychological therapy. Though it is an exaggeration to conclude that the wartime debate on neurasthenia was essentially between Freudian "analysts and anti-Freudian moralists," as some historians have suggested, the question of appropriate therapy was not easily resolved.[152] Among combat officers and certain medical experts there was a pronounced lack of sympathy for those whose *constitution anxieuse* predisposed them to the psychiatric wards, often combined with the suspicion that among them were a preponderance of patients whose symptoms were less than authentic. This view contributed to the common complaint that malingering and simulated symptoms were proliferating under the guise of neurasthenia. As Eric Leed points out in his study of the psychology of the war, often the distinction between malingering and war-related neuroses was vague. A French military physician, J. A. Secard, distinguished between two kinds of malingerers: the *simulateur de création*, who pretended the symptom to escape danger, and the *simulateur de fixation*, who manufactured and maintained an initial neurotic symptom long after the dangerous condition had disappeared. As in the case of *sinistrose de guerre*, even a genuine victim of shell shock was a potential *simulateur de fixation*, especially if he was met with pity and attention and became accustomed to the rewards of his ailment. Leed found that many hysterics and neurasthenics were treated to severe disciplinary therapy, including the "manière forte" and "torpillage" in the French army, the "quick cure" and "queen square" in the British, the "Kaufmann" technique in Austria, and "Überrumpelung" (hustling) in Germany—all of which combined pain administered by electrical current, with barked commands, isolation, and discipline.[153]

Against these harsh methods, psychoanalytically informed experts proposed short-term analysis to locate the traumatic sources of the disorder, while psychophysical exponents proposed a series of equally short-term methods to restore the energies of the afflicted—preferably baths and rest cures. Given the apparent reality of malingering, the techniques of ferreting out fraud developed in industrial medicine were extended to the wards and rear lines. Some physicians even found Mosso's ergograph a useful tool to distinguish the simulator from the sufferer. By measuring the fatigue of the neurasthenic, the dubious practices used by shirkers could be detected by military physicians.[154]

The war accelerated the application of the insights and techniques developed in the peacetime laboratory to the exigencies of combat, especially in the vocational rehabilitation and retraining of the wounded. It also hastened the rapprochement between Taylorism and the science of work in industry by pointing toward the postwar uses of psychophysical aptitude testing, the fatigue study, and it helped achieve public acceptance of neurasthenia as pathological fatigue. But, if the war demonstrated the utility of a scientific approach to fatigue, it also underscored its limitations. Long hours, overwork, and exhaustion were, it proved, entirely compatible with a rationalization of movement, nutrition, and the deployment of skill. The war, in short, was a laboratory for social hygiene. If it fostered the emergence of a scientific approach to work in the modern world, it also severely undermined one of the most utopian impulses of the European science of work: its commitment to a calculus of productivism and reform, carefully maintained and regulated by the state.

CHAPTER TEN

━━━━━━━━━━━━

The Science of Work Between
the Wars

PRODUCTIVISM BETWEEN THE WARS

THE efforts of European physiologists and liberal reformers to discover a scientific solution to the worker question in the objective laws of fatigue before the First World War foreshadowed the postwar politics of social modernity and social rationalization. The war hastened the acceptance of new ideas of industrial organization, including the many approaches developed by the science of work. Only after the war, however, did European governments, industrialists, and universities begin to recognize fully the potential benefits of these endeavors. During the interwar period a panoply of "social technologies" supported by European governments developed standardized forms of knowledge, became established professions, and became institutionalized as academic disciplines.

As the great wave of postwar "Americanism" swept Europe in the early 1920s, labor physiologists and psychologists applied the idea of conserving the nation's energy through the rationalization of the body in the military and industry. The methods and practices of the science of work were gradually introduced into vocational training programs, employment offices, and into industrial enterprises as components of the "rationalization" of German and, to a lesser extent, French industry. In the first decades of the 1920s psychotechnical aptitude testing, the ergonomic calculation of "optimal" energy expenditure, and the fatigue study became part of the standard arsenal of techniques available to trained professionals whose services were increasingly demanded by government and industry.

Though the application of the science of work in industry spread rapidly in this period, its vision narrowed as the fatigue experts apprenticed their craft to the more powerful American "science of work," Taylorism. The ideology of "productivism," the balancing of expanded output with social reform, to conserve energy was no longer the monopoly of a dedicated coterie of liberal scientists and reformers convinced that industrial power could be curbed by scientific knowledge. To be sure, post–World War I productivists still held that the energies of society could be deployed according to the basic principles of physics and biology, but their aims were no longer associated with extensive proposals for liberal reform. Productivism became the common coin of European industrial management and of the pro-Taylorist technocratic movements across the European political landscape between the wars. As Charles S. Maier has shown in a classic article, on all points of the political spectrum "Taylorism and technocracy" were the watchwords of a three-pronged idealism: the elimination of economic and social crisis; the expansion of productivity through science; and the reenchantment of technology.[1] The vision of a society in which social conflict was eliminated in favor of technological and scientific imperatives could embrace liberal, socialist, authoritarian, and even communist and fascist solutions. Productivism, in short, was politically promiscuous.

Taylorist management, technocratic initiatives, and planning were adopted in interwar France by nationalist radicals like Ernest Letailleur, as well as by socialists like Alphonse Merrheim and Albert Thomas. In Germany and Austria the productivist spectrum ranged from Prussian conservatives like Wichard von Moellendorf to liberals like Walther Rathenau, and included the leading German and Austrian social democrats Rudolf Hilferding and Otto Bauer.[2] Antonio Gramsci and Lenin gave the Communist imprimatur to Taylorism, and the famous institute of Alexei Gastev, the "Soviet Taylor," persisted well into the Stalin era. In the 1920s Soviet Taylorism became a mass movement with the proliferation of the "time-leagues," founded to teach young workers how to budget their daily activities according to rigidly prescribed schedules.[3] In Italy, Fascist leaders like Massimo Rocca and Mussolini paid homage to Taylor's achievements, and his ideas found expression in the fascist leisure-time movement, *dopolavoro*.[4] Even the antimodernist, cultural pessimist Oswald Spengler could find hope in a future controlled by the engineer—"the scientific priest of the machine."[5] In both the interwar liberal democracies and in the totalitarian states, the connection between labor time and political control became integral to the politics of productivism.[6]

THE POSTWAR RAPPROCHEMENT BETWEEN THE SCIENCE OF WORK AND TAYLORISM

After the war, critics and supporters of Taylorism generally conceded that many of the modifications introduced into scientific management, especially after Taylor's death in 1915, already paved the way for a combination of the two approaches.[7] Jules Amar, for example, maintained that the Taylor system did not measure fatigue scientifically, and emphasized that well-trained ergonomists were still indispensable.[8] Lahy too softened his criticisms, concentrating on the ancillary field of the selection of aptitudes and vocational counselling—psychotechnics —in the postwar organization of French industry. His articles in the 1920s and 1930s reveal none of the critical fire that motivated his assault on the Taylor system in 1913.

Postwar management science was a synthesis of Taylorism and the science of work, focusing on productivity while eliminating the psychological and physiological excesses of the American system. Even conservative promoters of Taylorism, engineers like Le Chatelier and Georg Schlesinger (the "German Taylor"), recognized that in Europe the tradition of workers' protection by the state had to be affirmed if scientific management was to be successful. In the cautiously worded passages of Josefa Ioteyko's 1919 textbook, *The Science of Labour and Its Organization*, a rapprochement between the science of work and scientific management was clearly evident. Though Ioteyko condemned Taylor for opposing the "progress of hygiene," and for guaranteeing only the "freedom of the individual to overwork," she called for an international commission of physiologists and engineers to examine charges of excessive fatigue among workers in industries that had adopted Taylor's methods.[9] Ioteyko admitted the value of the Taylor system for increasing output and rebuilding the European economy and acknowledged that it was not accidental that this system was "at the present hour claiming the attention of manufacturers."[10]

Finally, Ioteyko conceded that those, like Lahy and Imbert, who rejected Taylorism, may have been too hasty in their judgments about its negative consequences. Rather its cautious supporters, like Amar, may have been right to see that those who "follow Taylor may correct the errors of his system and perfect it."[11] Moreover, Ioteyko argued that final judgment had to be withheld: "Only after inquiry, and in the event of a favorable answer, will Taylor's system deserve the name 'scientific' and may be considered free from all defect."[12]

For Ioteyko, the fusion of Taylorized factories and psychotechnical testing promised to create a postwar society free of the arbitrariness and irrationality of free-market capitalism. The "principle of the most apt" would regulate the society by the rational deployment of personnel throughout society:

> What should we say of a society wherein everyone should have followed the line of his tastes, of his leanings and aptitudes, where each would occupy the place best suited to him, and wherein the various occupations were allotted to the "most apt"? Such a society would be reformed from top to bottom, in the sense of greater equity, greater productivity, and greater happiness.[13]

Ioteyko's vision marked a distinct shift in the emphasis of the science of work in the postwar era: from the fatigue study to the psychophysics of aptitude, from deployment to selection, from physiology to psychology. The state, not private employers, would bring about the social rationalization necessitated by, and parallel with, the economic and technical rationalization of industry. According to a report on "Scientific Management in Europe" issued in 1927 by the Geneva-based International Labor Organization, by 1919 European industrial psychologists and physiologists appeared "to have abandoned their attitude of opposition to the spread of scientific management."[14]

Along with "industrial medicine," the experimental testing of aptitudes, or industrial psychology, became subdisciplines of a broader management science after the war. Despite Ioteyko's utopian evocation of a future governed by the rational mechanisms of testing and placement, psychophysical aptitude testing and fatigue research defined a narrow area of competence within a broader science of industrial organization. At the Psychotechnical Congress that convened in Moscow in September 1931, a unified terminology for the international psychotechnical movement was established.[15] But twentieth-century European managers could draw on many different traditions—industrial paternalism, socially concerned liberalism, Christian social doctrine, and scientific management. Increasingly they saw "the enterprise as a social organization and not merely as a productive unit."[16] Having won its place as an adjunct of Taylorist management in the postwar era, the science of work fragmented into a bewildering array of academic subdisciplines: psychotechnical aptitude testing; industrial physiology and psychology; plant sociology; personnel management, and the like.

By the late 1920s, the gloss had worn off the psychotechnical movement, which fell on hard times: the need for rapid selection of workers diminished as management sought more efficient ways to "sort out" unproductive or unreliable militant workers. By the 1930s conflicts over the role of the science of work during the Depression divided practitioners along political and ideological lines. Especially in Germany, by the mid-1930s several explicitly political approaches to the workplace emerged to challenge the claims of the ergonomists to neutrality by using methods similar to theirs to undermine the hegemony of unions and foster authoritarian versions of plant "harmony." Hostile to Taylorism and Americanism, these authoritarian approaches politicized the "science of work" in the 1930s, combining politics and professionalization in the era of National Socialism.

THE INSTITUTIONALIZATION OF THE SCIENCE OF WORK

Psychotechnical aptitude testing was the principal innovation to emerge from the wartime uses of the science of work. In the later phases of the war, governments and employers collaborated with trade unions to demand greater output from an often untrained—and largely young and female—labor force. In November 1917, the German Women's Vocational Agency (Frauenberufsamt) promoted occupational counseling, which used psychotechnical aptitude testing, as did other vocational agencies set up for younger workers.[17] Wartime economic demands, long hours of work, and increasingly limited food supplies called attention to the deteriorating health of the labor force.

In Britain there had been little development of a science of work before 1914, a deficiency corrected during the war when a large-scale effort soon made Continental observers envious. In September 1915, Lloyd George, minister of munitions, appointed a committee of scientists to investigate the condition of British workers. They issued a report demonstrating that a reduction in worktime from 75 to 67 or even 55 hours per week did not significantly reduce output and in some cases even increased it.[18] As a result the fatigue and health experts on the Health of Munitions Workers Committee exercised some influence over British munitions plants, reducing hours, introducing rest pauses and holidays, and identifying health risks.[19] By 1918, under the leadership of H. M. Vernon and Stanley Kent, Britain became a leader in the development of industrial psychology and physiology in Europe.[20] At the end of the war the committee was converted into the Industrial Fatigue Research Board, which was expanded to include physiologists, academic psychologists, and industrial experts who produced an impressive series of reports on fatigue in various occupations. Despite its strong publication record, the Board's work was confined to recommending minor technical adjustments in plants; its wartime emphasis on a progressive reduction in the hours of work was quietly abandoned.[21]

The success of Taylorism in postwar Europe gave engineers a new role in the organization and administration of larger industrial firms, opening the way for the professionalization of the social hygienist, the fatigue expert, and eventually the industrial psychologist. At the Siemens-Halske electrical works in Berlin, studied by Heidrun Homburg, the firm first began calculating the performance of machine tools and carrying out systematic time, motion, and fatigue studies only after 1919.[22] Although engineers controlled the practice of psychotechnical testing in industry, psychologists and physiologists provided training for the plant engineers. The newly achieved academic legitimacy for its concerns was another key factor in the postwar rapprochement between Taylorism and the science of work. Teaching positions, professorships, and chairs were founded in numerous polytechnical colleges and universities, along with professional journals of industrial sociology and psychology.

The European experience of scientific management in the 1920s—and to a large extent the Anglo-American—was a significant departure from the narrow application of the principles laid down by Taylor. In the United States Henry L. Gantt and Frank B. Gilbreth borrowed heavily from Taylor, but they stressed the combination of efficiency and enlightened management in the application of his methods. For the next generation of Taylorist managers, emphasis on the "human factor" became an important corrective to Taylor's limited interest in the social consequences of scientific management. The sophisticated system of industrial administration developed by Henri Fayol in France, the Bédaux system of incentive payments in Britain and the United States, and the widespread introduction of the mechanized "Fordist" assembly line in the mid-1920s—all altered or transformed the original conception of Taylorism by extending the principles of rational management to the administration, technology, and social policies of the enterprise. According to the 1927 ILO report, postwar scientific management was redefined to include industrial psychology and physiology (fatigue study, industrial hygiene, studies of monotony and repetitive work),

FIGURE 15: Charles Fremont, Chronophotograph, hammerer, 1895. (*Le Monde Moderne* [Paris, 1895].)

FIGURE 16: Charles Fremont, Chronophotograph. Hammerers striking a blow, with body blurred to show trajectory of motion, 1895. (*Le Monde Moderne* [Paris, 1895].)

FIGURE 17: Angelo Mosso, Ergograph, 1884. (Angelo Mosso, *La Fatica* [Milan, 1921].)

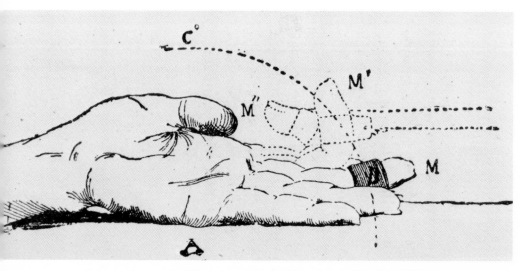

FIGURE 18: Angelo Mosso, Successive positions assumed by the middle finger in raising the weight of the Ergograph. (Angelo Mosso, *La Fatica* [Milan, 1921].)

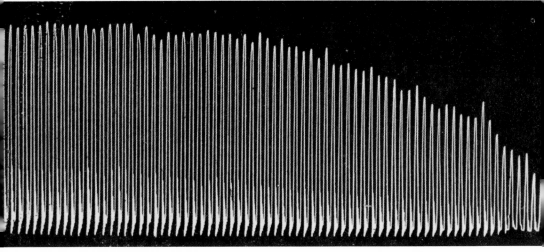

FIGURE 19: Angelo Mosso, Ergographic fatigue tracing, demonstrating the beneficial effect of exercise on muscular force, ca. 1890. (Angelo Mosso, *La Fatica* [Milan, 1921].)

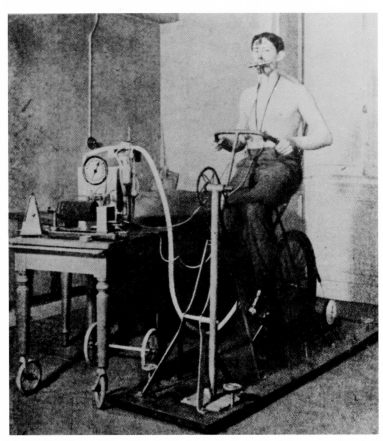

FIGURE 20: Jules Amar, Ergometric Monocycle, 1912. (Jules Amar, *Les Lois scientifiques de l'éducation respiratoire* [Paris, 1920].)

FIGURE 21: Jules Amar, Filer with respiratory apparatus, ca. 1912. (Jules Amar, *Le Moteur humain et les bases scientifiques du travail professionnel* [Paris, 1914].)

FIGURE 22: Jules Amar, Chronophotographic studies of the filer, correct *(left)* and false *(right)* positions, ca. 1913. (Fritz Giese, *Psychologie der Arbeitshand, Handbuch der biologischen Arbeitsmethoden* [Stuttgart, 1923].)

FIGURE 23: Nathan Zuntz, Portable Respirator, military subject, 1900. (Nathan Zuntz and Wilhelm Schumburg, *Studien zu einer Physiologie des Marches* [Berlin, 1901].)

FIGURE 24: Nathan Zuntz, Portable Respirator, military subject, 1900. (Jules Amar, *Le Moteur humain et les bases scientifiques du travail professionnel* [*Paris, 1914*].)

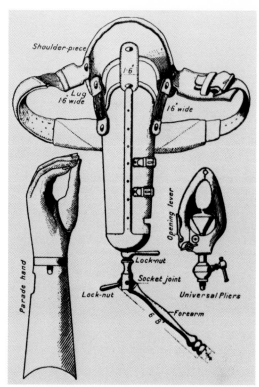

FIGURE 25: Jules Amar, Working Arm, industrial prosthesis, 1916. (Jules Amar, *The Physiology of Industrial Organization and the Re-employment of the Disabled* [New York, 1919)].)

FIGURE 26: Jean-Marie Lahy, Measuring reaction time in his laboratory at the front, Somme, 1916. (*La Science et la Vie* [Paris, 1917/18].)

psychotechnics, aptitude testing, management training, training for shop-floor personnel, vocational counselling and guidance.[23]

During the heyday of the rationalization movement in German industry, there was a broad consensus on the right and on the left that the introduction of "human factor" considerations, including the results of fatigue research and psychotechnical testing in industry, represented a necessary corrective to the dangers of "mis-rationalization" *(Fehlrationalisierung).*[24] In its 1931 manual, the Reichskuratorium für Wirtschaftlichkeit (RKW), the influential umbrella agency of the German rationalization movement, stressed that in addition to organizational and technological aspects of labor, physiological conditions were "an essential task" of the scientific investigation of work. The national railways and postal system adopted guidelines for the workers' health, safety, and diet and prescribed extensive psychological testing.[25] Rubner's successor, Edgar Atzler, director of the rebaptized Arbeitsphysiologische Institut which relocated in Dortmund in October 1929, distinguished between Taylorism, based on a *maximum* productivity, and *optimum* productivity, which required a physiological calculation of energy expenditure.[26] The Berlin psychologist Otto Lipmann proposed that "human efficiency" replace Taylor's method of "economic efficiency"; expenditure of energy, not expenditure of time, was paramount in the standardization of work processes and the selection of workers.[27]

Although most of the postwar pioneers of industrial psychology, for example, H. C. Link and H. E. Burtt in the United States, Walther Moede and Curt Piorkowski in Germany, and E. P. Cathcart and H. M. Vernon in Great Britain were oriented toward more efficient management, a minority, like Lahy and Lipmann remained sympathetic to workers' concerns, and even to socialism. Moreover, in the early postwar years industrial psychology's effort to find suitable workers seemed to eliminate Taylor's negative method of removing unsuitable workers from the production lines. In the early 1920s both managers and trade union leaders looked to experts who could establish "rational" physiological and psychological standards for selection, thereby winning the support and cooperation of rank and file unionists, as well as social democratic activists. Social hygienists were often called upon by the German factory councils *(Betriebsräte)* to attest to the health and safety benefits of shorter hours, lengthier pauses, and rational guidelines for work.[28] Such efforts were clearly undertaken with the hope that along with technical progress, the opportunity for greater conservation of "human economy" could also be achieved. Ludwig Teleky, an influential exponent of the social hygiene movement in the Weimar Republic, proposed that psychotechnical examinations be used to determine whether firms demanded too much of workers, not simply to improve their competitiveness.[29]

THE ERA OF PSYCHOTECHNICS

During the early 1920s psychotechnical institutes were developed in England, France, Germany, Italy, Poland, Russia, and Japan. In 1924 Lahy founded the Psychotechnical Laboratory of the Paris Public Transport System, and in 1927, organized the Paris Congress of the International Association of Psychotechnics.[30] Lahy's influential journals, the *Revue de Sciences du Travail* (1929–30) and *Le Travail Humain* (1933–present), promoted the professionalization and internationalization of the psychotechnical movement.[31] In Belgium, Omer Buyse developed a comprehensive approach to aptitude testing, also called *ergologie*, a term used to describe the science of work in the French-speaking world after the war.[32]

In Germany, under the slogan "the right man for the right job," psychotechnical aptitude testing became a virtual craze in the immediate postwar period. Two distinct trends of pre–World War I academic psychology fused in the psychotechnical movement: the testing of the capacities of the human motor—fatigue, trainability, and performance (via Kraepelin and Münsterberg)—and William Stern's "differential psychology," the classification of occupational aptitudes and inclinations.[33] These methods were extensively applied to reintegrating returning soldiers into the peacetime economy. Adapting techniques developed during wartime, Moede and Piorkowski attracted the support of private industry and extensively promoted the use of psychotechnical examinations to retrain veterans for various civil occupations, such as street-car conductors, typesetters, typists, and clerks (their privately funded Berliner Begabtenschule trained plant engineers). Moede developed a wide array of tests to determine physical and intellectual aptitude, and later in his career he elaborated psychological tests of "character" and motivation.[34]

The psychotechnical movement in Germany experienced its greatest influence in the early 1920s. In 1919, Prussian law decreed that occupational counseling and employment agencies be set up with the aid of psychologists.[35] By 1922 no less than 170 psychotechnical "testing stations" were operating in Germany and a special agency to deal solely with psychotechnical problems created by the Ministry of Labor.[36] Private employers also found psychotechnics an efficient method of deploying labor according to vocational skills (as well as screening out workers considered unreliable politically). According to the journal *Industrielle Psychotechnik,* sixty-three firms established their own testing stations in the early 1920s.[37] Moreover, psychotechnical methods were adopted by the German railway and postal system, and by nearly all large metalworking, engineering, and electrical firms.[38] Psychotechnical aptitude tests were also used to select candidates for occupations as diverse as office work and hairdressing.

The success of psychotechnics was in part due to the massive restructuring of the labor market immediately after the war, but it was also a consequence of the generally positive reception afforded it by the trade unions, who viewed aptitude testing as an equitable and objective basis for "achieving the best relation between labor power and the labor process."[39] The use of psychotechnical aptitude testing for choosing officers' candidates in the Army and Navy was also welcomed by Social Democratic party officials as a way of breaking the traditional hegemony of aristocratic elites in those careers.

The practical achievements of psychotechnics in the military and industry during and after the war paved the way for academic acceptance. The early 1920s were the glory days of academic psychology in Germany and the chief means by which the new applied science was most effectively disseminated. New psychotechnical journals—*Industrielle Psychotechnik, Praktische Psychologie, Psychotechnische Zeitschrift*—were testimonies to newfound academic legitimacy (and professional differences).[40] Between 1918 and 1927, psychotechnical professorships were established at six polytechnical colleges (Technische Hochschulen) throughout Germany, and psychotechnical institutes were created at several major universities in cities such as Hamburg, Munich, and Berlin.[41] Most prominent among the new psychotechnical institutes was the showcase Institut für Psychotechnik at the Technische Hochschule in Berlin-Charlottenberg, founded by Georg Schlesinger (and later directed by Moede). Engineering students in German universities were required to complete a course in scientific management and in scientific factory organization—courses that included the latest developments in applied psychology.[42] Intense rivalry developed between engineers and academic psychologists over which profession was best qualified to develop and administer the new testing methods in industry.[43] By 1922 Prussian state officials complained that there were no less than "five psychological and psychotechnical institutes" competing for financial support.[44]

Despite its initial popularity, the psychotechnical craze was short-lived. By the middle of the decade the German had economy stabilized, and the need for extensive aptitude testing and vocational counseling reached a saturation point. Technological and social improvements, which scarcely compensated for heavy costs, were curtailed, by firms and public enthusiasm for psychotechnics waned considerably.

Increasingly, during the 1930s, critics pointed to the limits of these methods in industry. Fatigue testing, though it might determine the point at which a particular muscle or limb could no longer perform at a single movement, was unable to set relevant guidelines for any trade or occupation. As one observer pointed out, in practice muscular fatigue could not be isolated from any other factors—skill, intelligence, conditions of work—that might impinge on performance. Fatigue seemed to mean far less than was assumed: it did not refer to one thing, but rather to a complex of physiological and psychological "causes" that resulted in less work.[45]

Psychotechnics was also justly accused of readily sacrificing its scientific neutrality and objectivity when it came to the interests of the firm. The industrial psychologist, Wladimir Eliasberg, for example, charged that "for psychotechnics the human being is only a factor of the firm [Betriebsfaktor]."[46] In 1930 a public controversy erupted when Otto Lipmann accused Walther Moede of misusing the methods of psychotechnics, of violating its neutrality, and of "discrediting" industrial psychology by advocating its use to remove "undesirable" employees.[47] Even among psychotechnical prophets, like Fritz Giese, director of the psychotechnical laboratory at the Polytechnical University in Stuttgart, it was becoming evident that the hour of what he called "subject psychotechnics"—the adaptation of individuals to economic circumstances—had passed. For Giese, the new economic situation demanded "object psychotechnics"—the adaptation of the conditions of work and technologies to the psychological needs of individuals. Giese looked beyond the workplace to expand psychotechnics to the study of crime, of revolution, and—anticipating his later conversion to National Socialism—of "spiritual epidemics."[48]

INDUSTRIAL PSYCHOLOGY AND THE PATHOLOGY OF WORK

By the mid-1920s, growing dissatisfaction with psychotechnics, even among its staunchest supporters, led to the development of several new psychological approaches to work in Germany. Experimentally oriented industrial psychologists promoted an approach that regarded "task" and "motivation" as the chief components of a new research strategy that took into account the character, or "total personality" of the worker, as well as social relations in the workplace. In his 1924 study of labor time, Lipmann pointed out that a reduction of labor time and more rest pauses might indeed improve work performance, but without a corresponding increase in the will to work such spurts in productivity proved ephemeral. "Capacity for work" was not identical with "ability to work."[49]

In the 1920s, monotony became an issue as psychologists like Lipmann began to recognize that the standardization of movements à la Taylor could inhibit efficiency if the worker's "bio-psychological" interests were not considered.[50] Lipmann also acknowledged the pleasure that a well-organized pattern of work could evoke. The problem of worker motivation gave rise to other psychological approaches—all of which stressed that no occupation could be sustained entirely out of "spiritless activity."[51]

By the end of the 1920s "characterology" *(characterologie)* came into vogue as a method of diagnosing workers' temperament, "reliability," "fortitude," devotion to duty, and other dimensions of "the working personality." Its reliance on essentially military criteria—discipline, subordination to superiors—to evaluate workers' abilities and personality characteristics explains why this authoritarian approach proved more suited to military use. Indeed, after 1927 it was adopted as the chief method of selecting candidates for officers' training.[52] Some psychologists believed that "sabotage" or "radicalism" could be studied as modern political pathologies of work. For Eliasberg, the "social therapist had to become not merely a social engineer, but a politician."[53]

Eliasberg's prognosis proved more than a rhetorical flourish. Throughout the 1920s various solutions to the labor question were put forward in Germany, including schemes to further "joy in work" *(arbeitsfreude)* or to introduce "social harmony" into the plant. These ideas initially emanated from the intellectual firmament of "yellow," or Catholic, unions but by the end of the decade a nationalist trend was in the ascendancy. Conservative proposals that prescribed a romanticized return to a preindustrial work community proliferated in Germany, for example, the "Werkgemeinschaft," or "plant community," promoted by Richard Ehrenbert and Josef Windschuh in the early 1920s. The workers' community they envisioned eliminated "obsolete" individualism and the "lordliness of the entrepreneur" *(unternehmerherrlichkeit)* through paternalistic programs within the firm, such as gardens, clubs, and newspapers.[54] Willy Hellpach, director of the Institut für Sozialpsychologie in Karlsruhe, and Richard Lang, an engineer at the Daimler-Benz works, conceived the idea of "group production" *(gruppenfabrikation)* modeled on a return to the methods of the small workshop within a larger plant.[55] Still more radical were the ideas of Eugen Rosenstock, whose "Werkstattaussiedlung" (workshop-relocation) advocated transferring workers from large urban plants to smaller groups in the country, where they would then create methods and rules of work in their own "living space."[56]

Although these nostalgic alloys of modern industry and preindustrial community were rarely adopted, by the mid-1920s, another American import, the "Fordist" assembly line, offered Europeans the promise of a more practical and successful work ideology. Unlike Taylorism, Fordism was based on predetermined work norms, higher wages, and on well-organized plant policies designed to discourage class solidarity and encourage loyalty to the firm. The Fordists emphasized the superiority of their methods of increasing mass consumption while subordinating the rationalization of the firm to the "larger rhythm of the flowing production process."[57] Above all, Fordism divorced social "happiness" from the workplace by promoting consumerism and leisure-time activities. Many European critics viewed Fordism, with its emphasis on the compensations of leisure and consumption, as the direct antithesis of Taylorism, with its "work-centered approach." In Germany political conservatives, like Friedrich von Gottl-Ottlilienfeld, could applaud the supersession of Taylor's ideas by Ford's synthesis of "economic and technical reason" and wax eloquently over Ford's claims to have eliminated class conflict while increasing productivity. For von Gottl-Ottlilienfeld, the Fordist firm was an "oasis of social peace," compared with the Taylorized firm, which he called a "paradise of the unskilled," where the skilled workman *(facharbeiter)* was "squeezed to death."[58]

SOCIAL POLITICS IN THE PLANT AND THE ROMANTIC PHILOSOPHY OF WORK

In the early phase of the Weimar Republic, industrialists, experts in fatigue, and Social Democratic trade unionists generally shared a positive view of the science of work as compensation for the negative effects of Taylorism. By the end of the 1920s, however, a spate of right-wing industrial ideologies appeared, which aimed at defusing political tensions and creating politically reliable workers and workplaces by authoritarian policies, by heavy-handed appeals, and by direct intervention. One novel approach was introduced by the sociologist Goetz Briefs early in 1929 at the Institut für Betriebssoziologie und soziale

Betriebslehre at the Technische Hochschule. The institute was oriented toward confronting social and political realities in the plant, as opposed to applying physiological and psychological insights to workers separated from the work process. Management could not leave the shop floor to the workers, but instead had to participate by fostering loyalty to the firm, joy in work, and a sense of well-being. Briefs' main efforts were directed at providing "future or already trained plant engineers" with "an extensive knowledge of the human and intra-human basis of a plant."[59] Prior to entering the "work and 'life sphere' of the industrial masses," engineers were given a general orientation in wage, work, and labor policies, and educated in how to "avoid tensions, conflicts, and idling." The purpose of Briefs' "social plant politics" was to overcome what he called the obsolete paternal organization of the plant, and to replace class hatred and worker solidarity with "industrial dissipation, decentralization and division."[60]

By the 1930s, the economic crisis and the radicalization of the political situation in Germany was accompanied by a crisis in the science of work. The "dismantling of social policy" during the Depression undercut the most optimistic promoters of social reform, though "rationalization" retained some of its tarnished aura. Taylorism and Americanism were now blamed for creating massive unemployment and for the rapid decline in Germany's world position. In the 1930s rhetoric of authoritarian and fascist movements, "work" was infused with a mystical, heroic, and transcendental meaning while its declining availability was attributed to a misplaced faith in the false promises of anti-German systems like Taylorism or socialism. Economic nationalism combined with spiritualization of work was evident in such popular treatises as Fritz Giese's *Philosophie der Arbeit* (1932), which showed a marked disinterest in economic or industrial problems and signaled a metaphysical or, romantic "turn," in the science of work by one of its leading proponents.[61] Giese attempted to fuse the late nineteenth-century energeticist approach and psychotechnics with a grand metaphysical doctrine, criticizing liberals like Lipmann, whose definition of work as an "economic activity" Giese ridiculed. Instead, Giese glorified labor as "an epochal phenomenon . . . corresponding to a purposive activity by individuals and society, and directed toward occupationally defined cultural goals *(kulturziele),* arising on the basis of biological and technological energy, but following a teleological impulse."[62]

Giese was inspired by Oswald Spengler's *Decline of the West,* but he was also influenced by Martin Heidegger's philosophic reflections on industrial labor and technology as decisive for the despiritualized condition of the West. During his year as the National Socialist Rector of

Freiburg University (1934), Heidegger frequently reiterated his view that labor is "not simply the production of goods for others," nor "the occasion and means to earn a living," but *as work,* something spiritual *(geistiges).* "[63] Giese shared Heidegger's view that restoring labor to its noble state required an end to class conflict and the restoration of a society based on order and authority. These ideas were not restricted to conservative circles. No less astute a Marxist critic of this neoromantic trend than Herbert Marcuse (in a 1933 essay entitled "On the Philosophical Foundation of the Concept of Labor in Economics") agreed with Giese's diagnosis, if not with his solution:

> Whenever it [the science of work] goes beyond the economic-technical dimension . . . [labor] is seen essentially as a psychological problem. Psychology, however, cannot adequately deal with the problem of labor since . . . labor is an ontological concept.[64]

Marcuse, a former student of Heidegger, was by then a Marxist and argued for the emancipation of labor as the basis for an end to human alienation. His depiction of work, however, in essentially the same spiritual terms as those of Giese (whom he credits), owed much to the mood of romanticizing labor evident in Germany in the late 1930s.

THE NATIONAL SOCIALIST SCIENCE OF WORK: DINTA AND BEAUTY OF LABOR

Giese's romantic metaphysics of labor pointed the way to his and many other German "psychotechnicians' " subsequent embrace of National Socialism.[65] Politically more significant was the rise of the Deutsche Institut für technische Arbeitsschulung (DINTA) founded in the fall of 1925 by the right-wing engineer Dr. Karl Arnhold. From the outset the DINTA received substantial support from politically conservative (DNVP) circles among Ruhr industrialists.[66] Distinguished by its unique combination of explicitly ideological appeals and its practical programs, the DINTA was directed equally at engineers, workers, and management. From the outset, Arnhold conceived of DINTA's mission in anti-socialist and nationalist terms: to create a "new industrial species, [the] carrier of a German ethos from the old germanic epoch."[67] The DINTA also distinguished itself from other branches of the science of work by its aggressive anti-Americanism and by its criticism of both Taylorist and Fordist methods of rationalization.

The DINTA's expansion as a "plant-pedagogical center" in the late

1920s and early 1930s was impressive. With considerable financial means at its disposal, it became the leading school of work science in the later phase of the Weimar Republic. Through its vocational counseling, adult education, and comprehensive "social plant policy," the DINTA's anti-Americanism and antisocialism found support among workers, while engineers and managers approved of its authoritarian approach to plant politics. By 1928 the DINTA built seventy-one training centers and completed the education of almost one hundred plant engineers. In 1930 more than 300 firms were operating according to DINTA guide-lines.[68] According to a sympathetic study of the DINTA conducted in 1931, participation in its various programs was especially high among younger, previously unemployed male workers.[69] DINTA engineers often trumped trade unions by more effectively criticizing speedup, by condemning obsolete work methods, and by improving poorly orga-nized and shabby workplaces while raising productivity.[70]

With the National Socialist seizure of power, Arnhold rose to prom-inence. The DINTA was absorbed into the German Labor Front (DAF), the Nazi Labor Organization, and in 1935 Arnhold became head of the Amt für Berufserziehung und Betriebsführung, the first national insti-tute of work psychology.[71] The April 1933 civil service laws directed against Jewish and socialist professionals forced as many as one-third of the (Ordinarius) professors of psychology out of their posts.[72] But it also afforded new opportunities for others. National Socialism demanded "Aryan descent," political reliability, and professional conformity: DINTA offered all three. The DINTA became the organizational center of industrial psychology in Nazi Germany, while competing institutes—Lipmann's and Briefs,' for example—were closed down.

Arnhold subsequently became an indefatigable advocate of the National Socialist organic "work idea": National Socialism had demon-strated that "it is possible to unify man and machine, that they must not assault each other." Along with mundane suggestions for improved lighting and for educating the plant "following" *(Gefolgschaft)* to or-derliness, cleanliness, and reliability, Arnhold's speeches and articles were generously laced with anti-Semitic slurs: "There is a heaven of difference between a working Jew and a working German."[73]

For numerous industrial psychologists, the National Socialist sei-zure of power called forth strong affirmations of the usefulness of their relatively young discipline, in both political and practical terms.[74] At the first Congress of the German Psychological Association (Deutsche Gemeinschaft für Psychologie) in October 1933, politicization and pro-fessional expertise were wedded to each other in appeals to eliminate Jewish influences and to draw on the deep psychological insights of

Mein Kampf for the new movement.[75] The psychologist, Walther Jaensch, for example, justified Nazi racial policies by contrasting the "Nordic integrated type" with the "Jewish-liberal dissolute type," in an effort to create a psychological anthropology based on race.[76] Under National Socialism, industrial psychology increasingly looked to the state, rather than to industry for financial support. With the labor shortage of the late 1930s, and with the militarization of the German economy, however, applied psychology and psychotechnics experienced an upsurge. The military utilized psychologists to select candidates for aviation training and for officers' candidate schools. The first permanent posts for psychologists in German firms were created under National Socialism, especially after the war economy went into full throttle after 1938. The use of psychotechnical tests was even extended to deploy the masses of Polish laborers deported into Germany for work after 1941.

The methods of the DINTA, psychophysical examinations, and many of the techniques developed by Briefs, Windschuh, and Hellpach, were employed in the German Labor Front's most successful industrial program, the Bureau of Beauty of Labor (Amt Schönheit der Arbeit), founded on 27 November 1933. During the 1930s Beauty of Labor developed extensive plans to encourage German plant managers to beautify and remodel factories and workrooms, offering tax incentives for the work. These efforts were accompanied by well-orchestrated propaganda campaigns ("Good Light—Good Work"; "Clean People in Clean Plants"), competitions for model workers and plants, and aesthetic and leisure-time programs, such as model designs for plant interiors, plant "evenings," and even "Beauty of Labor" tableware and furnishings. By 1938, annual expenditures by German employers for Beauty-of-Labor-sponsored projects had reached 200 million Reichsmarks: 67,000 plants were inspected, 24,000 had new washrooms and wardrobes, 17,000 had park and recreation areas, and 3,000 had new sports facilities.[77]

Beauty of Labor's "social aesthetic" combined a rationalizing and modernizing ideology with the anti-liberal and anti-capitalist propaganda of the National Socialist "plant community." After 1936 the Bureau of Beauty of Labor became the center for the modernizing thrust of National Socialism, incorporating its technocratic aesthetic in a *Sachlichkeit*. But Beauty of Labor also epitomized the extent to which National Socialism fused the styles of industrial modernism with conservative politics and antimodernist motifs. Carefully balancing management needs with its politically sophisticated programs, Beauty of Labor combined economic and technological rationality with the ideals of the "people's community." The glorification of technology was entirely compatible with antiliberal and anticapitalist rhetoric, just as industrial psychology could be used to justify and achieve immediate political objectives, like the "combing out" of unnecessary skills and "superfluous" occupations. As was the case with the similarly successful Labor Front leisure organization, "Kraft durch Freude" (Strength through Joy), which promoted modern high-speed ocean travel to palm-lined beaches and motorized outings through Medieval land-scapes, Beauty of Labor promoted a uniquely National Socialist synthe-sis of modernity and antimodernity, culture and civilization.[78]

The DINTA was incorporated into the German Labor Front in 1935, but not without friction. Arnhold and the DINTA often came into conflict with Beauty of Labor officials because of the Bureau's policy of not directly challenging reluctant industrialists and because of its em-phasis on aesthetics and architecture, as opposed to the hardcore politi-cal indoctrination advocated by Arnhold. Arnhold promoted his style of National Socialist work ideology by asserting that "the tempo of the machine be brought into harmony with the rhythm of the blood" through the "organic formation of the plant and the militarization of the leadership.[79] By the late 1930s, Arnhold's program of "mobilizing the performance reserves of industry" through quasi-military training and indoctrination was passed over. The technocratic direction of the Bureau after 1936 found him largely eclipsed. Arnhold's criticisms ironi-cally underscored the extent to which Beauty of Labor was more sophis-ticated in its use of the techniques of industrial psychology than was Arnhold's DINTA.

By the outbreak of the Second World War, industrial physiology and industrial psychology were no longer radical or utopian proposals drafted by idealistic reformers with little or no experience in industry. Success in the First World War made the techniques of ergonomics and fatigue research pioneered by Marey, Mosso, Imbert, and Amar, and the psychotechnics pioneered by Kraepelin and Münsterberg standard equipment of modern industrial management. Professionalization, however, was not the only consequence of the widespread acceptance achieved after the war. The realities of industrial conflict, the impact of the Depression, and rise of fascism and National Socialism under-mined the fundamental assumption of the European science of work: a calculus of efficiency and social improvements in industrial conditions (hours, wages, and restructuring the workplace) could be effected by government.

During the 1930s there were a few isolated figures like Lahy and Lipmann who believed that industrial psychology could still serve as a means of resolving conflicts between workers and employees. But it was the authoritarian anti-union approaches of Briefs, Giese, Windschuh, and the DINTA that predominated in Germany. The early 1920s saw a rapprochement between the most useful aspects of Taylorism and the humanitarian concerns of the industrial physiologists and psychologists, but the Depression brought to the fore political techniques of industrial organization and romantic ideologies of work that had little in common with the origins of the European science of work in the laboratories of reform-oriented scientists. Only after the Second World War was there a brief revival of the science of work among groups of academic sociologists, like Georges Friedmann and Pierre Naville in France, who turned to that tradition to counter the postwar resurgence of management ideologies focusing exclusively on productivity. By then, however, as these same sociologists began to admit, the productivist focus of the science of work on the body and its adaptation to the technology of the factory was itself becoming obsolete: the human motor was no longer the center of the industrial or the social cosmos.

Conclusion: The End of the Work-centered Society?

IN the mental life of the nineteenth century, work was at the center not only of society but of the universe itself. Social modernity, the project of superceding class conflict and social disorganization through the rationalization of the body, emerged at the intersection of two broad developments: the thermodynamic "model" of nature as labor power, and the concentration of human labor power and technology of the second industrial revolution. The metaphor of the human motor united these developments in the single idea that the working body is a productive force capable of transforming universal natural energy into mechanical work and integrating the human organism into highly specialized and technical work processes.

In the metaphor of the human motor Faustian idealism was embedded in a materialist substratum; conservation of energy was put in the service of industry. Scientific materialism accorded to the body and to energy a "trans-sensual" character: the manifest forms of energy belonged to the world of the senses, but energy itself was a hidden, invisible, and secret substance embedded in all these forms, "like the Platonic Idea over things."[1] The old mechanical materialism of the eighteenth century, as Engels never tired of pointing out, was based on the principle that there had to be something beyond matter to explain its movement. Why, if this were not so, had so many great thinkers been consumed by the chimerical search for perpetual motion? With the doctrine of energy, finally there was nothing beyond matter to explain its motion, no homunculus beneath the table. There was only matter in motion, with energy claiming true universality as the reservoir of perpetual motion of the universe, the source of its labor power. The old dualism of spirit and matter was overcome by a transcendental materialism with its own monadic conception of reality—the unity of matter and energy—as an energumen.

Late nineteenth-century materialism regarded energy as the basic element of all experience. The quantification of the body's energies, the isolation and determination of the economies of force in the physical and nervous systems, and ultimately the establishment of a system of equivalences between technology and physiological energy, expressed a Promethean idolization of productive force. Not only did science display an "unshakeable confidence" in its ability to bring the forces of nature under human control, but it also claimed the ability to bring human nature under the control of science. This "idealism" was at the metaphysical core of late nineteenth-century materialism.

Nature and society were assimilated to a single vision of the unity of all work. The universal laws of energy applied equally to the movements of the planets, the forces of nature, the mechanical work of machines, and the work of the body. The Copernican revolution of the nineteenth century was the discovery of this "laborcentric" universe. However different from one another in intent and ideology, these conceptions—Helmholtz's characterization of the universe as a reservoir of unlimited labor power, Marx's image of modernity as the relentless transformation of human labor power into the powerful engine of capital, Taylor's utopia of the body subordinated to the engineer at the helm of industry, and the European science of work—all placed the working body at the juncture of nature and society. The cosmos, the factory, and the worker were all extensions of "the human motor."

The nineteenth-century preoccupation with fatigue and exhaustion was the underside of this productivist obsession with the conquest and harnessing of every conceivable source of energy. The traditional Western proscription on idleness, which spiritualized and consecrated labor, was displaced onto the working body and recast in the language of scientific materialism. There is a clear line of ascent from the "cumulative" disorder of idleness, as Max Weber diagnosed it, to the cumulative disorder of fatigue—from the infirmity of the spirit to the infirmity of the body. Yet in this new image of fatigue as the universal form of resistance to work, the body, with its sensual substance of muscles and nerves, also constituted the chief mechanism of defense against the excessive demands of industry, against its own exploitation. The materialism of the discourse of fatigue and energy altered the political and intellectual framework in which work was situated: it redefined the *summum bonum* of industrial society as the elimination of the unreasonable demands on the energy system that *laissez-faire* economics and unbridled capitalism took for granted. The moralism of *industria* and the ancient proscription on sloth were insufficient to encompass the mechanized work of the industrial age.

Consequently, in the second half of the nineteenth century, liberal and socialist thinkers alike emphasized the importance of fatigue for identifying the limits of productivism, for demarcating the horizon of exploitation, and for affirming the legitimacy of reform. As opposed to traditional liberalism, which considered efficiency as the product of want and poverty (the work ethic), the new social liberalism of the nineteenth century substituted a calculus of energetics: the rational deployment of the body's forces in the interest of productivity. In this calculus fatigue was paramount: hence the search for its laws, its etiology, its chemical composition, and a vaccine to eliminate its negative effects. Fatigue, in this sense, was more than a subjective state: it represented the objective limit or, "optimal," point of exertion of the body, the outermost boundary of the human motor.

For the physiologists and reformers who shared this belief, the economies of motion and the conservation of the energies of this working body held the key to greater productivity, progress, and social justice. The energetic calculus was not restricted to any single political doctrine or ideology; it could be applied on different sides of the political spectrum by liberal reformers, by socialists, and by Marxists like Edouard Vaillant. According to the energeticist calculus, shorter workdays, rest pauses, ergonomic studies, and chronophotographic analysis of the body's most discrete motions—all contributed to a reduction of fatigue and to the proper use of the productive energies of the body. In global terms, enhanced conservation of social energy replenished the energy supply of the nation without a corresponding loss in the performance and productivity of labor. The optimistic correlation between the reduction, or perhaps elimination of fatigue, and the rationalization of the body's productive powers, was also a new politics of modernity, of a scientific and "objective" solution to the "social question."

Grounded in physiology, the European science of work was socially neutral and objective; the working body was subject only to the objective laws of motion of the physical and biological universe. As such it provided nineteenth-century reformers with a new vision of the working body that promised to transcend the moral and political dissension that plagued the modern workplace. Standing above the contest of labor and capital, the rational deployment of labor power in the interest of the whole would supersede traditional enmities and ideological camps, and resolve conflicts over wages, accidents, and industrial discipline.

To be sure, this vision of modernity also displayed a distinctly negative side in its preoccupation with the more severe manifestations of fatigue, mental torpor, neurasthenia, and the other "pathologies" of work. The persistent inertia and the chimerical ideas produced by "primitives," students, and other chronically fatigued groups were a constant reminder that the productivist ethos was threatened by internal and external inhibitions and impassable limits. Modernity was simultaneously a threat and a promise. And yet, as physiologists and psychologists discovered, fatigue was also a prophylaxis against this danger, a first line of defense against the irrational shocks and invasions of modernity, defining the corporal limits of production and overwork, and determining the point at which the energy supply demands restoration. As idleness defined the positive image of work on which the traditional spiritual discourse of ennobling labor insisted, fatigue represented the negative side of the well-regulated, energy-conserving materialist body.

With its arsenal of metaphors drawn from thermodynamics, social energetics conceived of society as a calculus of energy. Society could be constructed according to the rational deployment of the forces of the body: productivism could be harmonized with the need to conserve energy and preserve the productive power of society. Even after the Second World War this calculus has continued to exercise a strong attraction for those who saw in the diminution of fatigue the basis for a rational solution to social conflict and the "unfettering" of the power of production.

THE LEGACY OF THE HUMAN MOTOR

What is the long-term legacy of the "human motor"? Despite the declining intellectual significance of the science of work after the Second World War, it has proven to be a durable metaphor. A recent study of ergonomics in France claims that the "classic ideal of the "man-machine" remains the leitmotif of what has long since become a subordinate technology of industry and management: "The ergonomists are concerned in the first and second instance with the mass of considerable data that has been accumulated about the 'human machine' and its limits."[2]

But a more important legacy of the human motor is its role as a "paradigm" of social modernity. The subordination of the working body to the scientific calculus of reform and productivity, of energy and fatigue, has survived in another, much larger context. What has clearly been retained from the metaphor of the human motor is the vision of a society whose productive potential is ultimately linked to a calculus about the limits of society's ability to reduce social risks without damaging industrial growth and restricting freedoms. For nineteenth-century liberals, the vision of a scientifically regulated worker and workplace bridged productivism with reform. The elective affinity between expanded output, greater work performance, and more energetic workers was coupled with nineteenth-century efforts to shorten the working day, eliminate industrial risk, and secure health and safety. Reform was predicated on the intellectual vision of social positivism—on society's economically deploying the productive energies available to it while relying on scientific principles to conserve those energies by reducing social risk and human wear and tear.

The modern welfare state is in many respects the outgrowth of this dynamic, which was set in motion by the early efforts at health and accident protection legislation of the 1880s and 1890s. From a global standpoint, the much-discussed contemporary crisis of "Fordism"—the class compromise between welfare systems and capitalist organization plus high levels of mass consumption—is a crisis of the institutionalized form of this energeticist "calculus" in the contemporary West. The attempt to replace social conflict with the calculated management of risk is still at the core of the contemporary welfare state. It is precisely this element—the future of the risk-free society—that has most recently come into question in contemporary European and American debates on the future of the social state.[3] These debates are not so much concerned with the extension of social rights or with the long-term economic prospects of productivity and reform, but with the social and cultural costs of protecting individuals against social risks, with the extension of the compensatory model of risk protection from the workplace to other domains of social life. The cost-risk calculus now stretches beyond the workplace, to the economy, to the home, to the environment, to international conflicts, and to biotechnology.[4]

The contemporary crisis of the welfare state may, in certain respects, be traced to the gradual eclipse of the energeticist calculus. To be sure, advocates of the free market have always applied a different calculus, one in which inequality acts as a spur to efficiency, innovation, and productivity. But the unlimited faith of nineteenth-century reformers in scientific solutions to social problems and their linking of social rights to social policies administered by states attests to the attractiveness of the energeticist calculus. The principles articulated by the British liberal William Beveridge in his famous proclamation that "freedom from idleness" and "freedom from want" are inalienable social rights are no longer self-evident. But it is not just liberal naïveté and wartime social détente that accounts for this change.[5] Nor can the contemporary crisis of the welfare state be explained entirely by the overburdening of revenues, economic waste, or the moral and political debacle of "welfare." The problem is also with the *ethos* behind the ideal of social rights, the implied link between conservation of energy and social efficiency.

As the political theorist Alan Wolfe argued in a provocative article entitled "The Death of Social Democracy," "Without a sense that progress leads inevitably toward a more rational organization of society, politics becomes either a struggle for personal or group advantage in the present or a nostalgic longing for a better world presumed to have existed in the past."[6] Deprived of the Enlightenment's belief that society can be ordered according to the precepts of science, reform is reduced to self-interest, or worse, competitive advantage. For neoconservatives, of course, this crisis is to be welcomed, since the elimination of social risk contributes to the complacent expectation of social protection throughout life, undermining initiative and the work ethic. The high costs of reform contribute not only to permanent budgetary deficits, but to long-term economic weakness. Neoconservatives in Europe and America claim that such efforts ultimately reduce competitiveness and national efficiency and that the price of the risk-free society is an economy at risk. Yet their solution—a return to markets in all spheres of social life—has not, at least at this point, effectively undermined the basic premise of social rights in European societies.

In France, yet another approach has been articulated by neoliberal theorists of "the crisis of the providential state" like François Ewald, Jacques Donzelot, and Pierre Rosanvallon. In their view the society "invented" by the positivist calculus of deferred risk—in the interests of productivity, health and well-being, and domestic tranquility—has severely undermined the capacities of individuals to participate in democratic politics or even resist the incursions of the state into family life. Social modernity, embodied in the great transformation of social "rights" before the First World War and immediate postwar period, along with the "Fordist" compromises of the 1920s and 1930s, was the product of a social rapprochement between productivist trade unions and reform-oriented capitalism. Social democracy, however, has undermined democracy. The real crisis is a matter not only of budgets or ideologies, but of the growing "disqualification" of individuals from political life—from the trend toward greater reliance on a "hygienic" state that organizes social life in the interests of reducing the threat of dislocation.

Unlike American or British neoconservatives, the French neoliberals, like Donzelot, argue that it is not as much the economic costs of specific government programs and policies, as it is the *logic* of the historical rapprochement between productivism and reform that has created a cultural crisis. The greater cost of these measures is not a loss of competitive advantage, but the triumph of a dangerous social ethos, the long-term pursuit of "pleasure and work" at the expense of civic responsibility, public participation, and individual autonomy.[7]

Such grim assessments only underscore the utopianism of an earlier era in the development of social liberalism when the norm of social rights and productivism were first linked to the neutral ideal of an interventionist state and the unlimited possibilities of scientific progress. The discourse of energy and fatigue represented a faith in science, in nature, and in industrial regulation. But, as the example of industrial accidents demonstrates from the outset, competing discourses of labor, of management, and of the insurance companies and physicians attempted to undercut that utopianism. If they are correct, these contemporary assessments offer another explanation for the decline of the "science of work" in the postwar era.

To be sure, progressive trade unionists and academics continue to point out that in many respects the health and safety risks of the workplace have worsened and that environmental dangers—the dangers of radiation and toxic emissions—are a new and serious concern of work hygienists.[8] But even in the European welfare states, where job enrichment, worker representation in management, and workplace "group democracy" (the famous example of Volvo teams) are routine aspects of industrial democracy, communication is the slogan of the day, not the rationalization of the body. Physical work no longer occupies the position in social thought and practice that it once did in the perceptual universe of the nineteenth century.

THE OBSOLESCENCE OF THE BODY

To what extent has the eclipse of the image of the body as a human motor, and consequently the decline of this intellectual affinity between reform and productivism, generated another major transformation: the end of the work-centered society? Almost a decade ago, the sociologist

Ralf Dahrendorf addressed this question in a controversial article entitled, "The Disappearance of the Work-Society." According to Dahrendorf a salient characteristic of modernity has always been the pivotal role of labor as the core of social life: "Occupation is at the center of the social identity of the individual."[9] In contemporary society, Dahrendorf observed, the development of technology has increasingly dissolved this source of identity, displacing work from its core. The "work society has begun to transcend work."[10] Dahrendorf is hardly alone in this view: Daniel Bell in the United States, Anthony Giddens in England, Jürgen Habermas and Claus Offe in Germany, Alain Touraine and André Gorz in France, and a host of other observers have all defined the essential change occurring in postmodern society in terms of a departure from labor-centeredness, either in the sense of a declining interest in work, increasing leisure and consumption, or new forms of sociability and communication.[11]

Historians have added to this analysis a description of the long-term changes in the working class and the concomitant decline of the working-class movement. They point to the disappearance of the skilled (largely male) worker, the emergence of a racially and culturally segmented and professionalized workforce, and the fact that the proportion of a lifetime spent working is declining because of greater longevity and earlier retirement.[12] As traditional patterns of social identity and voting behavior no longer apply, working-class movements are becoming less and less prominent in national affairs, at least in Western Europe. Trade unions have also faced eroding membership and influence. In an article aptly titled "The Forward March of Labor Halted?" Eric Hobsbawm has forcefully described the decline of the manual worker, of class consciousness, and of a common workers' culture in contemporary Britain.[13] Apart from isolated efforts to revive movements for greater workers' power in France and Italy during the late 1960s, contemporary political and social movements in Western Europe have not by and large been defined by work-related issues or by working-class politics. Nonwork-related aspects of social experience have been in the forefront of "new social movements." The expansion of individual freedoms, the extension of social and political rights to the disenfranchised, and issues of gender and race have all taken precedence over work.

To be sure, there is less of a consensus about the social costs of this development. Daniel Bell has argued that having abandoned work as a source of self-fulfillment, the modern individual is vexed with a Faustian restlessness of spirit that offers little consolation. Culture, and certainly not mass culture, has proven no substitute for the gratifications of work. As we witness the transformation from an industrial society in which "energy and machines transform the nature of work" to one in which "social participation" replaces production, traditional social bonds no longer hold.[14] Even for those contemporary social theorists who do not share Bell's pessimism, there is little disagreement that production no longer defines or encompasses the totality of social being. Claus Offe, a German sociologist, rejects the proposition that "consumer-centered hedonism" is sufficient to explain the decline of the work ethic and the centrality of work. Rather, it is work itself that can be held accountable: "In so far as they are modelled on the pattern of 'Taylorization,' processes of technical and organizational rationalization result in the elimination of the 'human factor' and its moral capacities from industrial production."[15] Meaning has disappeared from work; consequently, work has disappeared as a source of meaning.

The argument that work was disappearing, or at least losing its centrality, originated during the controversy over automation—computerization, cybernetics, and robotics—during the late 1950s and early 1960s. This debate, carried on by influential economists, sociologists, and social theorists in Europe and America, underscored the rapidly changing work environment and proclaimed a qualitatively new stage in the development of industry. For the optimists, the worker liberated from manual labor would regain a universal "intelligence" while for the dystopians, the disappearance of the skilled worker meant long periods of unemployment and universal alienation.[16] Yet, opposing scenarios aside, there was little controversy over the impact of automation on work: technology had made possible the gradual elimination of the physical production of material objects. Physical work was being replaced by images, by communication, and by cybernetic systems of self-regulation. Organized around the substantiality of the muscles and nerves, and around the material objects that the body in tandem with machines transformed into products, nineteenth-century labor was being rapidly superseded by the work of technologies operating on and proliferating through abstract systems of knowledge. Technology was making corporal work obsolete.

In the 1950s, the influential Groupe de Sociologie du Travail, whose most prominent figures were Georges Friedmann and Pierre Naville, secured the support of the Conseil Supérieur de la Recherche Scientifique to produce a major study of automation in French industry. The result was a collaborative volume that defined the terrain of industrial sociology for more than a decade, the *Traité de Sociologie du travail* (1961).[17] For Naville and his co-researchers the arrival of automation was entirely consistent with Marx's prediction of the replacement of human labor power with the productivity of technology: "The extension of automatic systems signifies, today, the replacement and displacement of manual labor."[18] The potential application of technology extended beyond the domain "reserved for the brain, the eye, and the hand." What is new, wrote Naville, is the element of mechanical "supervision" *(asservissement)* over complex operations and over longer periods of time.[19]

For the authors of the *Traité,* the transformation of work could be grasped as a transition from "work of the laborer to the work of communication," from work centering on the physiology of muscles and nerves, to work of a "cognitive or semiotic" nature.[20] If work is no longer defined by the effects of the "productive force" of labor power and is no longer subordinated to the goal of the production of things, if the world of signs and communications belongs to human products and productive capacity, work can no longer be understood in terms of material output and performance. The appearance of the cerebral worker whose material and product is "information" is emblematic of the vast distance traversed between the worker who surveys complex technologies of communication and the "man-beef of Taylor."[21] Technical evolution and automation resemble the traditional forms of production only as "an informational residue, divested of muscular force."[22] Modern work was dematerialized work, work without the body.

In his speculative *L'automation et le travail humain* (1961), Naville foresaw the decline of traditional skill and of the moral strictures attached to work, and he predicted the coming of a mechanized society that would render the worker entirely superfluous, displaced by self-regulating, cybernetic production systems.[23] The society of the future, wrote Naville, would be largely "experimental" rather than permanent, rendering the habits and routines of traditional work obsolete. Naville prophesied that a new "cybernetic union" of man and technology would be forged that would abolish the sphere of production *tout court.* [24]

The argument that the body is no longer the "productive force" of industry was an invention not only of French Marxist sociology. In Germany, a similar claim was put forward by the conservative sociologist Helmut Schelsky, who argued that "the modern industrial worker is relatively less burdened in relation to machinery and machine production." For Schelsky, the sciences of work *(Arbeitswissenschaften)* were entirely successful insofar as they had become sciences of "the conscious organized planning of the adaptation of human beings to modern technology." By eliminating atavistic conflicts new technologies were assimilated "to the organic and spiritual needs of human beings."[25] Critics of industrial sociology in both Germany and France argued that the traditional emphasis on the skilled male worker in heavy industry was becoming anachronistic.[26] After the Second World War, industrial relations, industrial psychology, and industrial sociology were restricted to increasing motivation and job satisfaction within the tight economic and organizational constraints of modern firms.[27] By the end of the 1950s, industrial sociologists bemoaned the inevitability of alienation and joined the plentiful prophets of leisure.[28]

Even if much of this discussion appears today as little more than "pamphleteering on the grand scale," the European automation debate was significant for introducing the idea of the disappearing body into sociological discourse.[29] The automation controversy was a discourse of displacement: industrial work, in its traditional forms, faced extinction; the working body was no longer the human motor. The conservative Schelsky, the Marxist Naville, and the socialist Friedmann—all agreed that whatever the future might hold, it was not to be dominated by human labor power. The *Traité*, in fact, subverted its own legitimation. Friedmann cautioned that although industrial work was still "the motor which explains the evolution or revolution of social structures," he conceded that in the twentieth century, the working human being "is less and less a *homo faber* in the classic sense of the term."[30]

This diagnosis is also apparent in the great midtwentieth-century transformation in social theory, the shift from a Marxist or post-Marxist discourse to one that emphasizes language, meaning, and symbolic communication evident in structuralism, in poststructuralism, and in the work of the contemporary Frankfurt School. Jürgen Habermas, for example, has stressed that Marx's focus on the mastery of nature as a technical or instrumental process devalued the sphere of symbolic interaction and moral or cognitive interests in the constitution of society.[31] What Marx did not perceive is that work itself is "ideological" in the sense that it exclusively emphasizes the corporal and technical domination over nature to the exclusion of other social needs and values. To assert that the material reality of the work-centered society has been supplanted by the "information-" or "image-centered" society is, however, to ignore the ideological and cultural dimension of the "materialism" and the centrality of the body in nineteenth-century thought.

Can the disappearance of the work-centered society be fully explained by the decline of traditional labor intensive and technologically "primitive" factory work typical of the nineteenth century? This book proposes an explanation that reads these developments from a different perspective. The end of the work-centered model of society cannot be attributed to sociological factors alone. The work-centeredness of nine-teenth-century society was largely a phenomenon of the work-cen-teredness of the metaphor of the human motor. To the extent that this metaphor, which focused exclusively on the working body and on labor power as the organizing principle of nature and society, has disap-peared, the work-centered society has also begun to wane.

The disappearance of work is a phenomenon of mental life and of culture: a consequence of the declining power of an intellectual dis-course that placed energy and fatigue at the center of social perception. With the declining significance of industrial work as a paradigm of human activity and modernity, the body no longer represents the tri-umph of an order of productivism. The goal of emancipating labor from the constraints of tradition or unscientific practices no longer resonates with utopian anticipation as it once did for nineteenth-century reform-ers, scientists, and social thinkers. The displacement of work from the center to the periphery of late twentieth-century thought can thus be understood by the disappearance of the system of representations that placed the working body at the juncture of nature and society—by the disappearance of the "human motor."

Notes

INTRODUCTION

1. René Descartes, *Discourse on Method* [1637], trans. F. E. Sutcliffe (Harmondsworth, 1968), p. 73.
2. Hermann von Helmholtz, "Über die Wechselwirkung der Naturkräfte und die darauf Bezüglichen neuesten Ermittelungen der Physik," [1854] *Populäre wissenschaftliche Vorträge,* 2d ed., vol. 1 (Braunschweig, 1876), p. 115.
3. Rudolf Clausius, *Annalan der Physik* 125 (1865): p. 400. Cited in P. M. Harmon, *Energy, Force, and Matter: The Conceptual Development of Nineteenth-Century Physics* (Cambridge, 1982), p. 65.
4. Jean-Pierre Vernant, Pierre Vidal-Naquet, *Travail et esclavage en Grèce ancienne* (Paris, 1985), p. 3.
5. See Gaston Bachelard, *The New Scientific Spirit,* trans. Arthur Goldhammer (Boston, 1984), p. 65.
6. Agnes Heller, "Paradigm of Production: Paradigm of Work," *Dialectical Anthropology* 6 (1981): 71–79.
7. Jules Amar, "L'organisation scientifique du travail humain," *La Technique Moderne* 7, no. 4 (15 August 1913): 113.
8. See, for example, Joan W. Scott, "L'ouvrière! Mot impie, sordide . . .'; Women Workers in the Discourse of French Political Economy, 1840–1860," *The Historical Meanings of Work,* ed. Patrick Joyce (Cambridge, 1987), pp. 119–142.
9. On the "storing" of scientific traditions in literature, see Wolf

Lepenies, *Between Literature and Science: The Rise of Sociology* (Cambridge, 1988).

10. Fernand Lagrange, "La Fatigue et l'entraînement physique," *La Revue de Deux Mondes* 15, no. 7 (1892): 345, 346.

11. The most important analysis of the cultural consequences of the doctrine of entropy is Stephen G. Brush, *The Temperature of History; Phases of Science and Culture in the Nineteenth Century* (New York, 1978), especially pp. 61–76.

12. "Art" in Diderot, *Encyclopédie, ou dictionnaire raisonné des sciences, des arts et des métiers*, 17 vols. (Paris, 1751–72), vol. 1, p. 717. Cited in William H. Sewell, Jr., "Visions of Labor: Illustrations of the Mechanical Arts before, in, and after Diderot's *Encyclopédie*," in *Work in France: Representation, Meaning, Organization and Practice*, ed. Steven L. Kaplan and Cynthia J. Koepp (Ithaca, 1986), pp. 275, 276.

13. Auguste Comte, *Positivism and the Essential Writings*, ed. Gertrude Lenzer (New York, 1975), p. 425.

14. Cited in Jacques Donzelot, *L'invention du social: Essai sur le déclin des passions politiques* (Paris, 1984), p. 128.

15. Louis Querton, *L'augmentation du rendement de la machine humaine* (Brussels, 1905), p. 156.

16. Laurent Deschesne, "La productivité du travail et les salaires," *Revue d'Économie Politique* 13 (1899): 467.

17. Paul Louis, *L'ouvrier devant l'Etat: Histoire comparée des lois du travail dans les deux mondes* (Paris, 1904), p. 206.

18. Max Weber, *The Protestant Ethic and the Spirit of Capitalism* [1904/5], trans. Talcott Parsons (New York, 1958), p. 182.

19. Friedrich Nietzsche, *The Genealogy of Morals* [1887], trans. Francis Golfing (New York, 1956), p. 288.

20. For a general survey, see Michel Löwy, *Georg Lukács: From Romanticism to Bolshevism*, trans. Patrick Camiller (London, 1979), chap. 1.

21. Max Weber, "Energetische Kulturtheorien," *Gesammelte Aufsätze zur Wissenschaftslehre*, ed. Johannes Winckelmann (Tübingen, 1968), p. 401.

22. See, for example, the essays in *The Authority of Experts: Studies in History and Theory*, ed. Thomas L. Haskell (Bloomington, 1984) and Theodore M. Porter, *The Rise of Statistical Thinking 1820–1900* (Princeton, 1986).

23. Some significant exceptions are Robert A. Nye, *Crime, Madness and Politics in Modern France: The Medical Concept of National Decline* (Princeton, 1984); Karen Offen, "Depopulation, Nationalism and Feminism in Fin-de-Siècle France," *American Historical Review* 89, no. 3 (June 1984): 648–76.

24. See Dominick LaCapra, *Rethinking Intellectual History: Texts, Contexts, Language* (Ithaca, 1983), pp. 23–72; also Joan Wallach Scott, "On Language, Gender, and Working-Class History," *International Labor and Working-Class History* 31 (Spring 1987): 1–14, and the responses by Bryan D. Palmer, Christine Stansell, and Anson Rabinbach.

25. For a critical discussion, see John E. Toews, "Intellectual History after the Linguistic Turn: The Autonomy of Meaning and the Irreducibility of Experience," *American Historical Review* 32, no. 4 (October 1987): 879–907; Martin Jay, "Should Intellectual History Take a Linguistic Turn? Reflections on the Habermas-Gadamer Debate," *Fin-de-Siècle Socialism* (New York, 1988), pp. 17–36.

26. An excellent formulation of this approach is Joan Wallach Scott, *Gender and the Politics of History* (New York, 1988), pp. 4–6. The essays in *The New Cultural History* ed. Lynn Hunt (Berkeley, 1989) provide a good overview of the issues and of differing approaches, including some examples of how these might be fruitful for research.

27. Max Horkheimer and Theodor W. Adorno, *Dialectic of Enlightenment*, trans. John Cumming (New York, 1968).

28. See Jürgen Habermas, *Knowledge and Human Interests,* trans. Jeremy J. Schapiro (Boston, 1971); *The Philosophical Discourse of Modernity: Twelve Lectures,* trans. Frederick Lawrence (Cambridge, Mass., 1987); Jean Baudrillard, *The Mirror of Production,* trans. Mark Poster (New York, 1975); Martin Jay, *Marxism and Totality: The Adventures of a Concept* (Berkeley, 1984).

29. And, I should add, some criticism of his methods. See Allan Megill, "The Reception of Foucault by Historians," *Journal of the History of Ideas* 48 (1987): 117–41; Mark Poster, "The Future According to Foucault: The Archeology of Knowledge and Intellectual History," in *Modern European Intellectual History: Reappraisals and New Perspectives*, ed. Dominick LaCapra and Steven Kaplan (Ithaca, 1982); Patricia O'Brien, "Foucault's History of Culture," in *The New Cultural History*, pp. 25–46; *L'impossible prison*, ed. Michelle Perrot (Paris, 1980); Paul Veyne, *Foucault révolutionne l'histoire* (Paris, 1978).

30. On Foucault's thought, see John Rajchman, *Michel Foucault: The Freedom of Philosophy* (New York, 1985); Hubert L. Dreyfus and Paul Rabinow, *Michel Foucault: Beyond Structuralism and Hermeneutics* (Chicago, 1982).

31. Like Weber, Foucault focused on law and legal institutions (also on medicine and psychiatry) to "examine how forms of rationality inscribe themselves in practices, or in systems of practices, and what role they play in them." However, Foucault did not consider rationality to be an inescapable "iron cage" nor an "anthropological invariant"; nor did he employ "ideal types" in describing the mechanisms of power and rational procedures. For Foucault's view of Weber, see "Questions of Method: An Interview with Michel Foucault," *Ideology and Consciousness. Power and Desire: Diagrams of the Social* 8 (Spring 1981): 8–11.

32. Rajchman, *Foucault,* p. 85.

33. Michel Foucault, *Discipline and Punish: The Birth of the Prison,* trans. Alan Sheridan (New York, 1977), p. 80.

34. Alfred Fouilée. *La Science sociale contemporaine,* 2d ed., 2 vols. (Paris, 1885).

35. The crucial point of transition between these two differing conceptions is Foucault's "The Discourse on Language" (1970), in *The Archeology of Knowledge,* trans. A. M. Sheridan Smith (New York, 1972).

36. Michel Foucault, "Nietzsche, Genealogy, History," in *Language, Counter-Memory, Practice: Selected Essays and Interviews,* ed. Donald F. Bouchard (Ithaca, 1977), p. 150.

37. Walter Benjamin, *Das Passagen-Werk. Gesammelte Schriften,* vol. 5:2 (Frankfurt-am-Main, 1982), p. 962.

CHAPTER 1. FROM IDLENESS TO FATIGUE

1. Friedrich Nietzsche, *The Will to Power,* trans. Walter Kaufmann, Walter Kaufmann and R. J. Hollingdale, eds. (New York, 1967), pp. 48, 134.

2. Karl Marx and Friedrich Engels, *The Communist Manifesto* [1848] (New York, 1964), p. 7.

3. George Steiner, "The Great Ennui," in *Bluebeard's Castle: Some Notes Towards the Redefinition of Culture* (New Haven, 1971), p. 11.

4. Nietzsche, *Will to Power,* p. 40.

5. Theodore Zeldin, *France 1848–1945,* vol. 2, *Intelligence, Taste and Anxiety* (Oxford, 1977), pp. 831, 832.

6. See Stephen G. Brush, *The Temperature of History: Phases of Science and Culture in the Nineteenth Century* (New York, 1978), p. 61 and passim.

7. Saul Friedländer, "Themes of Decline and End," in *Visions of Apocalypse: End or Rebirth?,* ed. Saul Friedländer, Gerald Holton, et al. (New York, 1985), p. 65.

8. Friedrich Nietzsche, *The Twilight of the Idols* in *The Portable Nietzsche,* ed. and trans. Walter Kaufmann (New York, 1968), p. 547.

9. *Index Catalogue of the Library of the Surgeon General's Office, United States Army,* 2d ser. (Washington, D.C., 1900), pp. 481, 482.

10. Among the first articles on muscle fatigue is an 1853 notice concerned with "fatigue pains" resulting from gymnastics, or study, and "fatigue cramps," which occur in writers, shoemakers, and flutists. See *Schmidt's Jahrbücher* (Leipzig, 1853), p. 31.

11. Nietzsche, *Will to Power,* p. 34.

12. Zeldin, *France,* vol. 2, pp. 822–75.

13. Robert A. Nye, "Degeneration, Neurasthenia, and the Culture of Sport in *Belle Époque* France," *Journal of Contemporary History* 17, no. 1 (January 1982): 51–69; Théodule Ribot, *Les Maladies de la volonté* (Paris, 1891); *Diseases of the Will,* trans. Merwin-Marie Snell (Chicago, 1896); Albert Deschamps, *Les Maladies de l'énergie, therapéutique génerale* (Paris, 1908).

14. Théodule Ribot, *La Psychologie des sentiments* (Paris, 1896); English trans. *The Psychology of the Emotions* (London, 1897), p. 147.

15. Eugen Weber, "Introduction: Decadence on a Private Income," *Journal*

of Contemporary History (Beverly Hills) 17 (1982): 16. Brush also draws a parallel between thermodynamics and theories of degeneration in his *Temperature of History*, pp. 103–20.

16. Georges Canguilhem, *On the Normal and the Pathological*, trans. Carolyn R. Fawcett (Dordrecht, 1978), pp. 13–18.

17. Ibid.

18. Claude Digeon, *La Crise allemande de la pensée française* (Paris, 1959), p. 431.

19. See Robert A. Nye, *Crime, Madness and Politics in Modern France: The Medical Concept of National Decline* (Princeton, N.J., 1984), especially chaps. 2 and 9.

20. Philippe Tissié, *La Fatigue et l'entraînement physique* (Paris, 1897), p. 85.

21. Weber, "Introduction," p. 16.

22. Louis Querton, *L'augmentation du rendement de la machine humaine* (Brussels, 1905), pp. 144, 145.

23. Ludwig Teleky, *Vorlesungen über soziale Medizin* (Jena, 1914), p. 2. Also see Dietrich Milles, "Prävention und Technikgestaltung. Arbeitsmedizin und angewandte Arbeitswissenschaft in historischer Sicht," *'Gestalten': Eine neue gesellschaftliche Praxis*, ed. Felix Rauner (Bonn, 1988), 41–71.

24. Ernst Haeckel, *Die Welträtsel* (Stuttgart, 1899), p. 87.

25. See Dolf Sternberger, "Natural/Artificial," *Panorama of the Nineteenth Century*, trans. Joachim Neugroschl (New York, 1977), p. 36.

26. Charles Fremont, "Les mouvements de l'ouvrier dans le travail profession-nel," *Le Monde Moderne* (February 1895): p. 193.

27. Siegfried Wenzel, *The Sin of Sloth: Acedia in Medieval Thought and Literature* (Chapel Hill, 1960), p. 31.

28. Cited in Ibid., pp. 30, 46.

29. Ibid., p. 36.

30. Jacques Le Goff, "Labor Time in the 'Crisis' of the Fourteenth Century: From Medieval Time to Modern Time," *Time, Work and Culture in the Middle Ages*, trans. Arthur Goldhammer (Chicago, 1980), pp. 44, 51, 52.

31. Wenzel, *Sin of Sloth*, p. 92.

32. Roland Barthes, "Dare To Be Lazy," in *The Grain of the Voice: Interviews 1962–1980*, trans. Linda Coverdale (New York, 1985), pp. 338.

33. Max Weber, *The Protestant Ethic and the Spirit of Capitalism*, trans. Talcott Parsons (New York, 1958), p. 264n.

34. Michel Foucault, *Madness and Civilization: A History of Insanity in the Age of Reason*, trans. Richard Howard (New York, 1973), p. 56.

35. E. P. Thompson, *The Making of the English Working Class* (New York, 1963), p. 357.

36. "Paresse," *Encyclopédie ou dictionnaire raisonné des sciences, des arts et des métiers*, vol. 11, p. 939.

37. Jean-Jacques Rousseau, *Émile* [1762], trans. Barbara Foxley (New York, 1974), p. 158.

38. Barthes, "Dare To Be Lazy," p. 340.

39. Gotthold Lessing, cited in Paul Lafargue, *The Right To Be Lazy,* trans. Charles H. Kerr (Chicago, 1967), p. 5.

40. Walter Benjamin, "Müssigang," *Das Passagen-Werk,* (Frankfurt-am-Main, 1982), p. 961.

41. Wilhelm Heinrich Riehl, *Die deutsche Arbeit* (Stuttgart, 1861), p. 6.

42. "Paresse," *Encylopédie des jeunes étudiants et des gens du monde, au dictionnaire raisonné des connaissances humaines, des moeurs, et des passions,* 2d ed., vol. 2 (Paris, 1835), pp. 187, 188.

43. Thompson, *Making of the English Working Class,* p. 357.

44. E. P. Thompson, "Time, Work-Discipline, and Industrial Capitalism," *Past and Present* 38 (1967): 69.

45. Napoleon Bonaparte, "Décision" Osterode, 5 March 1807, *Correspondance de Napoléon Ier* (Paris, 1863), pp. 374–6. Napoleon wrote: "I am the authority . . . and I should be disposed to order that on Sunday after the hour of service be past, the shops be reopened and the laborers return to their work."

46. Leopold von Ranke, *Sämtliche Werke,* vol. 48 (Leipzig, 1867–90), pp. 361, 444.

47. C. L. Bergery, *Économie industrielle ou science de l'industrie,* vol. 1, *Économie de l'ouvrier* (Metz, 1833), p. 32.

48. Guillaume Ferrero, "Les formes primitives du travail," *Revue Scientifique* 5, no. 11 (7 March 1896): 331.

49. Friedrich Ratzel, *Völkerkunde,* vol. 2 (Leipzig, 1888), p. 120.

50. George Sand, *Un hiver a Majorique, Oeuvres Complètes,* vol. 51 (Paris, 1869), p. 143.

51. Francois Auguste Péron, *Voyage de découvertes aux terres australes executé sur les corvettes, le géographe, le naturaliste . . . pendant les années 1800, 1801, 1802, 1803, et 1804* (Paris, 1815, 1816), pp. 461, 462. Péron planned to write a "philosophical history of diverse peoples considered according to their physical and moral relations" but died in 1810 before he could complete it. The first naturalist to use the dynamometer invented expressly for him by Regnier was Henri Buffon, who measured the muscular force of individuals of different ages and occupations. For a description of Regnier's device, see Edme Regnier, *Mémoire explicatif du dynamomètre et autres machines inventées par lui* (Paris, 1900).

52. Riehl, *Die deutsche Arbeit,* pp. 112, 113.

53. Ibid., p. 114.

54. Hyppolyte Adolphe Taine, *Note sur l'Angleterre* (Paris, 1872), p. 82.

55. Karl Marx, "Die Preussische Revolution," *Aus dem literarischen Nachlass von Karl Marx und Friedrich Engels und Ferdinand Lassalle, Gesammelte Schriften,* vol. 3, ed. Franz Mehring (Stuttgart, 1902), p. 211.

56. "Paresse," *Encyclopédie des jeunes,* pp. 187, 188.

57. Ibid., p. 187.

58. Ibid.

59. Bergery, *Économie industrielle,* p. 37.

60. See the pertinent remarks by Gary Cross in "Worktime and Industrialization: An Introduction," *Worktime and Industrialization: An International
History,* ed. Gary Cross (Philadelphia, 1988), pp. 6, 7.

61. Bergery, *Économie industrielle;* See Michelle Perrot, "Travailler et pro-
duire: Claude-Lucien Bergery et les débuts du management en France,"
Mélanges d'histoire sociale offerts à Jean Maitron (Paris, 1876), pp. 177–90.

62. Meeting of 28 July 1868, *The General Council of the First International
1866–1868. Minutes. Documents of the First International* (Moscow, 1964),
p. 232. Marx's exact words are "I do not say that it is wrong that women
and children should participate in our social production. I think every
child above the age of nine ought to be employed at productive labour a
portion of its time, but the way in which they are made to work under
existing circumstances is abominable."

63. Hubert Treiber, Heinz Steinert, *Die Fabrikation des zuverlässigen
Menschen: Über die "Wahlverwandschaft" von Kloster-und Fabrik-
disziplin* (Munich, 1980), p. 29.

64. Edouard Foucaud, *Paris inventeur: Physiologie de l'industrie française*
(Paris, 1844), p. 222. Cited in Walter Benjamin, *Charles Baudelaire: A Lyric
Poet in the Era of High Capitalism,* trans. Harry Zohn (London, 1873),
p. 38.

65. Jeffrey Kaplow, "La fin de Saint-Lundi: Étude sur le Paris ouvrier aux XIXe
siècle," *Temps Libre* 2 (Paris, 1981), 107–18; Douglas A. Reid, "Der Kampf
gegen den 'Blauen Montag' 1766 bis 1876," in *Wahrnehmungsformen und Protestverhalten: Studien zur Lage der Unterschichten im 18.
und 19. Jahrhundert,* ed. D. Puls and E. P. Thompson (Frankfurt-am-Main,
1979), pp. 265–95.

66. Moreau de Jonnès, *Statistique de l'industrie de la France* (Paris, 1851),
p. 332.

67. Friedrich Harkort, "Offener Brief an die Arbeiter," (May, 1849), cited in
W. Schulte, *Volk und Staat. Westfalen im Vormärz und in der Revolution
1848/1849* (Münster, 1954), p. 321.

68. Louis R. Villermé, *Tableau de l'état physique et moral des ouvriers employés dans les manufactures de coton, de laine, et de soie,* 2 vols. (Paris,
1840); Gérard Leclerc, *L'observation de l'homme: Une histoire des en-
quêtes sociales* (Paris, 1979).

69. Joan Scott, "Statistical Representations of Work: The Politics of the Cham-
ber of Commerce's *Statistique de l'Industrie à Paris, 1847–48 Work in
France: Representations, Meaning, Organization, and Practice,* ed. Steven
L. Kaplan and Cynthia J. Koepp (Ithaca, 1986), pp. 342, 359, 360.

70. Cited in Pierre Foissac, *Hygiène philosophique de l'âme,* 2d ed. (Paris,
1863), p. 198.

71. Ibid., pp. 194–200.

72. Denis Poulot, *Le Sublime: Ou le travailleur comme il est en 1870, et ce qu'il
peut être,* ed. Alain Cottereau (Paris, 1980), p. 161.

73. Jules Simon, *L'ouvrière* (Paris, 1871), p. vii.

74. Alain Cottereau, "Usure au travail, destins masculins et destins féminins dans les cultures ouvrières, en France, au XIXe siècle," *Le Mouvement Social* 124 (July–September 1983): 85.

75. Charles Fourier, *Design for Utopia: Selected Writings of Charles Fourier*, ed. Charles Gide, trans. Julia Franklin (New York, 1971), pp. 166, 168.

76. See Pierre Rolle, "Proudhonism and Marxism in the Origins of the Sociology of Work," *Industrial Sociology: Work in the French Tradition*, ed. Michael Rose, trans. Alan Reybould (London, 1987), p. 102.

77. William Sewell, *Work and Revolution in France: The Language of Labor from the Old Regime to 1848* (Cambridge, Mass., 1980), pp. 236–42. The influence of the ideology of artisanal labor on social theory, as well as on how contemporary labor historians nostalgically view nineteenth-century work and workers is criticized by Jacques Rancière, "The Myth of the Artisan: Critical Reflections on a Category of Social History," in *Work in France: Representations, Meaning, Organization, and Practice*, ed. Steven L. Kaplan and Cynthia J. Koepp (Ithaca, 1986), pp. 317–34.

78. Lafargue, *The Right To Be Lazy*, p. 5.

79. Lafargue's pamphlet was republished in America by Paul Buhle and Franklin Rosement, the midwestern surrealist faction of *Radical America* in 1969.

80. Lafargue, *Right To Be Lazy* pp. 8, 9.

81. Ibid., p. 19.

82. Ibid., p. 35.

83. Georges Ribeill, "De l'oisiveté au surmenage: Les figures critiques du travail au XIXe siècle," *VRBI: Arts, histoire, ethnologie des villes* 2 (December 1979): 50. I am indebted to Mr. Ribeill for the comments he made on an earlier draft of this work and have used his suggestions liberally in this section.

84. Apollinaire Bourchardat, *Le Travail et son influence sur la santé* (Paris, 1863), pp. 5–6. Cited in Ribeill, ibid., p. 51.

85. Ibid., p. 29.

86. Jules Simon, *Le Travail* (Paris, 1870), p. 6.

87. Martin Méliton, *Le Travail humain: Son analyse, ses lois, son évolution* (Paris, 1878), p. 3.

88. Ibid., pp. 40, 42, 43.

89. George V. Poore, "On Fatigue," *The Lancet* (31 July 1875): 163, 164.

90. M. Carrieu, *De la fatigue et de son influence pathogénique* (Paris, 1878), p. 3.

91. Ibid.

92. C. Ribaud, *Le Travail, ses lois et ses fruits* (Paris, 1864), p. 32.

93. James Boswell, *The Life of Samuel Johnson* (28 July 1782), vol. 4 (Oxford, 1934), p. 153.

94. "Fatigue," *Encyclopédie ou dictionnaire raisonné des sciences, des arts et des métiers*, vol. 6, p. 429.

95. *Middle English Dictionary*, Hans Kurath, ed. (Ann Arbor, 1952), p. 422.

96. Jules and Edmund Goncourt, *Journal des Goncourt—Mémoires de la vie littéraire*, vol. 1 (Paris, 1887), pp. 219, 220.

97. Maurice Keim, *De la fatigue et du surmenage au point de vue de l'hygiène et de la médecine légale* (Lyon, 1886), p. 2.

98. Pierre Révilliod, *De la fatigue* (Lausanne, 1880), p. 13.

99. Keim, *De la fatigue,* pp. 24, 25.

100. Joris K. Huysmans, "Preface—Written 20 Years after the Novel," in *Against the Grain* (New York, 1969), p. xi.

101. Cited in Zeldin, *France,* vol. 2, p. 834.

102. "Destruction," Charles Baudelaire, *Selected Poems,* trans. Johanna Richardson (Harmondsworth, 1976), p. 189.

103. Arthur Schopenhauer, "Physiognomy," *The Essays of Arthur Schopenhauer,* trans. T. Bailey Saunders (New York, 1902), p. 70.

104. Huysmans, *Against the Grain,* p. 2.

105. Ibid., p. 7.

106. Ibid., p. 16.

107. Ibid., p. 14.

108. Dolf Sternberger, "Inside the Home," *Panorama of the 19th Century,* p. 151.

109. Adolf Göller, "Was ist die Ursache der immerwährenden Stilveränderung in der Architektur?" in *Zur Ästhetik der Architektur* (Stuttgart, 1887), p. 20.

110. Vladimir Jankélévitch, *L'aventure, l'ennui, le sérieux* (Aubier, 1963), p. 71.

111. Walter Benjamin, "Paris—the Capital of the Nineteenth Century," in *Charles Baudelaire, p. 169.*

112. Émile Tardieu, *L'ennui: Études psychologiques* (Paris, 1903), p. 1.

113. Ibid., p. 10.

114. Charles Féré, "Épuisement et criminalité," *Dégénérescence et criminalité: Essai physiologique* (Paris, 1888), p. 89.

115. Ibid., p. 91.

116. Théodule Ribot, "Le moindre effort en psychologie," *Revue Philosophique* 70 (July 1910): 376.

117. Angelo Mosso, *Fatigue,* trans. M. and B. Drummond, 1st. ed. (New York, 1904), p. 238.

118. Ribot, *Diseases of the Will,* p. 58.

CHAPTER 2. TRANSCENDENTAL MATERIALISM: THE PRIMACY OF *ARBEITSKRAFT* (LABOR POWER)

1. In an address to the British Association in 1854. Cited in P. M. Harmon, *Energy, Force, and Matter: The Conceptual Development of Nineteenth-Century Physics* (Cambridge, England, 1982), p. 58.

2. Cited in Charles Coulston Gillispie, *The Edge of Objectivity: An Essay in the History of Scientific Ideas* (Princeton, 1960), p. 357.

3. Peter T. Manicas, *A History and Philosophy of the Social Sciences* (New York, 1987), p. 83.

4. For a discussion of German philosophical and poetic visions of nature, M. H. Abrams, *Natural Supernaturalism: Tradition and Revolution in Romantic Literature* (New York, 1971), remains unsurpassed.

5. Thomas Kuhn, "Energy Conservation as an Example of Simultaneous Discovery," in *The Essential Tension; Selected Studies in Scientific Tradition and Change* (Chicago, 1977), pp. 66–104. The idea of bodies capable of conserving natural force was anticipated by Descartes and Leibniz, who argued that it was a natural property of living force, or *vis viva,* to be conserved and that "the forces are not destroyed, but dissipated amongst the small parts," as money can be divided into smaller increments. Yet these conceptions only vaguely approximated the demonstration of the unity and equivalence of all natural forces that was the achievement of the nineteenth-century doctrine of energy conservation. See Émile Meyerson, *Identity and Reality,* trans. Kate Loewenberg (New York, 1962), p. 191.

6. The sociologist Axel Honneth is right to see the cleansing of the "traditional normative contents from the concept of work" as a historical process, but he is wrong to attribute this development solely to the Taylorization of industrial work. See Axel Honneth, "Work and Instrumental Action," *New German Critique* 26 (Spring/Summer 1982): 38.

7. See the discussion in Harmon, *Energy, Force, and Matter,* pp. 65–69; and in the context of evolutionary theory, Peter Bowler, *Evolution: The History of an Idea* (Berkeley, 1989), pp. 205–8; Joe D. Burchfield, *Lord Kelvin and the Age of the Earth* (New York, 1975).

8. Brush, *The Temperature of History,* pp. 61–76; Bowler, *Evolution,* pp. 206, 207.

9. Gillispie, *Edge of Objectivity,* p. 400.

10. Jules Amar, *The Human Motor or the Scientific Foundations of Labour and Industry,* trans. Elsie P. Butterworth and George E. Wright (London, 1920), p. 206.

11. Gaston Bachelard, *The New Scientific Spirit,* trans. Arthur Goldhammer (Boston, 1984), pp. 68, 69.

12. On the political program of scientific materialism, see Mendelsohn, "Revolution and Reduction: The Sociology of Methodological and Philosophical Concerns in Nineteenth-Century Biology," in *The Interaction Between Science and Philosophy,* ed. Y. Elkana (Atlantic Highlands, N.J., 1974), pp. 407–26; Frederick Gregory, *Scientific Materialism in Nineteenth-Century Germany* (Dordrecht, Boston, 1977), especially chap. 2, and Oswei Temkin, "Materialism in French and German Physiology of the Early Nineteenth Century," *Bulletin of the History of Medicine* 20 (1946): 322–27.

13. Mendelsohn, "Revolution and Reduction," pp. 407–26.

14. Jacob Moleschott, *Der Kreislauf des Lebens,* 5th ed., vol. 2 (Giessen, 1887), p. 155.

15. Friedrich Albert Lange, *The History of Materialism and Criticism of its*

Present Importance, trans. Ernest C. Thomas, 3d English ed., vol. 2 (New York, 1950), p. 306. It is interesting that Lange uses Büchner as a foil with which to exalt the views of Helmholtz, who he believes reconciles both idealism and materialism in a neo-Kantian epistemology. For a detailed analysis see Gregory, *Scientific Materialism*, chap. 5.

16. Ludwig Büchner, *Kraft und Stoff: Empirisch-naturphilosophische Studien* (Frankfurt-am-Main, 1855), p. xi.

17. Ibid. The English translation of the tenth German edition is Louis Büchner, *Force and Matter: Empirico-Philosophical Studies Intelligibly Rendered*, trans. J. Frederick Collingwood (London, 1870), p. 2. On the belated integration of conservation of energy in Büchner see Gregory, *Scientific Materialism*, pp. 160, 161; and Lange, *History of Materialism*, vol. 2, pp. 274, 275.

18. Mendelsohn, "Revolution and Reduction," p. 421; Gregory, *Scientific Materialism*, p. 149.

19. Georges Canguilhem, "Machine et organisme," *La Conaissance de la vie* (Paris, 1965), p. 113.

20. The Paris Academy abandoned the search for a perpetuum mobile in 1775. "Histoire de l'Académie Royale des Sciences," Année 1775 (Paris, 1777), pp. 61, 65, cited in Meyerson, *Identity and Reality*, p. 213. Also see Walter Dirks, *Perpetuum Mobile: Search for Self-Motive Power During the 17th, 18th and 19th Century* (London, 1861), pp. 117–32; Georg Helm, *Die Lehre von der Energie historisch-kritisch entwickelt* (Leipzig, 1887), p. 81.

21. Georges Canguilhem, "The Role of Analogies and Models in Biological Discovery," *Scientific Change: Historical Studies in the Intellectual, Social and Technical Conditions for Scientific Discovery and Technical Invention from Antiquity to the Present. Symposium on the History of Science, University of Oxford, 9–15 July, 1961* (New York, 1963), pp. 510, 511. On the eighteenth-century automata and the application of biomechanics to the "animal economy," see A. Doyon and L. Liaigre, "Méthodologie comparée du biomécanisme et de la mécanique comparée," *Dialectica* 10, no. 4 (December 1956): 319.

22. Voltaire, *Registre contenant le Journal des Conférences de l'Académie de Lyon*, quoted in A. Doyon and L. Liaigre, *Jacques Vaucanson, mécanicien de génie* (Paris, 1966), p. 148. An English translation appears in Jean-Claude Beaune, "The Classical Age of Automata: An Impressionistic Survey from the Sixteenth to the Nineteenth Century," in *Fragments for a History of the Human Body*, ed. Michel Feher with Ramona Naddaff and Nadia Tazi (New York, 1989), p. 457.

23. Martin S. Staum, *Cabanis: Enlightenment and Medical Philosophy in the French Revolution* (Princeton, 1980), pp. 86, 89.

24. Michel Serres, *Hermes: Literature, Science, Philosophy*, ed. Josué V. Harari and David F. Bell (Baltimore, 1982), p. 71. This theme is also elaborated in Serres, *Hermes: La traduction*, vol. 3 (Paris, 1974), p. 258.

25. Meyerson, *Identity and Reality*, p. 194.

26. Cited in Leo Köenigsberger, *Hermann von Helmholtz*, trans. Frances A. Welby (New York, 1965), p. 46.
27. On Helmholtz's influence on modern thought see Manicas, *A History and Philosophy of the Social Sciences*, pp. 185–88.
28. Hermann von Helmholtz, "The Conservation of Force: A Physical Memoir [1847]," in *Selected Writings of Hermann von Helmholtz*, ed. Russel Kahl (Middletown, Conn., 1971), p. 4.
29. Ibid. For a discussion of the distinctions between the scientific materialists and Helmholtz, compare Gregory, *Scientific Materialism*, pp. 160, 161 and F. A. Lange, *History of Materialism*, vol. 2, especially p. 307.
30. In the 1847 lecture, Helmholtz prefaces his remarks with a brief discussion of the epistemological significance of Kant's *Critique of Pure Reason* (a preface that Helmholtz decided to retain after throwing "everything over-board that savoured of philosophy"). The "fundamental principle" of science, he argued, was "that every change in nature must have a sufficient cause, to which we refer natural phenomena."

These considerations, however, only reaffirmed Helmholtz's belief that all knowledge of nature presupposed the concept of *Kraft*, since only energy, or "force," could account for the effects produced in nature. Hence, in order to gain knowledge of matter, "we can do so only by ascribing force to it, that is, by adding a second abstraction, the capacity to produce effects." This was not, however, pure idealism, but neo-Kantian materialism with a psychological dimension: "Objects in nature are not, however, inert. Indeed, we come to knowledge of them only through their effects upon our sense organs." Helmholtz tried to account for our knowledge of nature by a kind of physiological or sensualist psychology. Matter and energy were both abstractions that attained the explanatory power of natural laws. The external world is not merely imprinted or "reflected" in consciousness, but mediated through the sense organs or sensations. When we perceive phenomena it is because we have sensations of them first, and interpret them secondarily. Such ideas as cause and effect are a consequence of perception, not of sensation, and there was, in Helmholtz's view, nothing metaphysical about "causes as in principle unwitnessable productive powers," or "Ursache," as he put it. On Helmholtz's psychology see Manicas, *History and Philosophy of the Social Sciences*, p. 176; Köenigsberger, *Hermann von Helmholtz*, p. 37.
31. Köenigsberger, *Hermann von Helmholtz*, p. 8.
32. Hermann von Helmholtz, "Über die Wechselwirkung de Naturkräfte und die darauf Bezüglichen neuesten Ermittelungen der Physik," *Populäre wissenschaftliche Vorträge*, 2d ed., vol. 1 (Braunschweig, 1876), p. 115.
33. Cited in Köenigsberger, *Hermann von Helmholtz*, pp. 44, 45.
34. Kuhn, *Essential Tension*, pp. 66–104. Helmholtz himself acknowledged that the essential principles of energy conservation were first articulated by the physician Julius Robert Mayer as early as 1842. See Helmholtz's 1881 appendix to "Erhaltung der Kraft," in Helmholtz, *Selected Writings*, pp.

52–54. Nor was Mayer alone in this insight. According to Kuhn, "Between 1837 and 1844, C. F. Mohr, William Grove, Faraday and Liebig all described the world of phenomena as manifesting a single 'force,' on which could appear in electrical, thermal, dynamical, and many other forms." In addition to Joule's experimental studies of heat and mechanical work, and Faraday's work on batteries, Kuhn emphasizes two other important "external" sources of the theory, "the relevance of engines" and the German tradition of romantic natural philosophy, which posited a primordial force in nature, or "Urkraft," denoting the motive power of the cosmos. Émile Meyerson and Yehuda Elkana have confirmed some aspects of Helmholtz's claim to priority, though Kuhn emphasizes the "simultaneous discovery of conservation of energy" in the 1840s. See also Meyerson, *Identity and Reality*, p. 202; Elkana, *The Discovery of the Conservation of Energy* (London, 1974).

35. Gillispie, *Edge of Objectivity*, p. 381.
36. Kuhn, *Essential Tension*, p. 87.
37. Hermann von Helmholtz, "Goethe's Anticipation of Subsequent Philosophical Ideas," in *Selected Writings*, p. 499.
38. Kuhn, *Essential Tension*, p. 88; According to Kuhn, Helmholtz cites a translation of Clapeyron's version of Sadi Carnot's memoir, which renders "puissance motrice" as "bewegende Kraft." "To this extent," Kuhn concludes, "the tie to the engineering tradition is explicit."
39. Ibid., pp. 86–90.
40. John T. Merz, *A History of European Scientific Thought in the Nineteenth Century* (London, 1904–1912), vol. 2, p. 101.
41. Cited in Koenigsberger, *Hermann von Helmholtz*, p. 142.
42. Gillispie, *Edge of Objectivity*, p. 393.
43. Helmholtz, "The Endeavor to Popularize Science [1874]," *Selected Writings*, p. 331.
44. Elkana, *Discovery of the Conservation of Energy*, p. 137.
45. Kuhn, *Essential Tension*, p. 88.
46. Hermann von Helmholtz, "Wechselwirkung der Naturkräfte," p. 101. An English translation can be found in "On the Interaction of Natural Forces," Hermann von Helmholtz, *Popular Lectures on Scientific Subjects*, trans. E. Atkinson (London, 1908), pp. 137–74. I have consulted both the original and the English translation, which is dated and somewhat inaccurate. Translations are my own. (Page numbers refer to the original and the translation, respectively, for example, 101; 137.)
47. Ibid., pp. 101; 138.
48. Ibid., pp. 102; 138.
49. Ibid.
50. Ibid., pp. 103; 139.
51. Ibid., pp. 103; 139.
52. Ibid., 125; 162.
53. Ibid., pp. 104; 140.

54. Hermann von Helmholtz, "Über die Erhaltung der Kraft," [1862–63] *Populäre wissenschaftliche Vorträge*, vol. 1, 2d ed. (Braunschweig, 1876), p. 142; "On the Conservation of Force," Hermann von Helmholtz, *Popular Lectures on Scientific Subjects*, p. 280.

55. Helmholtz, "Wechselwirkung der Naturkräfte," pp. 104; 140.

56. Helmholtz, "Über die Erhaltung der Kraft," pp. 142; 280.

57. Ibid., pp. 143; 281.

58. Ibid.

59. Ibid., pp. 176; 316.

60. Helmholtz, "Wechselwirkung der Naturkräfte," pp. 125; 163.

61. Ibid., pp. 117; 153–54. Also see P. M. Harmon, *Energy, Force, and Matter*, p. 67.

62. For an illuminating discussion, see Brush, *Temperature of History*, pp. 73–75.

63. Helmholtz, "Erhaltung der Kraft," pp. 120; 157.

64. I am indebted to Dolf Sternberger's brilliant essay, "Natural / Artificial," for this, and many other insights in this section. Sternberger, *Panorama of the Nineteenth Century*, p. 38.

65. Serres, *Hermes*, p. 36.

66. See Brush, *Temperature of History*, pp. 66–69.

67. Serres, *Hermes*, p. 58.

68. Paul Ricoeur, "A Philosophical Interpretation of Freud," *The Conflict of Interpretations: Essays in Hermeneutics*, ed. Don Ihde (Evanston, 1974), p. 167.

69. Though his idea of the "death instinct" has sometimes been equated with entropy (and parallels the "final-state" thesis), Freud's overriding emphasis on biological regression or devolution was—largely because of its vitalism—distinct from entropy. Freud especially cited the neo-Darwinist Auguste Weissmann's "germ plasm" theory to argue that one "part" of the living substance—"the body," or "the soma"—morphologically tended towards death, while the "immortal portion," or "germ plasm," tended towards life. Frank Sulloway has argued this case by pointing out that Freud's chief sources for the scientific validity of the death instinct were derived from contemporary vitalistic biology. See Frank J. Sulloway, *Freud, Biologist of the Mind* (New York, 1979), p. 406. Sigmund Freud, *Beyond the Pleasure Principle*, [1920], *Standard Edition*, vol. 18, pp. 45, 56.

70. Cited in Arno Baruzzi, *Mensch und Maschine: Das Denken sub specie machinae* (Munich, 1973), p. 63.

71. Cited in Wolf Lepenies, *Das ende der Naturgeschichte: Wandel kultureller Selbstverständlichkeiten in den Wissenschaften des 18. und 19. Jahrhunderts* (Frankfurt-am-Main, 1978), p. 83.

72. See the discussion of Lavoisier in William Coleman, *Biology in the Nineteenth Century: Problems of Form, Function, and Transformation* (New York, 1979), pp. 124–27. In his *Mémoires sur la respiration et la transpiration des animaux* (1789–90), Lavoisier introduced the concept of the "regu-

lator" and the comparison of the animal "machine with a motor." Also, the concept of physiological work was equated with the mechanical work of the machine. It was not however, until the second half of the nineteenth century that this analogy was linguistically universalized. Compare Georges Canguilhem, "Die Herausbildung des Konzeptes der biologischen Regulation im 18. und 19. Jahrhundert," *Wissenschaftsgeschichte und Epistemologie: Gesammelte Aufsätze,* ed. Wolf Lepenies (Frankfurt-am-Main, 1979), pp. 97, 98.

73. See *La Mettrie's L'Homme machine: A Study in the Origins of an Idea,* ed. Adam Vartanian (Princeton, 1960), p. 129.

74. Canguilhem, "The Role of Analogies and Models," p. 510.

75. Cited in Alfred Rupert Hall, *Scientific Revolution, 1500–1800: The Transformation of the Modern Scientific Attitude* (New York, 1954), p. 247.

76. Everett Mendelsohn, *Heat and Life: The Development of the Theory of Animal Heat* (Cambridge, 1964), p. 179.

77. A brief study of the mechanist-vitalist controversy is G. J. Goodfield, *The Growth of Scientific Physiology: Physiological Method and the Mechanist-Vitalist Controversy, Illustrated by the Problems of Respiration and Animal Heat* (London, 1960), p. 125.

78. Emil Du Bois-Reymond, "Über die Lebenskraft [March 1848]," in *Reden,* vol. 2 (Leipzig, 1886), p. 14.

79. Mendelsohn, "Revolution and Reduction," pp. 407–23; Everett Mendelsohn, "Physical Models and Physiological Concepts: Explanation in 19th Century Biology," *British Journal for the History of Science* 2 (1965): 201–13; Temkin, "Materialism in French and German Physiology," pp. 322–27.

80. Cited in Georges Canguilhem, "Les physiologistes," *Études d'histoire et de philosophie des sciences* (Paris, 1968), p. 251.

81. Emil Du Bois-Reymond, "Über die Lebenskraft," pp. 1–23.

82. Ibid., p. 16.

83. Alfred Noll, *Die "Lebenskraft" in den Schriften der Vitalisten und ihrer Gegner* (Leipzig, n.d.), p. 83.

84. Stanley Joel Reiser, *Medicine and the Reign of Technology* (Cambridge, 1978), pp. 100, 101. Also see chapter 4.

85. Emil Du Bois-Reymond, "Über thierische Bewegung," pp. 29–52.

86. M. A. Herzen, "L'activité musculaire et l'equivalence des forces," *Revue Scientifique* 24, no. 8 (19 February 1887): 237.

87. Coleman, *Biology in the Nineteenth Century,* p. 158.

88. Ibid., pp. 157, 158.

89. Richard Kremer, "From Stoffwechsel to Kraftwechsel: Voit, Rubner and the Study of Nutrition in the 1880s," History of Science Society. October, 1983.

90. Wilhelm Wundt, *Grundriss der Psychologie* (Leipzig, 1914), p. 400.

91. Théodule Ribot, *Diseases of the Will,* trans. M. Snell (Chicago, 1896), p. 134.

92. Jean Starobinski, "A Short History of Body Consciousness," in *Humanities in Review,* vol. 1 (New York, 1982), pp. 22–39.

93. Ribot, *Diseases of the Will,* p. 2.

94. G. H. Lewis, "De l'énergie spécifique des nerfs," *Revue Philosophique* 1 (1876): 161–69.

95. Charles François-Franck, "Nerveux," *Dictionnaire encyclopédique des idées médicales* (Paris, 1878), p. 572.

96. Charles Rouget, "La conservation de l'énergie," *Revue Scientifique* 9, no. 50 (12 June 1880): 1209.

97. Herzen, "L'activité musculaire et l'équivalence des forces," p. 237.

98. Helmholtz, "Über die Erhaltung der Kraft," pp. 143; 282.

CHAPTER 3. THE POLITICAL ECONOMY OF
LABOR POWER

1. Georg Helm, *Die Lehre von der Energie: Historisch-kritisch entwickelt* (Leipzig, 1887), p. 72.

2. Baron Charles Dupin, *Géométrie et mécanique des arts et métiers et les beaux-arts* (Paris, 1826), pp. 81–104, 108. Dupin refers to Coulomb's *Sur la force des hommes* (1785), but he also observed the use of the "treadmill" to calculate the working power of English prisoners. Several years later the engineer Navier used soldiers to measure the amount of labor that could be performed in a day, and J. -V. Poncelet drew up tables of measurements for simple tasks, like climbing a staircase or raising a weight. J. -V. Poncelet, *Introduction à la mécanique industrielle* (Metz, 1839), p. 230..

3. Pelligrino Rossi, *Cours d'économie politique* (Brussels, 1842), pp. 370–71, uses *puissance de travail* as capacity for labor. Sismondi also uses this phrase; compare Jean-Charles Sismondi, *Nouveaux principes d'économie politique,* vol. 1 (Paris, 1819), p. 113.

4. Baron Charles Dupin, *"Force productive des nations concurrentes, depuis 1800 jusqu'à 1851,"* in *Travaux de la Commission Française sur l'Industrie des Nations.* Exposition Universelle de 1851 (Paris, 1851).

5. On Gossen, see Charles Gide and Charles Rist, *Geschichte der volkswirtschaftlichen Lehrmeinungen* (Jena, 1921), p. 625; Leon Walras, "Un économiste inconnu, Hermann Henri Gossen," *Journal des Économistes: Revue Mensuelle de la Science Economique et de la Statistique* (1885):179.

6. On Walras, see *Études d'économie sociale: Théorie de la production de la richesse sociale* (Paris, 1936), p. 239.

7. Gide and Rist, *Geschichte der volkswirtschaftliche Lehrmeinungen,* p. 625.

8. Cited in Helm, *Die Lehre von der Energie,* p. 72.

9. Wilhelm Launhardt, *Mathematische Begründung der Volkswirtschaftslehre* (Leipzig, 1885). Cited in ibid., p. 72.

10. Friedrich Engels, "Review for the Fortnightly Review," in *Engels on Capital* (New York, 1937), p. 19. The review was written in English.

11. See, for example, G. A. Cohen, *Karl Marx's Theory of History: A Defence*

(Princeton, 1978); Sebastiano Timpanaro, *On Materialism,* trans. Lawrence Garner (London, 1975).

12. Karl Marx, *Capital: A Critique of Political Economy,* trans. Ben Fowkes (London, 1976). (All subsequent references to *Capital,* vol. 1, are to this edition.) For a lucid account of labor in the early Marx, see R. N. Berki, "On the Nature and Origins of Marx's Concept of Labor," *Political Theory* 15, no. 56 (February 1979): 35–56; also see Kostas Axelos, *Alienation, Praxis and Techné in the Thought of Karl Marx,* trans. Ronald Bruzina (Austin, 1976), pp. 53–66.

13. See Agnes Heller, *The Theory of Need in Marx* (London, 1976), p. 33.

14. Here I follow Agnes Heller's argument that Marx's work contains two distinct approaches to this question, or two models, which she calls the "paradigm of work" and the "paradigm of production." The first, characteristic of the *Paris Manuscripts* of 1844, conceives of work as "an anthropologically primary human universal," and a model for every kind of human activity. The second, characteristic of *Capital,* claims the primacy of instrumental activity or production "against other types of human action and communication." The shift from a "paradigm of work" to a "paradigm of production" in Marx's *oeuvre* occurred, Heller argues, with the abandonment of the idea of emancipation *through labor* and with the adoption of a historical philosophy of emancipation *from labor* predicated on the historically necessary expansion of the productive forces. See Agnes Heller, "Paradigm of Production: Paradigm of Work," *Dialectical Anthropology* 6 (1981): 71–79.

15. Marx, *Capital,* p. 667.

16. Ibid., p. 666.

17. Karl Marx, *Capital: A Critique of Political Economy,* vol. 3 (New York, 1967), p. 820. On this point also see Albrecht Wellmer, "The Latent Positivism of Marx's Philosophy of History," *Critical Theory of Society* (New York, 1971), p. 113.

18. Marx, *Capital,* p. 667.

19. Jean Baudrillard is typical of many post-Marxist critics when he says that Marx confounded "the liberation of productive forces" with "the liberation of man." Jean Baudrillard, *The Mirror of Production,* trans. Mark Poster (St. Louis, 1975), p. 21. However, Baudrillard pays little attention to the new rationale offered by the mature Marx for his conviction that the latter presupposed the former. Marx's most passionate defenders have done more to underscore his debt to nineteenth-century science and materialism, than his critics who have generally ignored these connections. See also Cohen, *Karl Marx's Theory of History;* Timpanaro, *On Materialism,* pp. 57, 58.

20. Marx, *Capital,* p. 548.

21. Ibid.

22. Friedrich Engels, Introduction to *Wage Labour and Capital* (New York, 1933), p. 10.

23. Marx, *Capital,* p. 129.
24. The conflict between defenders of the "physiological" and the "historical" schools of labor power has a long intellectual history. The best survey, from the standpoint of an "unorthodox" Soviet defender of the historical interpretation is Isaak Illich Rubin, *Essays on Marx's Theory of Value* (Moscow, Leningrad, 1928), trans. Milos Samardzija and Fredy Perlman (Detroit, 1972), pp. 130–38. The physiological mechanists were represented by P. Struve, a pre-Soviet Russian Marxist, who argued that "Marx accepted the mechanical-naturalistic point of view ... because labor which creates value is understood by Marx in a purely physical sense as an abstract expenditure of nervous and muscular energy." P. Struve, Foreword to the Russian edition of *Capital,* 1906, p. 28 (cited in Rubin, p. 132). The historical position was defended by Otto Gerlach, *Über die Bedingungen wirtschaftlicher Thätigkeit* (Jena, 1890).
25. This sentence stems from Struve's Foreword to the Russian edition of volume 1 of *Capital* (1906), p. 28. Cited in Rubin, *Essays on Marx's Theory of Value,* p. 132.
26. Marx, *Capital,* vol. 1, p. 323n. My translation differs slightly here. Compare Karl Marx, *Das Kapital: Kritik der politischen Ökonomie,* vol. 1, Karl Marx, Friedrich Engels, *Werke,* vol. 23 (Berlin, 1972), pp. 229, n. 27.
27. Marx, *Capital,* p. 272.
28. Ibid., p. 270.
29. Ibid., pp. 274, 276.
30. Ibid., p. 129.
31. Ibid., p. 134.
32. Ibid., p. 137.
33. For an extensive discussion of the origins of this concept see Martin Nicolaus, "Foreword" to the *Grundrisse: Foundations of the Critique of Political Economy (Rough Draft)* [1857–1858], trans. Martin Nicolaus (Harmondsworth, 1973), pp. 21, 44–52. In a little known study of the origins of this concept, Walter Tuchscheerer noted a one-time earlier appearance of "Arbeitskraft," in *Wage Labour and Capital* (1849) but attributes this usage to accident. "An einer Stelle spricht Marx zwar davon, das in Lohn ausgelegte Kapital werde 'gegen eine Arbeitskraft ausgetauscht,'" but concludes that "Es bleibt also kein anderer Schluss, als den einmaligen Gebrauch des Terminus 'Arbeitskraft' als zufällig anzusehen." See Walter Tuchscheerer, *Bevor Das Kapital Entstand: Die Herausbildung und Entwicklung der ökonomischen Theorie von Karl Marx in der Zeit von 1843 bis 1858* (Berlin [GDR], 1968), p. 314.
34. Marx, *Grundrisse,* p. 464.
35. Ibid., p. 307.
36. Ibid., pp. 282, 283. The labor objectified in use value is the physiological capacity for work, which includes the nourishment necessary to replenish the working body, "the objectified labor necessary bodily to maintain not

only the general substance in which his labor power exists, i.e., the worker himself, but also that required to modify this general substance so as to develop its particular capacity."

37. Ibid., p. 462.

38. Alfred Schmidt, *The Concept of Nature in Karl Marx*, trans. Ben Fowkes (London, 1971), pp. 86, 89. Schmidt's account suffers from a contradictory set of impulses. On the one hand Schmidt argues for a historical materialist Marx in the Feuerbachian mode, for example, he sees utopia in terms of the reintegration of man and nature; on the other, Schmidt defends Adorno and Horkheimer's *Dialectic of Enlightenment* (1947), a far more pessimistic indictment of the human domination of nature. Schmidt posits the concept of "metabolism of man and nature" as the quintessential concept of nature in Marx, young and old. Schmidt concedes, however, that there is a difference between Marx's use of this concept in 1844 and in 1863: "In the Paris Manuscripts, while under the influence of Feuerbach and Romanticism, Marx portrayed labor as a process of progressive humanization of nature, a process which coincided with the naturalization of man. . . . The later and more critical Marx of the economic analyses, took the view that the struggle of man with nature could be transformed but not abolished" (p. 76). Both of these views are evident in the *Grundrisse* (a text we should recall, not destined for publication). This accounts for the ambiguity of the concept of labor power in the *Grundrisse,* and also for the greater conceptual clarity of *Capital,* which largely abandons the older approach and self-consciously employs the vocabulary of energeticism.

39. In the 1840s Marx was familiar with the achievements of the scientific materialists: Feuerbach, Jacob Moleschott, Ludwig Büchner, Carl Vogt, and Justus von Liebig. This did not mean that he was on good terms with them. His legal struggle with Vogt is summarized in Marx's most boring work, *Herr Vogt* (which does not deal with natural science at all). For a detailed discussion of Marx's relations with the scientific materialists, see Frederick Gregory, *Scientific Materialism in Nineteenth-Century Germany* (Boston, 1977), pp. 200–204. When he was seeking a French publisher for his *Das Kapital,* Marx wrote to Büchner, "a complete stranger," to ask his assistance. In the letter Marx refers to Büchner's "schrift über 'Stoff und Kraft' (sic). Compare Letter to [Ludwig] Büchner, in Karl Marx and Friedrich Engels, *Briefe über Das Kapital* (Erlangen, 1972), pp. 134, 135. Schmidt argues that "The preparatory work for *Capital* took place in the decade between 1850 and 1860, a period in which there flourished in Germany the natural scientific materialism associated with Büchner, Vogt, and Moleschott. Marx and Engels repeatedly and severely criticized this dogmatic, and in general, crudely mechanical form of materialism. This does not, however, exclude the possibility that Marx owed certain insights to this materialism." Schmidt, *The Concept of*

Nature in Karl Marx, pp. 86, 92. In fact, Marx's earlier works reflect the "metabolic" metaphors of Moleschott, while *Capital* owes more to William Grove and the new physics.

40. Karl Marx, *A Contribution to the Critique of Political Economy,* trans. N. I. Stone (Chicago, 1904), p. 55.

41. Marx, *Grundrisse,* p. 461.

42. Ibid., p. 300.

43. Marx, *Capital,* vol. 1, p. 290. In the *Critique of Political Economy* (1859), Marx invokes the language of forces and substances: Labor time is the "vital substance of labor." The products of social life are "the result of human vital power, *materialized labor.*" For example, he conceives of social exchanges as chemical exchanges: "They [use values] are exchanged for one another in definite proportions, or form equivalents, just as chemical elements combine in certain proportions, forming chemical equivalents." See pp. 22, 23, 30. Though drafted somewhat earlier, the *Critique of Political Economy* was submitted for publication in January 1859.

44. Ibid., p. 284.

45. Marx cites the eighteenth-century political economist, Pietro Verri approvingly, "All the phenomena of the universe, whether produced by the hand of man or indeed by the universal laws of physics, are not to be conceived of as acts of creation but solely as a reordering of matter." Marx, *Capital,* p. 133n.

46. Ibid., p. 984.

47. Cohen, *Karl Marx's Theory of History,* p. 44.

48. Karl Marx, *Theories of Surplus Value,* pt. 1 (Moscow, 1956), p. 381.

49. Axelos, *Alienation, Praxis, Techné,* p. 84.

50. There is no clear-cut point where Marx registers his "turn" and comments on it. On the other hand, Marx does indicate that he relied on *both* the French engineering tradition and on the new physical sciences for this conceptual shift from the vocabulary of metabolism to the language of *Kraft.*

51. Marx, *Capital,* p. 277. This passage from the French political economist, Pelligrino Louis Edouard Rossi, *Cours d'économie politique,* année 1836–37, is part of a larger collected work, *Cours d'économie politique* (Brussels, 1843). It appears in Marx's notebooks to *Capital* (no. B33) to have been excerpted in Belgium from February and March of 1845. Marx's notebooks (A52-25/1) also contain another reference to "la puissance du travail et des autres instruments producteurs de la richesse peut être augmentée indéfiniment par l'emploi des produits de ce travail et de ces instruments, comme moyens d'une nouvelle production," cited in a French translation of Nassau W. Senior, *Principes fondamentaux de l'économie politique,* trans. Jean Arrivabene (Paris, 1836). Text and notes of Marx's notebooks, Karl Marx, *Exzerpte über Arbeitsteilung, Maschinerie und Industrie. His-*

torisch-kritische Ausgabe, transcribed and edited by Rainer Winkelmann (Frankfurt, 1982), pp. CLIX, 95, 230.

52. Marx, *Capital,* p. 664n.

53. According to Kuhn, Grove's book, which appeared in 1846, was reprinted at least six times in England, three times in America, twice in France, and once in Germany between 1850 and 1875. See Kuhn, *Essential Tension,* p. 82.

54. Marx, *Capital,* vol. 3, p. 820.

55. Ibid., p. 667.

56. Ibid., p. 820.

57. Louis Althusser was thus correct in claiming that the mature Marx abandoned "anthropology's theoretical pretensions." In the later Marx "the labor process as a material mechanism is dominated by the physical laws of nature and technology." See Louis Althusser and Étienne Balibar, *Reading Capital,* trans. Ben Brewster (New York, 1970), p. 171. Heller provides a similar argument, though from the opposing viewpoint, when she argues that Marx abandoned the "paradigm of work" as a model for nonalienated labor, for a different "paradigm of production" in *Capital.* Thus critics and defenders of Marx's scientific attitudes generally agree on the change in Marx's language, though neither have adequately explained it.

58. Marx, *Grundrisse,* p. 690.

59. Marx, *Capital,* p. 661.

60. Marx, *Grundrisse,* p. 705.

61. We need not dwell on the ways that Engels attempted to ground Marxism in scientific materialism and reduce experience to the unity of matter and motion. Commentators have already filled pages with criticisms of Engels' naturalistic reading of Marx, with the consequence that after Georg Lukács' *History and Class Consciousness* (1921), few Western Marxists appealed to natural science in their defense of Marxism. Lukács was among the first to argue—inaccurately, however—that Engels suppressed the side of Marx that constantly emphasized that social relations could not be raised to the level of objective laws, and that it was in Engels, not in Marx, that social relations were assimilated to the level of natural laws. Georg Lukács, *History and Class Consciousness: Studies in Marxist Dialectics,* trans. Rodney Livingstone (Cambridge, Mass., 1971), pp. 128–32.

62. See Frederick Engels, *Dialectics of Nature,* trans. Clemens Dutt (New York, 1940), p. 332.

63. Frederick Engels, *Herr Eugen Dühring's Revolution in Science (Anti-Dühring),* trans. Emile Burns (New York, 1934), p. 18.

64. Ibid., p. 21.

65. Ibid., p. 169.

66. Ibid., p. 18.

67. Cited in B. Hessen, "The Social and Economic Roots of Newton's 'Principia,'" *Science at the Crossroads* (London, 1931), p. 203.

CHAPTER 4. TIME AND MOTION: ETIENNE-JULES MAREY AND THE MECHANICS OF THE BODY

1. The best recent treatment of this subject is Stephen Kern, *The Culture of Space and Time 1880–1918* (Cambridge, Mass., 1983). Also insightful are: Marshall Berman, *All That Is Solid Melts into Air: The Experience of Modernity* (New York, 1982); David Gross, "Time, Space and Modern Culture," *Telos* 50 (Winter 1981–82): 59–78; Donald M. Lowe, *History of Bourgeois Perception* (Chicago, 1982); Eugene Lunn, *Marxism and Modernism: An Historical Study of Lukács, Brecht, Benjamin and Adorno* (Berkeley, 1982).

2. Edmund Husserl, *The Crisis of the European Sciences and Transcendental Phenomenology: An Introduction to Phenomenological Philosophy,* trans. David Carr (Evanston, 1970), p. 316.

3. Habermas, in contrast to Horkheimer and Adorno, reaffirms the Enlightenment's original scientific and ethical thrust. He emphasizes that classical modernism actually harbors a secret longing for tradition insofar as the "value placed on the transitory, the elusive, and the ephemeral," discloses "the longing for an undefiled, an immaculate and stable present." Jürgen Habermas, "Modernity versus Postmodernity," *New German Critique* 22 (Winter 1981): 5.

4. For three recent studies of this movement see Allan Megill, *Prophets of Extremity: Nietzsche, Heidegger, Foucault, Derrida* (Berkeley, 1985); Peter Dews, "Adorno vs. Post-Structuralism," *New Left Review* 157 (May–June 1986): 28–44. Jürgen Habermas, *The Philosophical Discourse of Modernity: Twelve Lectures,* trans. Frederick Lawrence (Cambridge, Mass., 1987).

5. A good example is the metahistorical view of science in Max Horkheimer and T. W. Adorno, *Dialectic of Enlightenment* (1947), trans. John Cumming (New York, 1972), which argues for the syncreticism of social rationalization and the Enlightenment's vision of nature dominated through the formulation of objective laws.

6. Michel Foucault's early works, especially *The Order of Things: An Archeology of the Human Sciences* (New York, 1973), attempted to rethink the ways in which the "configuration" of Western thought beginning with Descartes led inevitably to the positivist sciences of man with their accompanying conflation of self-knowledge and institutional power.

7. Anthony Giddens, "Modernism and Postmodernism," *New German Critique* 22 (Winter 1981): 16.

8. The scientific debates on the problem of space and their impact on the cubist revolution have been deftly analyzed by Linda Dalrymple Henderson, in *The Fourth Dimension and Non-Euclidean Geometry in Modern Art* (Princeton, 1983), a magnificent book that has not received the attention it deserves. Henderson provides an indispensable guide to the arcane debates on non-Euclidean geometry (the geometry of higher dimensions) in French intellectual life at the turn of the century.

9. Stephen Kern also acknowledges Marey's significance, juxtaposing his and Muybridge's contribution to the motion studies of Frank B. Gilbreth. See Kern, *Culture of Time and Space*, p. 21. This connection between Marey and Gilbreth was first made by Siegfried Gideon, *Mechanization Takes Command* (New York, 1948), p. 24. A recent study of Marey is François Dagognet, *Etienne-Jules Marey* (Paris, 1987).

10. Gideon, *Mechanization Takes Command*, p. 24.

11. Sources for Marey's childhood and personal life are scant. Information is contained in: H. A. Snellen, *E. J. Marey and Cardiology: Physiologist and Pioneer of Technology (1830–1904)* (Rotterdam, 1980), pp. 11–21; Henri Savonnet, "Flash sur la vie d'Etienne-Jules Marey," *Société d'Archéologie de Beaune* (Côte d'Or), *Mémoires, années 1973–1974*, 58 (Beaune, 1975); Charles François-Franck, *L'oeuvre de E. J. Marey* (Paris, 1905), pp. 5–21; *Hommage à M. Marey*. Institut Marey, Boulogne-Sur-Seine (Paris, 1902).

12. Snellen, *E. J. Marey and Cardiology*, p. 11.

13. Jacques Léonard, *La Médecine entre les pouvoirs et les savoirs: Histoire intellectuelle et politique de la médecine Française au XIXe siècle* (Paris, 1981), pp. 145, 146. During his internship in Paris with a champion of experimental physiology in the 1850s, Martin Magron, Marey became acquainted with the generation of young physicians who were to leave their mark on French medicine in the second half of the nineteenth century: C. Pierre Potain in cardiology; Léon Labbé in physical education; Paul Brouardel in hygiene and public health; Paul Joseph Lorain in experimental physiology. During the final year of his internship, 1858, Marey met Henri Milne-Edwards, who introduced the concept of physiological "organization," modeled on the division of labor in industry, into French physiology in 1827. Milne-Edwards became Marey's mentor, along with the famous cardiologist, Dr. Joseph Beau. On Marey's early medical career see Snellen, *E. J. Marey and Cardiology*, pp. 11–16; François-Franck, *L'ouevre*, pp. 6–11.

14. Claude Digeon, *La Crise allemande de la pensée française* (Paris, 1959) pp. 42–45.

15. Snellen, *E. J. Marey and Cardiology*, pp. 14, 15.

16. Marey's decision to abandon clinical practice and pursue a career in research, led to his establishing the first private laboratory for experimental physiology in 1864. His decision can be understood in terms of his generation's shared conviction that the laboratory was becoming the locus of medical knowledge, increasingly modeled on the physical sciences. The atmosphere of urgency in France contributed to Marey's prominence, since, as a result of his cardiological researches, he was already strongly identified with the French school of German physiology. René Quinton, "E. J. Marey," *La Revue des Idées* 1, no. 6 (1904): 485.

17. "Discours de M. Chauveau," *Inauguration du monument élevé à la mémoire de Etienne-Jules Marey au Parc des Princes, le mercredi 3 Juin 1914* (Paris, 1914), p. vii.

NOTES

18. Léonard, *La Médecine entre les pouvoirs et les savoirs*, p. 234. At the suggestion of Émile Alglave, director of the influential *Revue des Cours Scientifiques*, Duruy took an interest in Marey, observing him at work in the physics laboratory at the Collège where he was surrounded by the instruments that were his lifelong passion. On Duruy's support for Marey, see "Discours de M. Henri de Parville," *Hommage à M. Marey*, pp. 6, 22. François-Franck, *L'oeuvre*, pp. 10, 11; E. J. Marey, "The Work of the Physiological Station at Paris," *Smithsonian Institution. Annual Report, 1895* (Washington, D.C., 1896), p. 396 (Translation of "La station physiologique de Paris," *Revue Scientifique* 2 (29 December 1894): 802–8; 3 (8 January 1895): 2–12.

19. Stanley Joel Reisser, *Medicine and the Reign of Technology* (Cambridge, Mass., 1978), pp. 99–102. Marey's sphygmograph was a significant improvement on the earlier, clumsier efforts by Carl Ludwig (the kymograph) and Karl Vierordt. As Reisser notes, Marey's simple and easy to use device could make a clear tracing, "converting pulsations into visual form." Above all, "since Marey lucidly explained the phenomena his instrument uncovered, and had carefully designed and constructed his machine, it met with a better reception than its predecessors." Also see Snellen, *E. J. Marey and Cardiology*, p. 16; Edgar Holden, *The Sphygmograph and the Physiology of the Circulation* (New York, 1871) and Etienne-Jules Marey, *Recherches sur le pouls au moyen d'un nouvel appareil enregistreur, le sphygmographe* (Paris, 1860).

20. Édouard Toulouse, "Nécrologie—Marey," *Revue Scientifique* 1, no. 22 (28 May 1904): 673.

21. Quinton, "E. J. Marey," p. 484.

22. See chapter 2 and Georges Canguilhem, "The Role of Analogies and Models in Biological Discovery," *Scientific Change: Historical Studies in the Intellectual, Social and Technical Conditions for Scientific Discovery and Technical Invention From Antiquity to the Present. Symposium on the History of Science, University of Oxford, 9–15 July 1961* (New York, 1963), pp. 507–20; "Die epistemologische Funktion des 'Einzigartigen' in der Wissenschaft vom Leben," *Wissenschaftsgeschichte und Epistemologie. Gesammelte Aufsätze*, trans. Michael Bischoff und Walter Seitter (Frankfurt am Main, 1979), p. 70; A. Doyon and L. Liaigre, Méthodologie comparée du biomécanisme et de la mécanique comparée." *Dialectica* 10, no. 4 (December 1956): 304; Paul Delauney, *L'évolution philosophique et médicale du biomécanisme* (Clermont, 1927).

23. Etienne-Jules Marey, *Animal Mechanism: A Treatise on Terrestrial and Aerial Locomotion* (New York, 1874), p. 59. [English translation of *La Machine animale, locomotion terrestre et aérienne* (Paris, 1873).

24. Ibid., p. vii.

25. Etienne-Jules Marey, *Du mouvement dans les fonctions de la vie: Leçons faites au Collège de France* (Paris, 1868), p. 69.

26. Ibid., pp. 69, 70.

27. Claude Bernard, *Rapport sur les progrès et la marche de la physiologie générale en France* (Paris, 1867), p. 230; Marey indirectly takes up these arguments in *Du mouvement dans les fonctions de la vie*, p. 22. Also see Canguilhem, "The Role of Analogies and Models," and "Theorie und Technik des Experimentierens bei Claude Bernard," in Canguilhem, *Wissenschaftsgeschichte und Epistemologie*, pp. 75–88.

28. Canguilhem, *Wissenschaftsgeschichte und Epistemologie*, p. 77. Also see Joseph Schiller, *Claude Bernard et les problèmes scientifiques de son temps* (Paris, 1967), p. 215.

29. Marey, *Du mouvement dans les fonctions de la vie*, p. 22.

30. Ibid., p. vii.

31. Ibid., p. 203. This remarkable discussion in many ways anticipates Michel Foucault's schema of the development from taxonomy to morphology; from classification to organic structure, and from space to time in the human sciences. Foucault's analysis, which focuses on the epistemological constitution of the social and natural sciences, closes with the revolution in epistemology at the end of the eighteenth century brought about by the location of truth in organic structure, as opposed to spatial order. Georges Canguilhem, on whose work Foucault's argument rests, and Michel Serres also deserve credit for documenting the subsequent shift to "motion" and thermodynamics. See Michel Foucault, *The Order of Things*, pp. 264–79.

32. Marey, *Du mouvement dans les fonctions de la vie*, p. 55.

33. Ibid., p. 65.

34. Marey, *Animal Mechanism*, p. 9.

35. Ibid., p. 8.

36. Ibid., p. 15.

37. Ibid., p. 69.

38. Ibid. Marey, who was skeptical of Darwin, was a cautious defender of *transformisme*, the Lamarckian doctrine of the growing complexity of physiological systems in their aptitude to respond to the requirements of life. See Etienne-Jules Marey, "Le transformisme et la physiologie expérimentale," *Revue des Cours Scientifiques* 4 (1 March 1873): 813–22.

39. Marey, *Animal Mechanism*, p. 8.

40. Cited in Kern, *The Culture of Time and Space*, p. 33.

41. Marey, *Animal Mechanism*, pp. 44, 45

42. Etienne-Jules Marey, *La Méthode graphique dans les sciences expérimentales et principalement en physiologie et en médecine* (Paris, 1878). All citations from Etienne-Jules Marey, *La Méthode graphique dans les sciences expérimentales*, 2d ed. (Paris, 1885), p. 153.

43. Reisser, *Medicine and the Reign of Technology*, pp. 100–103; L. Campan, "Marey et la capture des temps physiologiques," *Agressologie* 19, no. 4 (1978): 233–38.

44. Marey, *La Méthode graphique*, p. xi.

45. Campan, "Marey et la capture des temps physiologiques," p. 234.

46. Etienne-Jules Marey, *Physiologie médicale de la circulation du sang* (Paris, 1865), p. 5.

47. Marey, *Du mouvement dans les fonctions de la vie,* p. 93.

48. Marey, *La Méthode graphique,* p. iv.

49. Ibid., p. 108.

50. Ibid., p. ii.

51. Ibid., p. iii.

52. Walter Benjamin, "On Language as Such, and on the Language of Man," *Reflections: Essays, Aphorisms, Autobiographical Writings,* ed. Peter Demetz, trans. Edmund Jephcott (New York, 1978), p. 330.

53. Marey, *Du mouvement dans les fonctions de la vie,* p. 93.

54. Ibid., pp. 83–85.

55. Marey, "The Work of the Physiological Station at Paris," p. 393. As opposed to Bernard, Marey abhorred vivisection, one of the most sharply and passionately contested issues in French medicine in the 1860s and 1870s. Vilified by the antivivisectionists, Bernard sat on a commission empowered by Napoleon III to decide the question. When a statue for Bernard was proposed, the antivivisectionists suggested that he be represented by "a decapitated dog on the operating table." Compare Schiller, *Claude Bernard,* p. 36; Hubert Bretschneider, *Der Streit um die Vivisektion im 19. Jahrhundert* (Stuttgart, 1962). Marey noted, "From Galen to Claude Bernard the scalpel brought about discoveries, but it brought to the functions of life only very preliminary sorts of notions, which in our day can be completed by the use of precision instruments." Etienne-Jules Marey, "La méthode graphique et les sciences expérimentales," *Revue Scientifique* 8, no. 6 (7 August 1897): 162.

56. Marey, *Du mouvement dans les fonctions de la vie,* p. 33.

57. Marey, *La Méthode graphique,* p. 23. On Quetelet see J. Lottin, *Quetelet, statisticien et sociologue* (Louvain and Paris, 1912); Maurice Halbwachs, *La Théorie de l'homme moyen: Essai sur le Quételet et la statistique morale* (Paris, 1912).

58. Marey's inscriptors were not merely an improvement over earlier models. By adapting a set of independent "lever drums" (tambours à levier) connected by air tubes to the tracing mechanisms, an almost perfect recording of the most minute alterations in movement could be achieved. Marey, *La Méthode graphique,* p. 125.

59. Marey, *La Méthode graphique,* pp. 113–16.

60. Reisser, *Medicine and the Role of Technology,* p. 100; Marey, *La Méthode graphique,* p. 110.

61. Reisser, p. 100; Marey, *Du mouvement dans les fonctions de la vie,* p. 135; Marey, *La Méthode graphique,* p. xiii.

62. Marey, *Du mouvement dans les fonctions de la vie,* p. 144. Marey's first efforts with water tubes were unsuccessful. The first inscription by air tubes was achieved by one of Marey's colleagues, Dr. Charles Buisson, in 1858. See Charles Buisson, "Premiers appareils pour l'inscription des

mouvements à distance au moyen de la transmission par l'air," cited in François-Franck, *L'oeuvre*, pp. 43, 44.

63. These are each described in detail in Marey, *La Méthode graphique*, chap. 4.

64. Marey, *La Méthode graphique*, pp. 193–202; Marey, *Du mouvement dans les fonctions de la vie*, p. 165; Etienne-Jules Marey, "De la production du mouvement chez les animaux," *Revue des Cours Scientifiques* 4, no. 14 (2 March 1867): 209–17; Marey, *Animal Mechanism*, p. 33.

65. *Hommage à M. Marey*, p. 14; H. Milne-Edwards, "Rapport de la Commission de l'Académie des Sciences chargée de décerner le prix Lacaze (physiologie). *Revue Scientifique* (16 January 1875): 682.

66. François-Franck, *L'oeuvre*, p. 39; H. Milne-Edwards, "Rapport de la Commission," p. 684.

67. Etienne-Jules Marey, "Travail de l'homme dans les professions manuelles," *Revue de la Société Scientifique d'Hygiène Alimentaire* 1 (1904): 194.

68. Marey, *Du mouvement dans les fonctions de la vie*, p. 103.

69. In the action of Marey's tracing mechanism there is a kind of repetition compulsion, ironically not unlike the charge of repetitiveness directed against him by less sympathetic colleagues. In a sense, all of his inventions did the same thing, inscribing all aspects of physical motion, human and animal. Marey was not bothered by such assaults (especially in a pre-Freudian age), remarking: "They say, when talking about my work, that it is always the same thing. Such appreciation, . . . would be, if I deserved it, the greatest recompense for my efforts. It demonstrates that I have achieved my goal." Marey, *La Méthode graphique*, p. xix.

70. Etienne-Jules Marey, "Étude de la locomotion animale par la chronophotographie," *Association Française pour l'Avancement des Sciences, Compte-Rendu du Congrès de Nancy 1886* (Nancy, 1887), p. 53.

71. Ibid.

72. Marey, "The Work of the Physiological Station at Paris," p. 395; Etienne-Jules Marey, *Le Vol des oiseaux* (Paris, 1890), p. viii.

73. Marey's debate with early nineteenth-century theorists of flight, for example, Chabrier, is contained in his "Mécanisme du vol chez les insectes. Comment se fait la propulsion," *Revue des Cours Scientifiques* 6 (20 March 1869): 252–56.

74. Marey, *Animal Mechanism*, p. 273.

75. Etienne-Jules Marey, "Les mouvements de l'aile chez les insectes," *Revue des Cours Scientifiques* 6 (13 February 1869): 174; Etienne-Jules Marey, *Movement*, trans. Eric Pritchard (New York, 1895), p. 243.

76. Marey, "Mécanisme du vol chez les insectes," p. 254; Marey, *Movement*, p. 246.

77. François-Franck, *L'oeuvre*, p. 11; Marey, *Animal Mechanism*, p. 229. 228; Marey, *Le Vol des oiseaux*.

78. Marey, *Animal Mechanism*, p. 241.

79. Ibid., p. 242.

80. Marey, "Mécanisme du vol chez les insectes," p. 254.

81. "Aéroplane à l'air comprimé," *Travaux du laboratoire de M. Marey* (Paris, 1878/79), p. 227; Victor Tatin, "Expériences sur le vol mécanique," *Travaux du laboratoire de M. Marey* (Paris, 1876), pp. 87–108; *E. J. Marey. 1830/1904: La Photographie du mouvement,* Centre National d'Art et de Culture Georges Pompidou, Musée National d'art Paris. Exhibition Catalogue (Paris, 1979), p. 99; Marey, "The Work of the Physiological Station at Paris," p. 396; *Le Vol des oiseaux,* p. viii.

82. Marey, *Le Vol des oiseaux,* p. viii; Louis Mouillard, *L'empire de l'air: Essai d'ornithologie appliquée à l'aviation* (Paris, 1881).

83. Etienne-Jules Marey, "Des allures de cheval, étudiées par la méthode graphique," *Compte-rendus des séances de l'Académie des Sciences* [hereafter cited as CRAS] 75 (Paris, 1872): 884.

84. Ibid. "The animal was ridden by M. Pellier, who held in one hand the inscriptive apparatus, and in the other, with the reigns, the india rubber ball connected to the tracing instrument." There appear to have been two types of inscriptors, one attached to the horse's ankle, and another affixed to the horseshoe. In the latter, when the foot strikes the ground, the India rubber ball is compressed, and drives air into the tubes, which carry the signal to registering instruments. A detailed description is in Marey, *Animal Mechanism,* pp. 138–77. See also Marey, *Movement,* pp. 8–12.

85. Equestrian experts had actually reached similar conclusions in the mid-1870s, but no conclusive proof existed until Marey's famous experiments. See Émile Duhousset, *Le Cheval: Études sur les allures, l'extérieur et les proportions du cheval* (Paris, 1874).

86. Marey, *Animal Mechanism,* pp. 125, 129.

87. Etienne-Jules Marey, "Moteurs animés: Expériences de physiologie graphique," *La Nature* 278 (28 September 1878): 273–78; 279 (5 October 1878): 289–95.

88. Etienne-Jules Marey, "Études sur la marche de l'homme au moyen de l'odographe," CRAS 99 (3 November 1884): 732–33.

89. Marey, "Moteurs animés," p. 273.

90. Ibid., p. 294.

91. Several years earlier Émile Duhousset, an equestrian expert and lieutenant colonel in the French cavalry, corrected some of the greatest painters—both ancient and modern—by realistically drawing the horse according to Marey's graphic notations. Émile Duhousset, *The Gaits, Exterior and Proportions of The Horse* (London, 1896), p. xi. In a letter to Duhousset, written after the first edition of the latter's *Le Cheval* appeared, (see note 85), Marey emphasized that science and art could collaborate through expert knowledge. Françoise Forster-Hahn, in her superb catalogue essay, notes that "Duhousset based his analytical drawings on Marey's chronographic notations." In fact, Duhousset was unaware of Marey's work when he completed the first edition of *Le Cheval,* and in the subsequent English edition explained that he had revised his publication based on Marey's

research. See Duhousset, *The Gaits, Exterior and Proportions of The Horse*, p. xi. In 1874 Duhousset published the first article in France on Muybridge's motion photographs of the horse, "Reproduction instantanée des allures du cheval, au moyen de l'électricité appliquée à la photographie," *L'Illustration* (25 January 1874): 58, 59. He also contributed the drawings for Marey's article "Moteurs animés" in *La Nature*, as well as for *Animal Mechanism*. See Françoise Forster-Hahn, "Marey, Muybridge and Meissonier: The Study of Movement in Science and Art," *Eadweard Muybridge. The Stanford Years 1872–1882. Stanford Museum of Art Catalogue* (Palo Alto, 1972), pp. 87, 107.

92. Forster-Hahn, "Marey, Muybridge and Meissonier," pp. 92, 117, 132, especially fn., 10.

93. See Ibid., pp. 92, 93, 97. Muybridge's *The Attitudes of Animals in Motion: A Series of Photographs Illustrating the Consecutive Positions Assumed by Animals in Performing Various Movements* was published in 1881.

94. Letter to *La Nature* 29 (28 December 1878): 54, English translation in Forster-Hahn, "Marey, Muybridge and Meissonier," p. 116.

95. Ibid.

96. Ibid., p. 117. Muybridge's letter, dated 17 February 1879 appeared in *La Nature* (22 March 1879).

97. Etienne-Jules Marey, "Le fusil photographique," *La Nature* (22 April 1882), p. 327. Several years later, in his *Le Mouvement* (1894), Marey juxtaposed Duhousset's drawings with Muybridge's photographs in order to show how the "comparison of the two is not to the advantage of the first." Marey, *Movement,* p. 196. In a review of his photographic achievements, written in 1899, however, Marey simply stated that Muybridge's photographs "confirmait ma chronographie." Etienne-Jules Marey, *La Chronophotographie* (Paris, 1899), p. 8. In January 1879, only a few weeks after Muybridge's photographs first appeared in France, Duhousset published a full account of them in the popular magazine, *L'illustration,* including eleven sketches from the photographs. He also indicated that Muybridge had provided his French agent with fifty-four poses. See Duhousset, "Reproduction instantanée des allures du cheval," p. 58.

98. *La Globe* (27 September 1881), cited in Forster-Hahn, "Marey, Muybridge and Meissonier," pp. 130, 131.

99. Ibid. Gordon Hendricks, *Eadweard Muybridge: The Father of The Motion Picture* (New York, 1975), p. 135.

100. Marey, *La Chronophotographie,* p. 23. Marey incorrectly dates the event one year later. In an article published in April 1882, Marey gave a more detailed summary of the history of synthetic motion, noting that "Parmi les auteurs qui ont réalisé des zootropes avec les photographies instantées, on doit citer M. Muybridge lui-même; en France, M. Mathias Ducal, professeur d'anatomie à l'École des Beaux-Arts, et le colonel Duhousset; en Hongrie, M. Ziekly, également professeur à l'École des Beaux-Arts; enfin, en Angleterre plusieurs industriels vendaient, l'an dernier, des zootropes

formés avec les figures que M. Muybridge a publiées." Marey, "Le fusil photographique," *La Nature* (22 April 1882), p. 327. It would be interesting to pursue the last remark to see whether or not the zoetrope with Muybridge's images also had reached a popular audience on the Continent. The origins of devices using revolving images to achieve the illusion of continual motion are, of course, much older. In 1830 Joseph Plateau, a Belgian physicist, constructed an instrument called the "phenakistoscope" (eye-deceiver). The phenakistoscope was a popular toy that created animation by means of a series of slightly differing pictures on a revolving disc. Plateau is also credited with having discovered the principle of optics which explains this effect, the confusion of images on the retina, which later became known as the "Talbot-Plateau law" or "Talbot's law of fusion." For an extensive discussion of this problem in the prehistory of cinema see Stephen Heath, "The Cinematic Apparatus: Technology as Historical and Cultural Form," *The Cinematic Apparatus,* Teresa de Laurentis and Stephen Heath, eds. (London, 1980), pp. 78, 79.

101. Cited in Hendricks, *Muybridge,* p. 137.

102. Forster-Hahn, "Marey, Muybridge and Meissonier," pp. 95, 96.

103. Marey, *La Chronophotographie,* p. 217. According to Marey's assistant, Georges Demeny, Meissonier's initial reaction to the photographs was a "cry of astonishment" and an accusation: "notre appareil de voir faux." Pointing to one of his sketches, he exclaimed: "Quand vous me donnerez un cheval galopant comme celui-ci . . . je serai satisfait de votre invention." Cited in Georges Demeny, *Les Origines de la cinématographie* (Paris, 1909), p. 15.

104. Marey, Muybridge and Meissonier conceived of an elaborate plan to collaborate on an exhaustive book "on the attitudes of animals in motion as illustrated by both ancient and modern artists," but the project never materialized. Muybridge, Letter of December 23, 1881, cited in Hendricks, *Eadweard Muybridge,* p. 139.

105. Etienne-Jules Marey, *Développement de la méthode graphique par l'emploi de la photographie* (Paris, 1884), p. 9. Marey also introduced Muybridge to the simpler rapid gelatin dry-plate photography, as opposed to more cumbersome and slower wet collodian plates.

106. Marey, "Le fusil photographique," p. 327; Etienne-Jules Marey, "Sur la reproduction, par la photographie, des diverses phases du vol des oiseaux," CRAS 94 (1882): 683–85; Etienne-Jules Marey, "Emploi de la photographie instantanée pour l'analyse des mouvements chez les animaux," CRAS 94 (1882):1013–20.

107. Commenting on Muybridge two decades later, Marey only reluctantly admitted him to the pantheon of chronophotography, and with the following qualification: "We place the experiments of Muybridge along with those of chronophotography, although this ingenious experimenter did not succeed in taking his instantaneous photographs at equal intervals of time. For the velocity of the horse not being quite uniform, the equidistant wires

were not reached at equal intervals of time. Besides, the wire was more or less stretched before rupture took place. From these causes there was a certain inequality in the rates of succession which Muybridge did not succeed in satisfactorily overcoming by letting off the shutters independently of the horse's motion." Marey, "The History of Chronophotography," *Smithsonian Institute, Annual Report 1901* (Washington, D.C., 1902) p. 319 n.

108. See Eadweard Muybridge, *Animal Locomotion: An Electro-Photographic Investigation of Consecutive Phases of Animal Movements.* Prospectus. (Philadelphia, 1887), p. 16.

109. Etienne-Jules Marey, "La Station Physiologique de Paris," *La Nature* (8 September 1883): 226.

110. Marey, *Animal Mechanism*, p. 3.

111. Marey, "The Work of the Physiological Station at Paris," p. 397.

112. Marey, "La Station Physiologique de Paris," p. 227.

113. Marey, "The Work of the Physiological Station at Paris," p. 317.

114. Marey, "La Station Physiologique de Paris," p. 227.

115. Letter of Marey to Demeny, 6 August 1881, in Demeny, *Les Origines de la cinématographie*, p. 36. Marey included among his "friends" on the municipal council, Hérédia and Hector Depasse.

116. Letter from Marey to Demeny, Naples, 27 December 1881, in ibid., p. 39. In 1882, the Minister of Education, Jules Ferry, argued for additional support before the Chamber of Deputies. Compare Marey, "La Station Physiologique de Paris," p. 227. On Ferry and science, see Louis Legrand, *L'influence du positivisme dans l'oeuvre scolaire de Jules Ferry: Les origines de la laïcité* (Paris, 1961), p. 100.

117. Demeny, *Les Origines de la cinématographie*, p. 32. Much of the construction of the new Physiological Station took place under Demeny's supervision, especially during Marey's habitually long absences from Paris (he spent a good part of the year at his villa in Naples).

118. On Janssen see Friedrich von Zglinicki, *Der Weg des Films: Geschichte der Kinematographie und ihrer Vorläufer* (Berlin, 1956), p. 169. Also see *Bulletin de la Société Française de Photographie* (1874): 15, 97.

119. Janssen's astronomical revolver produced a series of images at precise intervals of seventy seconds on a circular light-sensitive glass plate by means of a rotating disc with seventeen windows. With each displacement of a few degrees, the plate received an image at a different point on its surface. Pierre Jules Janssen, "Présentation d'un spécimen de photographies d'un passage artificiel de Vénus, obtenu avec le revolver photographique," CRAS 79 (1874), pp. 7, 8; Marey, *Méthode graphique par l'emploi de la photographie*, pp. 6, 7; Marey, *Movement*, pp. 103, 104.

120. Pierre Jules Janssen, "Mémoire," *Bulletin de la Société Française de Photographie* (14 December 1876), cited in Marey, *Méthode graphique par l'emploi de la photographie*, p. 8.

121. Cited in Marey, *Movement*, p. 105.

122. Etienne-Jules Marey, "Sur la reproduction par la photographie des diverses phases du vol des oiseaux," p. 684.

123. Ibid.

124. Marey claimed that in good sunlight he could reduce the exposure time to 1/720 of a second. Ibid. See Marey, "Le fusil photographique," pp. 326–30. Motion photography was predicated on the reduction of exposure time necessary to produce a photographic image. As late as 1866 Professor Herman Wilhelm Vogel, one of the great German photographic pioneers, responded to the question, what is the photographic "moment," with "three seconds." This duration was inadequate for instantaneous photography. By improving the chemical process and the shutter speed, photographers were able to gradually whittle down the exposure time from the thirty minutes required for a Daguerrotype in 1839 to the 1/200 of a second that Muybridge achieved in 1878. According to Marey, Muybridge's photographs crossed the 1/500th threshold by 1882. See Zglinicki, *Der Weg des Films*, p. 165. On the important, but relatively unsung history of exposure time see Josef Maria Eder, *Geschichte der Photographie* (Halle, 1932), p. 611.

125. *E. J. Marey, 1830/1904. La Photographie du mouvement*, p. 28; Etienne-Jules Marey, "Le vol des oiseaux," *La Nature* (16 June 1883), p. 35.

126. Marey, "The Work of the Physiological Station at Paris," p. 395.

127. Marey, "La Station Physiologique de Paris," p. 228.

128. Marey, *La Chronophotographie*, p. 11.

129. Marey, "La Station Physiologique de Paris," p. 229.

130. Marey, *La Chronophotographie*, p. 11. The theme of absolute blackness *(le noir absolu)* was discussed in French scientific circles at the time. See for example, E. Chevreul, *De la loi du contraste simultané des couleurs* (Paris, 1886).

131. Etienne-Jules Marey, "La photographie du mouvement," *La Nature* (22 July 1882), p. 116; Marey, *Méthode graphique par l'emploi de la photographie*, p. 23.

132. Marey, *Méthode graphique par l'emploi de la photographie*, p. 23.

133. Marey, "La photographie du mouvement," p. 116. Marey also conducted experiments with pigeons, a white horse, and an elephant. See François-Franck, *L'oeuvre*, p. 50.

134. Marey, "The History of Chronophotography," p. 317.

135. Marey, *Méthode graphique par l'emploi de la photographie*, pp. 31, 32.

136. Ibid., p. 35.

137. Ibid., p. 33. Speeding up the rotation of the disc only exacerbated the problem. In addition to Marey, several other contemporary photographers confronted this problem, and some found a solution in multiple objectives. One interesting experiment was designed by Albert Londe, chief photographer at the mental institution at Salpétrière. He aligned twelve objectives in rows of four on a single plate. Marey also attempted a similar solution with six lenses (1883–84). These odd devices generally did not

perform well, leaving images that were blurred and indistinct. See *E. J. Marey. 1830/1904. La Photographie du mouvement*, pp. 63, 81; Marey, "The History of Chronophotography," p. 323.

138. Marey, *Méthode graphique par l'emploi de la photographie*, p. 36.

139. Marey, "The History of Chronophotography," p. 323.

140. Ibid., p. 321; *E. J. Marey. 1830/1904. La Photographie du mouvement*, pp. 39, 40.

141. Ibid., p. 323. The original film consisted of long unperforated sheets of gelatino-bromide of silver paper, manufactured by Eastman, and only usable in a darkroom. The film itself was hard to obtain and uneven in sensitivity, producing "little miseries" and "lots of surprises." Demeny, *Les Origines de la cinématographie*, p. 16. The story of Marey's role in the history of the development of "synthetic motion" (or the motion picture camera/projector) is full of drama, tragedy, and betrayal, deserving of a more detailed treatment. Briefly, Marey and Demeny each developed several motion picture cameras and projectors in the early 1890s, but the devices were ultimately denied a patent according to a law prohibiting the patenting of inventions published in scientific journals. In 1892 Demeny did, however, obtain a patent for an apparatus which synthesized eighteen chronophotographic images on a rotating disc, the phonoscope. A commercial failure, Demeny eventually sold his patent to the far more successful Lumiére brothers in 1895 after a long and painful public schism with Marey. Marey, "The History of Chronophotography," p. 321; *E. J. Marey. 1830/1904. La Photographie du mouvement*, pp. 39, 40.

142. Marey, *La Méthode graphique*, p. xi.

143. Kern, *The Culture of Time and Space*, especially chaps. 1 and 6; Henderson, *The Fourth Dimension and Non-Euclidean Geometry in Modern Art*.

144. Georges Poulet, *Studies in Human Time* (Baltimore, 1956).

145. Henderson, *Fourth Dimension*, p. 13.

146. Ibid., pp. 307, 359–60.

147. Ibid., p. 17.

148. Marey, *Movement*, p. 30.

149. Henderson, *Fourth Dimension*, p. 17.

150. J.-M. Guyau, "L'évolution de l'idée de temps dans la conscience," *Revue Philosophique*, 19 (1885):353. Guyau's article, posthumously published in 1890 as *La Genèse de l'idée de temps*, and largely reedited by Alfred Fouillée, his stepfather, appeared at least on the surface to coincide with Bergson's ideas. Moreover, in his *La Psychologie des idées forces* (1893), and in some published letters to Guyau's son, Fouillée claimed that this influence had not been acknowledged. Compare vol. 2, p. 209. Bergson, however, published a highly critical review of Guyau's *La Genèse de l'idée de temps* in the *Revue Philosophique* 31 (1891): 185–90. Bergson also claimed that Fouillée had altered Guyau's work for publication, adding for example, the phrase "time as the fourth dimension of space," which does not appear in the original. This debate is summarized in Ben-Ami Scharf-

stein, *Roots of Bergson's Philosophy* (New York, 1943), pp. 37–43. For a detailed comparison of Bergson and Guyau see Vladimir Jankélévitch, "Deux philosophes de la vie, Bergson, Guyau," *Revue Philosophique* 97 (1924): 402–49.

151. Scharfstein, *Roots of Bergson's Philosophy*, chap. 2; Jankélévitch, "Deux philosophes de la vie"; Henderson, *The Fourth Dimension*, pp. 90–93.

152. Henri Bergson, *Essai sur les données immédiates de la conscience* (Paris, 1889); citations from the English translation, *Time and Free Will: An Essay on the Immediate Data of Consciousness*, trans. F. L. Pogson (New York, 1910), pp. 109, 110.

153. Compare Kern, *Culture of Space and Time*, pp. 22, 117. Henderson mentions Marey as an influence on the art of Frantisêk Kupka and Marcel Duchamp. Compare Henderson, *The Fourth Dimension*, pp. 105, 126, 127.

154. Henri Bergson, *Matière et mémoire* (Paris [1896] 1908); *Matter and Memory*, trans. Nancy M. Paul and W. Scott Palmer (New York, 1911), p. 277.

155. Henri Bergson, *L'évolution créatrice* (Paris, 1921); English translation *Creative Evolution*, trans. Arthur Mitchell (New York, 1911), pp. 272, 295.

156. Henri Bergson, *Mélanges*, ed. André Robinet (Paris, 1972), p. 510.

157. Bergson, *Creative Evolution*, pp. 332, 333.

158. Marey, "Moteurs animés," p. 295. In May 1904, according to François-Franck, an obituary for Marey appeared in *L'indépendance Belge* entitled "Marey et Phidias." The necrologist retold an anecdote in which a French cavalry officer, presumably Lt. Col. Duhousset, demonstrated in the Palais des Académies, how centuries before, the Greek Phidias had realized in the Parthenon frieze the principles later confirmed by Marey's photographs. François-Franck, *L'oeuvre*, p. 39. In subsequent editions of his book on the horse, Duhousset included a chapter on the "Bas-Reliefs of the Parthenon," praising Phidias for the "style and truth" of his horses. See Émile Duhousset, "The Bas-Reliefs of the Parthenon," in *The Horse*, pp. 112–17. Duhousset also refers to the work of an English hippologist, William Youatt, whose *The Horse* was critical of Phidias's work.

159. Bergson, *Creative Evolution*, p. 332.

160. Bergson, *Mélanges*, p. 573.

161. Ibid., pp. 573, 574.

162. Henri Bergson, *La Pensée et le mouvant* (Paris, 1934), p. 9, cited in Scharfstein, *Roots of Bergson's Philosophy*, p. 14.

163. Bergson, *Creative Evolution*, p. 306.

164. Bergson, *Matter and Memory*, p. 277.

165. Paul Valéry, *Variétés*, (Paris, 1944), p. 91.

166. Compare Georges Lechalas, *Étude sur l'espace et le temps* (Paris, 1896).

167. Bergson, *Creative Evolution*, p. 337.

168. Walter Benjamin, "Theses on the Philosophy of History," in *Illuminations*, ed. Hannah Arendt, trans. Harry Zohn (New York, 1968), p. 255.

169. Henri Poincaré, "Discours," *Inauguration du monument élevé à la mémoire de Etienne-Jules Marey au Parc des Princes, à Boulogne-Sur-Seine, le mercredi 3 juin 1914* (Paris, 1914), p. xvii.

170. Forster-Hahn, "Marey, Muybridge and Meissonier," p. 101.

171. This is a source of some controversy. See Aaron Scharf, "Marey and Chronophotography," *Artforum* (September 1976), pp. 62–70; and "Painting, photography and the image of movement," *The Burlington Magazine* (May 1962), p. 195.

172. Henry reviewed Marey's bicycle research in the *Revue Blanche* 7 (1894): 554–61. Also see José A. Arguelles, *Charles Henry and the Formation of a Psycho-Physical Aesthetic* (Chicago, 1972), pp. 106, 107.

173. Ibid., pp. 131–42.

174. Paul Valéry, *Degas, Manet, Morisot*, trans. D. Paul (New York, 1960), p. 41. Cited in Forster-Hahn, "Marey, Muybridge and Meissonier," p. 105.

175. Pierre Cabanne, *Entretiens avec Marcel Duchamp* (Paris, 1967), p. 57. Also see Henderson, *Fourth Dimension*, p. 127n; *E. J. Marey 1830/1904: La Photographie du mouvement*, pp. 103, 104.

176. For an extended discussion of Marey and Futurism, see J. Brun, "Le voyage dans le temps de la chronophotographie au futurisme," *Archivi di Filosofia* 2/3 (1975), pp. 355–64. Unfortunately this article is marred by numerous errors.

177. Cited in ibid., p. 360. Also see "Umberto Boccioni, peintre et sculpture futuriste 1914," in G. Lista, *Futuristie* (Lausanne, 1973), p. 196.

178. Gideon credits the American disciple of Taylor, Frank B. Gilbreth, who published his "cyclographs" in 1912 with this achievement. In fact, the first chronophotographs of work were completed in 1894. Gilbreth misleadingly claimed that Marey was not interested in work, or did not apply motion photography to it: "As for the motion study, Marey, with no thought of motion study in our present use of the term in his mind, developed as one of his multitudinous activities, a method of recording paths of motions, but never succeeded in his effort to record directions of motions photographically." Though technically Gilbreth did provide light photographs of the paths of motion, he ignored Marey's studies of work. Compare Frank B. Gilbreth, "Motion Study and Time Study: Instruments of Precision," in *The Writings of the Gilbreths*, ed. William R. Spriegel and Clark E. Myers (Homewood, Ill., 1953), p. 229.

179. Marey, "Étude de la locomotion animale par la chronophotographie," p. 66.

180. In the *La Nature* articles of 1878, Marey sketched how the portable odograph might be used to study "the forces of traction and animal energy." Two years later, Marey presented a study to the Académie des Sciences demonstrating how military recruits "of different sizes and energies, carrying loads of greater or lesser weight" might benefit from scientifically designed training methods. See Etienne-Jules Marey, "Études sur la

marche de l'homme," *Revue Militaire de Médecine et de Chirurgie* 1 (1881): 244–46; "Analyse cinématique de la marche," CRAS 98 (1884): 1218–20.

181. Charles Fremont, "Les mouvements de l'ouvrier dans le travail profession-nel," *Le Monde Moderne* (February 1895): 187–93.
182. Ibid., p. 189.
183. Ibid., p. 192.
184. Ibid.
185. Fremont, "Les mouvements de l'ouvrier," p. 193.
186. Etienne-Jules Marey, "L'économie de travail et l'élasticité," *La Revue des Idées* 1, no. 4 (1904): 161–77.
187. Etienne-Jules Marey, "Du moyen d'économiser le travail moteur de l'homme et des animaux," *Association Française pour l'Avancement des Sciences. Compte-rendu de la session de Lille 22 August 1874* (Paris, 1875), pp. 1157–65.
188. Ibid., p. 1156.
189. Marey, "L'économie de travail," p. 162.
190. Marey, "Travail de l'homme dans les professions manuelles," p. 196.

CHAPTER 5. THE LAWS OF THE HUMAN MOTOR

1. André Liesse, *Le Travail aux points de vue scientifique, industriel et social* (Paris, 1899), p. 16.
2. Armand Imbert, "Mode de fonctionnement économique de l'organisme," *Scientia: Exposé et développement des questions scientifiques à l'ordre du jour* (1902): 11.
3. Ibid., p. 6.
4. Ibid., p. 96.
5. Liesse, *Le Travail*, p. 19.
6. See Hermann von Helmholtz, "On the History of the Discovery of the Principle of Least Action," in Leo Köenigsberger, *Hermann von Helmholtz*, trans. Frances A. Welby (New York, 1965), pp. 352–56.
7. Josefa Ioteyko, *The Science of Labour and its Organization* (London, 1919), pp. 3, 4.
8. Jürgen Kocka, *Unternehmensverwaltung und Angestelltenschaft am Beispiel Siemens 1847–1914. Zum Verhältnis von Kapitalismus und Bürokratie in der deutschen Industrialisierung* (Stuttgart, 1969).
9. Patrick Fridenson, "France, États-Unis: Genèse de l'usine nouvelle," *Recherches* 32/33 (September 1978):375–88.
10. Michelle Perrot, "The Three Ages of Industrial Discipline in Nineteenth-Century France," in *Consciousness and Class Experience in Nineteenth-Century France*, ed. John M. Merriman (New York, 1979), pp. 160–63.
11. William Coleman, *Biology in the Nineteenth Century, Problems of Form, Function, and Transformation* (New York, 1979), p. 119.

12. Köenigsberger, *Helmholtz,* p. 33.
13. Everett Mendelsohn, *Heat and Life, The Development of the Theory of Animal Heat* (Cambridge, Mass., 1964), p. 182; Jaques Loeb, *The Mechanistic Conception of Life* (Chicago, 1912).
14. Robert Mayer, "Die organische Bewegung in ihrem Zusammenhänge mit dem Stoffwechsel: Ein Beitrag zur Naturkunde," *Die Mechanik der Wärme in Gesammelten Schriften von Robert Mayer,* ed. Jacob J. Weyrauch (Stuttgart, 1893), p. 88.
15. Coleman, *Biology in the Nineteenth Century,* p. 123.
16. Köenigsberger, *Helmholtz,* p. 33.
17. On electrophysiology, see Carlo Matteucci, "Electrophysiology: A Course of Lectures," *Smithsonian Institute Annual Report 1865* (Washington, D.C., 1866), pp. 291–345; Charles Richet, *Physiologie des muscles et des nerfs* (Paris, 1882).
18. See Adolf Fick, *Untersuchung über Muskel-Arbeit* (Basel, 1867), p. 56.
19. See Marey, *Du mouvement dans les fonctions de la vie: Leçons faites au Collège de France* (Paris, 1868), p. 279.
20. These experiments were widely known in France. See Charles Richet, *Physiologie des muscles,* pp. 40–43.
21. Adolf Fick, *Mechanische Arbeit und Wäremeentwicklung bei der Muskelthätigkeit* (Leipzig, 1882), p. 153.
22. Ibid., p. 146.
23. Ibid., p. 33.
24. "Ernährung," *Real Encyclopädie der gesamten Heilkunde: Medicinisch-Chirurgisches Handwörterbuch für praktische Ärtze,* ed. Albert Eulenburg, 5th ed. (Vienna, 1881), p. 76.
25. Coleman, *Nineteenth-Century Biology,* p. 135.
26. These experiments are summarized in Carl von Voit, "Physiologie des allgemeinen Stoffwechsels und der Ernährung," *Handbuch der Physiologie,* ed. Ludimar Hermann, vol. 6, no. 1 (Leipzig, 1881). See Frederick L. Holmes, "Carl von Voit," in the *Dictionary of the History of Science,* vol. 14, ed. Charles C. Gillipsie (New York, 1976), pp. 63–67.
27. Max Rubner, *Kraft und Stoff im Haushalte der Natur* (Leipzig, 1909), p. 53.
28. Rubner's calorimeter was originally designed for animals. An American, W. O. Atwater, who had studied with Voit while Rubner was an assistant there, and who had learned a great deal from Rubner, eventually designed a calorimeter for humans in 1892. See Coleman, *Nineteenth-Century Biology,* p. 142.
29. Max Rubner, *Die Gesetze des Energieverbrauchs bei der Ernährung* (Leipzig, 1902); *Nährungsmittel und Ernährungskunde* (Stuttgart, 1904); *Volksernährungsfragen* (Leipzig, 1908); *Unsere Ziele für die Zukunft* (Leipzig, 1910); *Wandlungen in der Volksernährung* (Leipzig, 1913); *Ernährung in der Kriegszeit: Ein Ratgeber für Behörden, Geistliche, Ärtze, Lehrer* (Braunschweig, 1915); *Deutschlands Volksernährung im Kriege* (Leipzig,

1916); *Deutschlands Volksernährung: Zeitgemässe Betrachtungen* (Berlin, 1930). On Rubner's role in the First World War, see chapter 9.

30. Cited in Max Weber, "Zur Psychophysik der industrielen Arbeit," *Gesammelte Aufsätze zur Soziologie und Sozialpolitik* (Tübingen, 1924), p. 76.

31. Marey, *Du mouvement dans les fonctions de la vie,* pp. 69, 70.

32. Auguste Chauveau, *Le Travail musculaire et l'énergie qu' il représente* (Paris, 1891), p. 10.

33. Cited in Imbert, "Mode de fonctionnement économique," p. 73.

34. Auguste Chauveau, Préface to F. Laulanie, *Énergétique musculaire* (Paris, 1898), p. 20; *Rapport scientifique sur le travail entrepris en 1905 au moyen des subventions de la caisse de recherches scientifiques* (Melun, 1906).

35. Gustave-Adolphe Hirn, *Conséquences philosophiques et métaphysiques de la thermodynamique* (Paris, 1868), p. 277.

36. Gustave-Adolphe Hirn, "La thermodynamique et le travail chez les êtres vivants," *Revue Scientifique* 22 (28 May 1887): 676.

37. Charles Rouget, "La conservation de l'énergie," *Revue Scientifique* 50 (12 June 1880): 1206–7.

38. Charles Richet, "La pensée et le travail chimique," *Revue Scientifique* 3 (15 January 1887):83–85. Richet concluded that "every time one compares the brain to the muscle, I found that the two organs behave according to the same reactions, and thus I conclude: muscular contraction is a phenomenon of chemical origins, and thought is also a phenomenon of chemical origins; therefore it is subject to the general and absolute law of the conservation of energy." See the response by Armand Gautier, "La pensée," *Revue Scientifique* 1 (1 January 1887): 14–18.

39. Napoleone Colajanni, untitled note, *Archives de l'anthropologie criminelle de Lyon* (Lyon, 1886): 481.

40. Liesse, *Le Travail,* p. 18.

41. "Ernährung," *Real Encyclopädie der gesamten Heilkunde,* p. 76.

42. Ibid., von Voit, *Physiologie des allgemeinen Stoffwechsels,* p. 162. For a lucid discussion see Richard Kremer, "From Stoffwechsel to Kraftwechsel: Voit, Rubner and the Study of Nutrition in the 1880s," History of Science Society Paper, October 1983. For Voit's study of the diet of public sector workers see *Untersuchungen der Kost in einigen öffentlichen Anstalten* (Munich, 1877). On the English versus German diet, also see chapter 8.

43. Gustave-Adolphe Hirn, *Recherches sur l'équivalent mécanique de la chaleur présentées a la société de physique de Berlin* (Colmar, 1858), pp. 45, 51.

44. Jules-Auguste Béclard, *De la contraction musculaire dans ses rapports avec la témperature animale, Mémoire présenté à l'Académie des Sciences* (Paris, 1860).

45. M. A. Herzen, "L'activité musculaire et l'équivalence des forces," *Revue Scientifique* 8 (19 February 1887): 237.

46. Paul Bert, *La Machine humaine,* 2 vols. (Paris, 1867–68).

47. Armand Gautier, *L'alimentation et les régimes chez l'homme sain et chez*

les malades (Paris, 1904); English translation: *Diet and Dietetics*, trans. A. J. Rice-Oxley (London, 1906), p. 83.

48. Ernest Lebon, *Armand Gautier: Biographie, bibliographie analytique des écrits* (Paris, 1912), p. 5.

49. Gautier, *Diet and Dietetics*, p. 49.

50. Ibid., p. 63. In the United States, W. O. Atwater experimented with immigrant groups in Chicago, California, Alabama and Virginia, comparing their albuminoid, fat, and carbohydrate intake. On Atwater, see Stephan Leibfried, Nutritional Minima and the State: "On the Institutionalization of Professional Knowledge in Social Policy in the U.S. and Germany." Center for Social Policy Research, Bremen University.

51. Liesse, *Le Travail*, p. 24.

52. Ibid.

53. Gautier, *Diet and Dietetics*, p. 94.

54. See the studies cited in ibid., p. 94.

55. "Kilogrammeter" is the standard measure of the force expended in raising a mass of one kilogram one meter, about 7.2 foot pounds. For example, an average worker provides between 80 to 100,000 kilogrammeters of work during 9 or 10 hours, Gautier, *Diet and Dietetics*, p. 87.

56. Armand Gautier, "À propos de la quantité minimum d'albuminoïde quotidiennement nécessaire à l'état de repos ou du travail," *Revue de la Société d'Hygiène Alimentaire et de l'Alimentation Rationnelle de l'Homme* 1 (August/September, 1904): 329–31 and Liesse, *Le Travail*, pp. 25, 26.

57. Liesse, *Le Travail*, p. 26.

58. Marey, *Du mouvement dans les fonctions de la vie*, p. 72.

59. Dragolioub Yovanovitch, *Le Rendement optimum du travail ouvrier: Étude sur les stimulants de l'activité ouvrière* (Paris, 1923), Thèse pour le doctorat des lettres, pp. 231–62.

60. Hugo Kronecker, "Über die Ermüdung und Erholung der quergestreiften Muskeln," *Berichte der Verhandlungen der sächsischen Gesellschaft der Wissenschaft zu Leipzig* (Leipzig, 1871), p. 718. According to Mosso, Wilhelm Wundt first proposed the idea of making use of the myograph invented by Helmholtz for the study of fatigue, but I have not been able to trace the origins of this remark. Angelo Mosso, *Fatigue*, trans. M. and W. B. Drummond (London, 1906), pp. 77, 82.

61. Mosso, *Fatigue*, p. 60.

62. Ibid., pp. 82–88.

63. Ibid., p. 122.

64. Ibid., p. 121.

65. Ibid., p. 92.

66. Cited in Tsuru Asai, *Mental Fatigue* (New York, 1912), p. 9.

67. Ibid., p. 154.

68. Ibid., pp. 154, 156.

69. Ibid.

70. Charles Henry, " 'À travers les sciences et l'industrie': La fatigue intellectuelle et physique d'après M. Mosso," *Revue Blanche* 4 (1894):170.

71. Review of Angelo Mosso, *La Fatica* by Théodule Ribot, *Revue Philosophique* 32 (1891):415, 416; Francesco S. Nitti, *Le Travail humain et ses lois*, trad. N. Politis (Paris, 1895), pp. 15–29; Ioteyko, *The Science of Labour*, p. 12.

72. For other versions of the ergograph see Zaccaria Treves, "Le travail, la fatigue, et l'effort," *Année Psychologique* 12 (1906):34.

73. Ibid.; Alfred Binet and Henri Vaschide, "Un nouvel ergographe," *Année Psychologique*, 2 (1898): 263–64; John Hough, "Ergographic Studies in Muscular Fatigue," *American Journal of Physiology* 4 (1900): 109–16; Zaccaria Treves, "Über den gegenwärtigen Stand unserer Kenntniss der Ergographie betreffend," *Archiv für den gesamte Physiologie* 78 (1899): 139–57.

74. Josefa Ioteyko, *Résumé des travaux scientifiques (1896–1906)* (Brussels, 1906).

75. Ernest Solvay, *Note sur des formules d'introduction à l'énergétique physio- et psycho-sociologique*, Institut Solvay. Travaux de l'Institut de Sociologie. Notes et Mémoires (Brussels, 1902), p. 14; Ernest Solvay, *Considérations sur l'énergétique des organismes au point de vue de la définition, de la genèse et de l'évolution de l'être vivant.* Travaux du Laboratoire de l'Institut Solvay (Brussels, 1901); and his programmatic *Principes d'orientation sociale* (Brussels, 1904).

76. Léopold Mayer, "Sur les modifications du chimisme respiratoire avec l'âge, en particulier chez le cobaye," CRAS 133 (13 July 1903); "Sur les modifications du chimisme respiratoire avec l'âge, en particulier chez le poulet et le canard," *Bulletin de la Société des Sciences Médicales et Naturelles de Bruxelles* (1904); Charles Henry and Louis Bastien, "Sur un critérium d'irréductibilité dans les ensembles statistiques," CRAS 133 (15 June 1903); Charles Henry, "Sur le travail statique du muscle, CRAS 134 (5 January 1904); Charles Henry and J. Ioteyko, "Sur une relation entre le travail et le travail dit statique, énergétiquement equivalents à l'ergographe," CRAS 133 (28 December, 1903).

77. Charles Henry, *Mesure des capacités intellectuelle et énergétique*, Institut Solvay, Travaux de l'Institut de Sociologie. Notes et Mémoires (Brussels and Leipzig, 1906), pp. 1–12.

78. Jules Amar, *The Physiology of Industrial Organisation and the Re-employment of the Disabled*, trans. Bernard Miall (Macmillan, 1919), p. 14.

79. Ibid., p. 14.

80. Imbert, "Mode de fonctionnement économique de l'organisme," p. 6.

81. Josefa Ioteyko, "Les lois de l'ergographie," *Bulletins de l'Académie Royale de Belgique* (May 1904):557. The fatigue quotient is the relation between the height of each contraction on the fatigue curve and the duration of the muscle contraction expressed in a simple formula ($n = H - at^3 + bt^2 - ct$) (n is the height of each contraction; H the maximum value of the effort expressed in millimeters, t the time or unity of time expended, a, b, c, the

contractions according to that rhythm. See Josefa Ioteyko, "L'équation de la courbe de fatigue et sa signification physiologique," *Communiqué au VIè Congrès International de Physiologie* (Brussels, 1904).

82. Ioteyko, "Les lois de l'ergographie," p. 562.

83. Ibid., p. 565.

84. Ioteyko, "Fatigue," *Dictionnaire de physiologie*, ed. Charles Richet (Paris, 1904), p. 171.

85. Charles Féré, "Influence du rythme sur le travail," *Année Psychologique* 6 (1902):45; Jules Amar, *Le Rendement de la machine humaine* (Paris, 1910), pp. 64–65.

86. Ioteyko, *La Fatigue* (Paris, 1920), p. 17.

87. See Jules Amar, *The Human Motor: Or the Scientific Foundations of Labor and Industry,* trans. Elsie P. Butterworth and George E. Wright (London, 1920), p. 210.

88. For a lengthy treatment of this debate Ioteyko, *La Fatigue,* pp. 30–34.

89. Jules Amar, "Observations sur la fatigue professionelle," *Journal de Physiologie* (March 1911):178–202.

90. A.-M. Bloch, "Enquête sur la fatigue musculaire professionnelle," *Société de Biologie,* Séance du 2. Mai 1903 (Paris, 1903): 548–50.

91. The example of the marathon runner is taken from a statue in the garden of the Tuileries by the sculptor, Cortot. On "Auto-intoxication," see Zaccaria Treves, "Sur les lois du travail musculaire," *Archives Italienne Biologique* 30 (1898):1–34. Charles Bouchard, *Leçons sur le auto-intoxication dans les maladies* (Paris, 1887).

92. Ioteyko, *La Fatigue,* p. 41.

93. Fernand Lagrange, "De l'essoufflement dans les exercices du corps." *Revue Scientifique* 23 (4 June 1887): 718–25.

94. E. Metschnikoff, *Études sur la nature humaine* (Paris, 1908), p. 379.

95. Ioteyko, *La Fatigue,* p. 40.

96. Josefa Ioteyko, "Les défenses psychiques: L'action défensive de la fatigue," *Revue Philosophique* 75 (1913):262. Also see Josefa Ioteyko, "La fatigue comme moyen de défense de l'organisme," *IVè Congrès International de Psychologie* (Paris, 1900), p. 230.

97. Ioteyko, "Les défenses psychiques," pp. 263, 264.

98. Ioteyko, *Science of Labour,* p. 4.

99. Ibid., pp. 20, 21.

100. Wilhelm Weichardt, *Über Ermüdungsstoffe* (Stuttgart, 1910), p. 2.

101. Ibid.

102. Ibid., p. 5.

103. Ibid., Josefa Ioteyko, "Fatigue," *Dictionnaire de physiologie,* ed. Charles Richet, p. 104.

104. Ibid., p. 37.

105. Ibid., p. 40.

106. Ibid., p. 41.

107. Ibid., p. 42. These results can of course be explained by the presence of a

group of adult men in lab coats scurrying about the room with their sprayers, stop-watches, and examination sheets.

108. The concept of "negative" fatigue was developed by Max Verworn who
conducted experiments on frogs inducing "fatigue without fatigue substances." Verworn also distinguished between fatigue, which produced
"positive" toxic substances, and exhaustion, which depleted the body of
oxygen, carbon, and sodium. This distinction was generally accepted by
fatigue researchers. Max Offener, *Die geistige Ermüdung* (Berlin, 1928), p.
9.

109. Arnold Durig, "Die Theorie der Ermüdung," *Körper und Arbeit: Handbuch der Physiologie,* ed. Edgar Ätzler (Leipzig, 1927), p. 255.

110. "Ermüdungsbekämpfung durch Antikenotoxin," *Deutsche militär-
ärtzliche Zeitschrift* 42, no. 1 (5 January 1913): 12–13.

111. Compare Hugo Münsterberg, *Grundzüge der Psychotechnik* (Leipzig,
1914), pp. 388–407.

CHAPTER 6. MENTAL FATIGUE, NEURASTHENIA, AND CIVILIZATION

1. Alfred Binet, *Les idées modernes sur les enfants* (Paris, 1899), p. 65.
2. Philippe Tissié, *La Fatigue et l'entraînement physique* (Paris, 1887), p. 177.
3. Gustave Lagneau, "Du surmenage intellectuel et de la sédantarité dans les
écoles," *Bulletin de l'Académie de Médecine* 15, Séance du 27 Avril, 1886
(Paris, 1886): 591–643.
4. Ibid., p. 593.
5. Ibid., p. 595.
6. Victor de Laprade, *L'éducation homicide* (Paris, 1868), p. 60.
7. For an important source of French views of the German educational system: See Michel Bréal, *Excursions pédagogiques: Un voyage scolaire en
Allemagne* (Paris, 1884), p. 133.
8. See *Rapport général sur les travaux du Conseil d'Hygiène et de Salubrité
du département de la Seine de 1878 à 1880* (Paris, 1884), pp. 152–56.
9. Lagneau, *Du surmenage intellectuel,* p. 612.
10. Ibid., p. 620.
11. For the older definition of see S. Arloing, "Surmenage," *Dictionnaire encyclopédique des sciences médicales* (Paris, 1884), p. 576.
12. Review of Lagneau in *Revue d'Hygiène et de Police Sanitaire* (1886): 431.
13. *Bulletin de l'Académie de Médecine,* Séance du 17 Mai 1887 (Paris, 1887).
14. Ibid.
15. Ibid.
16. Alfred Binet and Henri Vaschide, *La Fatigue intellectuelle* (Paris, 1898), p.
14.
17. Ibid., p. 21. Experimental studies of mental fatigue among schoolchildren
began as early as 1879, when J. Sikorski conducted a series of experiments

on the effect of mental fatigue on the number of errors that resulted in speech or in walking. J. Sikorski, "Sur les effets de la lassitude provoquée par les travaux intellectuels chez les enfants à l'âge scolaire, *Annales d'Hygiène Publique* 2 (1879): 458–67.

18. On fatigue in the German educational system see: Friedrich Kemsies, "Zur Frage der Überbürdung unser Schuljugend," *Deutsche Medizinische Wochenschrift* 22 (1896): 32, 33; Gabriel Anton, *Über geistige Ermüdung der Kinder im gesunden und kranken Zustande* (Halle, 1900); Emil Kraepelin, *Zur Überbürdungsfrage* (Jena, 1897).

19. For example, Rudolf Keller, "Pädagogisch-Psychometrische Studien," *Biologisches Centralblatt*, 14 (1894): 32, 33.

20. Kemsies, "Zur Frage der Überbürdung," pp. 32, 33.

21. Keller, "Pädagogisch-Psychometrische Studien," p. 32.

22. Théodore Vannod, "La fatigue intellectuelle et son influence sur la sensibilité cutanée," *Revue Médicale de la Suisse Romande,* 16 (1896): 712. A good summary of all the schools of mental fatigue research is Max Offener, *Mental Fatigue: A Comprehensive Exposition of the Nature of Mental Fatigue, of the Methods of its Measurement and of Their Results, with Special Reference to the Problems of Instruction,* trans. Guy Whipple (Baltimore, 1911), chaps. 2, 3.

23. Hermann Griesbach, "Communication," *Bericht über den XIV. Internationalen Kongress für Hygiene und Demographie,* Berlin, 23–29 September 1907 (Berlin, 1908):265.

24. Hermann Griesbach, *Energetik und Hygiene des Nervensystems in der Schule* (Munich, 1895).

25. Hermann Griesbach, "Untersuchungen über beziehungen zwischen geistiger Ermüdung und Hautsensibilität," *Archiv für Hygiene* 24, no. 2 (1895): 124–12.

26. Vannod, "La fatigue intellectuelle," p. 712. Vannod's algesiometer used perceptions of pain as indices of fatigue: "A prick-like pressure is applied to the skin by means of an 'algesiometer' which is an instrument closely similar to von Frey's hair aesthesiometer, and which consists essentially of a fine point and a scale that indicates the pressure of the hand upon the point, and consequently of the point upon the skin. In his experiments, Vannod found that at 8 A.M., before instruction began, pressure of 45 grams set up a pain sensation, whereas at 10 A.M., 39 grams, and at noon only 29 grams sufficed. Swift has carried on similar tests in American schools. . . . Binet, however, reached directly the opposite result, viz. that fatigue decreases, not increases, pain sensitivity." Offener, *Mental Fatigue,* p. 42.

27. Vannod, "La fatigue intellectuelle," p. 713.

28. Offener, *Mental Fatigue,* pp. 34–35.

29. Marx Lobsien, *Die experimentelle Ermüdungsforschung und Zeitschätzung,* (Langensalza, 1914), pp. 74–82.

30. Emil Kraepelin, *Über geistige Arbeit* (Jena, 1901), p. 3.

31. Ibid., p. 8.
32. Ibid., p. 12.
33. Ibid.
34. Georges Canguilhem, "What Is Psychology?" *Ideology and Consciousness: Technologies of Human Sciences* 7 (Autumn, 1980): 46.
35. Kraepelin, *Über geistige Arbeit*, p. 12.
36. Ibid., p. 28.
37. Ibid., p. 11.
38. On Kraepelin's views on rest pauses see Otto Amberg, "Über den Einfluss der Arbeitspausen auf die geistige Leistungsfähigkeit," *Psychologische Arbeiten* 1 (1896): 300–377.
39. Kraepelin, *Über geistige Arbeit*, p. 28.
40. A. Binet, V. Henri, *La Fatigue intellectuelle*, p. 1.
41. Ibid., p. 335.
42. Ibid.
43. George M. Beard, *A Practical Treatise on Nervous Exhaustion* (New York, 1869), p. vi. On Beard, see F. G. Gosling, *Before Freud: Neurasthenia and the Medical Community 1870–1910* (Urbana, 1988).
44. George M. Beard, *American Nervousness: Its Causes and Consequences* (New York, 1881).
45. On neurasthenia in Europe, see Robert A. Nye, *Crime, Madness, and Politics in Modern France: The Medical Concept of National Decline* (Princeton, 1984), pp. 148–54; "Degeneration, Neurasthenia, and the Culture of Sport in *Belle Epoque* France," *Journal of Contemporary History* 17, no. 1. (January, 1982):51–68; George Frederick Drinka, *The Birth of Neurosis: Myth, Malady and the Victorians* (New York, 1984).
46. M. Potel, "Neurasthénie," *La Grande encyclopédie* (1886), p. 986.
47. Ibid.
48. Cited in Drinka, *The Birth of Neurosis,* p. 109.
49. Georg Simmel, "The Metropolis and Mental Life," *The Sociology of Georg Simmel,* trans. ed. Kurt H. Wolff (New York, 1950), p. 410.
50. Émile Durkheim, *Suicide: A Study in Sociology,* trans. John A. Spaulding and George Simpson (New York, 1951), p. 68.
51. Ibid.
52. Pierre Janet, *Principles of Psychotherapy,* trans. H. M. and E. R. Guthrie (New York, 1924), p. 86.
53. Charles Féré, *La Famille névropathique. Théorie tératologique de l'hérédité et de la prédisposition morbides et de la dégénérescence* (Paris, 1894), p. 104.
54. Cited in Maurice de Fleury, *Introduction à la médecine de l'esprit* (Paris, 1898).
55. Léon Bouveret, *La Neurasthénie (épuisement nerveux)* (Paris, 1890), p. 4.
56. See Nye, *Crime, Madness and Politics,* p. 148.
57. Jean-Martin Charcot, "Leçons du mardi 1888–89," cited in Bouveret, *La Neurasthénie,* p. 28.

58. F. Levillain, *La Neurasthénie (maladie de Beard)* (Paris, 1891), p. 13.

59. On Féré's influence, see Nye, *Crime, Madness, & Politics,* pp. 125, 128.

60. Bouveret, *La Neurasthénie,* p. 9.

61. See Jan Goldstein, "The Wandering Jew and the Problem of Psychiatric Anti-Semitism in Fin-de-Siècle France," *Journal of Contemporary History* 20, no. 4 (1985): 521–52. Also see Potel, "Neurasthénie," p. 987.

62. Nicolas Vaschide and Claude Vurpas, "Contribution à l'étude de la fatigue mentale des neurasthéniques," *Société de Biologie,* Séance du 7 Mars (Paris, 1903):297.

63. Bouveret, *La Neurasthénie,* p. 10. Friedrich Ziemmsen, "Die Neurasthenie und ihre Behandlung," *Klinische Vorträge* (Leipzig, 1887).

64. Bouveret, *La Neurasthénie,* p. 10.

65. Arnold Kolwalewski, *Studien zur Psychologie des Pessimismus* (Wiesbaden, 1904), p. 121.

66. Bouveret, *La Neurasthénie,* p. 13.

67. J. M. Charcot, "Préface," in Levillain, *La Neurasthénie (maladie de Beard),* p. x.

68. Sigmund Freud, "A Case of Successful Treatment by Hypnotism (1892/3)," *The Standard Edition of the Complete Psychological Works of Sigmund Freud,* trans. James Strachey (London, 1966), vol. 1, p. 123.

69. See Bernard Straus, "Achille-Adrien Proust M.D.: Doctor to River Basins," *Bulletin of the New York Academy of Medicine* 50 (1974): 833–36; and his medical biography, *The Maladies of Marcel Proust: Doctors and Disease in His Life and Work* (New York, 1980), pp. 81–102.

70. Robert Le Masle, *Le professeur Adrien Proust (1834–1903)* (Paris, 1936), pp. 38, 39. For his interest in occupational diseases see Adrien Proust, *Traité d'hygiène,* 3d. ed. (Paris, 1902).

71. Straus, *The Maladies of Marcel Proust* (New York, 1980), p. 83.

72. Achille-Adrien Proust, Gilbert Ballet, *L'hygiène du neurasthénique* (Paris, 1887); English translation, *The Treatment of Neurasthenia* (New York, 1903), trans. Peter Campbell Smith (New York, 1903), p. 25.

73. Ibid., p. 6.

74. Ibid., p. 7.

75. Ibid., p. 16.

76. Ibid.

77. Ibid., pp. 21, 22.

78. Compare Charcot, "Préface," in Levillain, *La Neurasthénie (Maladie de Beard),* p. x; On traumatic neurasthenia, see Proust and Ballet, *Treatment of Neurasthenia,* p. 26 and the discussion of neurasthenia in industrial accidents in chap. 7.

79. Willy Helpach, *Nervenleben und Weltanschauung: Ihre Wechselbeziehungen im deutschen Leben von Heute* (Wiesbaden, 1906), p. 21.

80. Dr. Firmin Terrien, *L'hystérie et la neurasthénie chez le paysan* (Angers, 1906), cited in Theodore Zeldin, *France 1848–1945,* vol. 2 (Oxford, 1977), p. 843.

81. Proust and Ballet, *Treatment of Neurasthenia,* p. 32.
82. Ibid., p. 32.
83. Ibid., p. 3.
84. Ibid., p. 53.
85. Charles Féré, *The Pathology of Emotions: Physiological and Clinical Studies,* trans. Robert Park, M.D. (London, 1899), p. 89.
86. Proust and Ballet, *Treatment of Neurasthenia,* p. 30.
87. Sigmund Freud, "Hysteria" [1888] *Standard Edition,* vol. 1, p. 53.
88. Walter Benjamin, *The Origin of German Tragic Drama,* trans. John Osborne (London, 1977), p. 183. Here we might add that Benjamin alluded to a connection between allegory and melancholy when he noted that allegorical images quickly exhaust themselves precisely because they depend on the power of shock.
89. Potel, "Neurasthénie," p. 987.
90. Proust and Ballet, *Treatment of Neurasthenia,* p. 33.
91. Vaschide and Vurpas, "Fatigue mentale des neurasthéniques," p. 296.
92. Otto Binswanger, *Die Pathologie und Therapie der Neurasthenie. Vorlesungen für Studierende und Ärtze* (Jena, 1896), p. 110.
93. Edward Cowles, *The Mental Symptoms of Fatigue* (New York, 1893), p. 22.
94. Paul Dubois, *L'éducation de soi-même* (Self-Control and How to Achieve it) (New York, 1910). See the critical comments by Pierre Janet, *Principles of Psychotherapy,* p. 152.
95. P. E. Lévy, *L'éducation rationnelle de la volonté* (Paris, 1898).
96. Proust and Ballet, *Treatment of Neurasthenia,* pp. 33, 34.
97. Ibid., p. 30.
98. Ibid., pp. 96, 102.
99. Ibid., pp. 126, 127.
100. Alfred Binet and Charles Féré, "Recherche éxperimentale sur la physiologie des mouvements chez les hystériques," *Archive de Physiologie* 10 (1887): 320.
101. Albert Deschamps, *Les Maladies de l'énergie, thérapeutique générale* (Paris, 1908), p. 70.
102. Ibid.
103. Ibid., p. 46.
104. Ibid., p. 37.
105. Ibid., p. 47.
106. Angelo Mosso, *Fatigue,* trans. M. and W. B. Drummond (London, 1906), p. 209.
107. On Ribot see, M. Reuchlin, "The Historical Background of National Trends in Psychology: France," *Journal of the History of the Behavioral Sciences* 1, no. 2 (1965):115–22.
108. Théodule Ribot, *The Diseases of the Will,* trans. Merwin-Marie Snell (Chicago, 1896), p. 2.

109. Ibid., p. 4.
110. Marie Manacéine, *Le Surmenage mental dans la civilisation moderne: effets, causes, remèdes,* [Préface de Charles Richet] (Paris, 1890), p. 186.
111. Ribot, *Diseases of the Will,* p. 6.
112. Ibid., p. 8.
113. Ibid., pp. 10, 14. The problem of the laws of distribution of nervous fluid was subsequently taken up by Marey's student and associate, Charles François Franck, "Nerveux," *Dictionnaire encyclopédique des idées sciences médicales* (Paris, 1878), p. 572.
114. Ribot, *Diseases of the Will,* p. 14.
115. Ibid., pp. 18, 23.
116. Ibid., p. 27.
117. Ibid., p. 29. Compare Etienne Esquirol, *Des maladies mentales* (Paris, 1838), vol. 1, p. 421.
118. Ibid., p. 38.
119. The connection between inhibition to action *(Handlungshemmung)* in melancholia, and its antithesis, labor, was to my knowledge, first explored by the sociologist, Wolf Lepenies in his *Melancholie und Gesellschaft* (Frankfurt-am-Main, 1972), pp. 207–13.
120. Theodor Dunin, *Grundsätze der Behandlung der Neurasthenie und Hysterie* (Berlin 1902), p. 32.
121. Ibid., p. 33.
122. Proust and Ballet, *Treatment of Neurasthenia,* p. 72.
123. Ribot, *Diseases of the Will,* p. 40.
124. Ibid., pp. 51, 52.
125. See the superb analysis in Jan Goldstein, "The Hysteria Diagnosis in Nineteenth Century France," *Journal of Modern History* 54 (June 1982): 209–39.
126. Ribot, *Diseases of the Will,* p. 94.
127. There is an extensive literature on suggestion, hypnosis, and somnambulism. See Hyppolite Bernheim, *Hypnotisme, suggestion, psychothérapie, nouvelles études* (Paris, 1891); Jules Liégeois, *De la suggestion et du somnambulisme;* Georges Gilles de la Tourette, *L'hypnotisme et les états analogues au point de vue médico-legale* (Paris, 1887).
128. Ribot, *Diseases of the Will,* p. 115.
129. Sigmund Freud, "Extracts from the Fliess Papers, Draft B: The Aetiology of the Neuroses" *Standard Edition,* vol. 1, p. 180. Freud also believed that neurasthenia occurred primarily in males, and that "neurasthenia in women is a direct consequence of neurasthenia in men, through the agency of this reduction in their [males] potency."
130. See the discussion of this issue in Frank J. Sulloway, *Freud, Biologist of the Mind: Beyond the Psychoanalytic Legend* (New York, 1979), pp. 65–69.
131. For Brücke's influence, see Siegfried Bernfeld, "Freud's Earliest Theories

and the School of Helmholtz," *The Psychoanalytic Quarterly* 13 (1944):341–62.

132. Sigmund Freud, "Project for a Scientific Psychology," [1895] *Standard Edition,* vol. 2, p. 312.

133. Ibid., p. 158.

134. Jean Starobinski, "A Short History of Body Consciousness," *Humanities in Review* 1 (New York, 1982):22–39.

135. Zeldin, *France,* vol. 2, p. 845.

136. Janet, *Principles of Psychotherapy,* p. 161.

137. Ibid., p. 163.

138. Josefa Ioteyko, XIIIe Congrès international d'hygiène et de démographie, tenu à Bruxelles du 3 au 8 Septembre 1903, Compte-rendus du Congrès, 5 sec. 4 (Brussels, 1903), p. 72.

139. Charles Féré, *Travail et plaisir: nouvelles études expérimentales de psycho-mécanique* (Paris, 1904), pp. 20–26.

140. Ibid., p. 21.

141. Ibid.

142. For example, Guillaume Ferrero, "Les formes primitives du travail," *Revue Scientifique* 5, no. 11 (7 March 1896): 331–35; Karl Bücher, *Arbeit und Rhythmus* (Leipzig, 1897).

143. The *Revue Philosophique* carried on a veritable campaign against spiritualist conjectures about these states. Also see, Henri-Étienne Beaunis, *Le somnambulisme provoqué: Études physiologiques et psychologiques* (Paris, 1886).

144. Materialists like Richet believed that all movement was a function of "irritability," and that cellular irritability might be considered "elementary psychic life." On the theory of sensationalism, see Starobinski, "Short History of Body Consciousness," pp. 22–39; and Charles Richet, *Essai de psychologie générale* (Paris, 1891), p. 25.

145. Ferrero, "Le formes primitives du travail," p. 331.

146. Ibid., p. 333.

147. Ibid.

148. Ibid., p. 334.

149. Karl Bücher, *Arbeit und Rhythmus,* 3d ed. (Leipzig, 1902).

150. By 1924, Bücher's *Arbeit und Rhythmus* had reached six editions.

151. Bücher, *Arbeit und Rhythmus,* pp. 8–10.

152. Ibid., p. 25.

153. Ibid., p. 358.

154. Ibid., p. 375.

155. Ibid., p. 366.

156. Théodule Ribot, "Le moindre effort in psychologie," *Revue Philosophique* 70 (1910):364, 365.

157. Pierre Janet, *Les obsessions et la psychasthénie,* vol. 1 (Paris, 1911), p. 335. According to Ribot, the term *misonéisme* originated with the founder of criminal anthropology, Cesare Lombroso.

158. Ribot, "Le moindre effort," p. 374.
159. Ribot's source is Max Müller's *Lois de l'altération phonétique,* pp. 374–75.
160. Ibid., p. 376.
161. Ibid., p. 386.

CHAPTER 7. THE EUROPEAN SCIENCE OF WORK

1. Greg Myers, "Nineteenth-Century Popularizations of Thermodynamics and the Rhetoric of Social Prophecy," in *Energy and Entropy: Science and Culture in Victorian Britain,* ed. Patrick Brantlinger (Bloomington, 1989), pp. 332–34.
2. Cited in Georges Barnich, *Essai de politique positive basée sur l'énergétique sociale de Solvay* (Brussels, 1919), p. 373.
3. On Solvay's early life and work see: Louis Bertrand, *Ernest Solvay: Réformateur Social* (Brussels, 1918); Jacques Bolle, *Solvay: L'homme, la découverte, l'entreprise industrielle* (Brussels, 1968); Armand Detillieux, *La Philosophie sociale de M. Ernest Solvay* (Brussels, 1918); "Ernest Solvay, Soda King," *Scientific American Supplement* (6 December 1913): 364; Daniel Warnotte, *Ernest Solvay et l'Institut de Sociologie: Contribution à l'histoire de l'énergétique sociale,* 2 vols. (Brussels, 1946).
4. "Ernest Solvay, Soda King," p. 364.
5. Ernest Solvay, *Science contre religion* (Brussels, 1879), p. 21.
6. Bolle, *Solvay,* p. 141.
7. Ernest Solvay, *Notes sur le productivisme et le comptabilisme* (Brussels, 1900), vol. 2, p. 323.
8. Ernest Solvay, *Principes d'orientation sociale: Résumé des études de M. Ernest Solvay sur le productivisme et le comptabilisme* (Brussels, 1904), p. 29.
9. Ibid., pp. 33, 34.
10. Under this scheme, each individual would register his fortune with the national comptabilist office supervised by the state. All transactions would then be acquitted by using a "carnet" issued by the office, in which all debits and credits are recorded. Money, noted Solvay, "would be replaced by the mechanism of writing, pure and simple." This system, which more closely approaches the European postal savings system (modeled on the Austrian) rather than American bank-checking, can also be seen as the nineteenth-century forerunner of the credit card (without credit). Ernest Solvay, *Social Comptabilism: Its Principle and Ground of Existence* (Brussels, 1897).
11. Warnotte, *Ernest Solvay,* vol. 2, p. 521; Henry H. Frost, *The Functional Sociology of Émile Waxweiler and the Institut de Sociologie Solvay,* Publications of the Royal Academy of Belgium, 2d. ser. 53, no. 5 (Brussels, 1959) pp. 24–26. Solvay's decision to give Waxweiler a free hand in formulating the program of a new institute was in part motivated by his disapproval of his Belgian socialist collaborators, Emile Vandervelde, Guillaume De

Greef, and Hector Denis, whose convictions he considered incompatible with his own.

12. Wilhelm Ostwald, *Die energetischen Grundlagen der Kulturwissenschaft* (Leipzig, 1909). On the reception of this work and Ostwald's relations with Solvay, see Wilhelm Ostwald, *Lebenslinien: Eine Selbstbiographie,* vol. 3 (Leipzig, 1927), p. 322–29.

13. Ostwald, *Die energetischen Grundlagen,* p. 132; *Der energetische Imperativ* (Leipzig, 1912), pp. 81–97; *Lebenslinien,* vol. 2, pp. 180–86; and the discussion of the meeting and its repercussions in Erwin N. Hiebert, "The Energetics Controversy and the New Thermodynamics," *Perspectives in the History of Science and Technology,* ed. Duane H. D. Roller (Norman, Okla., 1971), pp. 67–86.

14. Ostwald, *Die energetischen Grundlagen,* p. 9.

15. Ibid., p. 120.

16. Ostwald, *Der energetische Imperativ,* p. 85.

17. Mosso and Solvay were in frequent contact and Solvay had worked at Carl Ludwig's laboratory with Kronecker. Ioteyko and Henry were both supported and sponsored by Solvay. Marey was widely recognized as the intellectual "father" of all of them, and was frequently eulogized as such. It would be tedious to detail the numerous threads linking the science of work in different countries, but in this early stage, at least until 1910, there is an international movement of advocates of the science of the "human motor." For this reason, any description of the development of the science of labor in purely national terms is limited and distorting. For some of the connections see the remarks of Mosso and Kronecker at the VIè Congrès International de Physiologie, Brussels, August/September 1914, in *Archives Internationales de Physiologie* 2 (1904/5): 21.

18. R. Lépine, "L'évolution de la médecine à la fin du XIXè siècle," *Revue du Mois,* 12 (1906):714.

19. Josefa Ioteyko, XIIIè Congrès international d'hygiène et de démographie, tenu à Bruxelles du 3 au 8 Septembre 1903, Compte-rendus du Congrès, 5, sec. 4 (Brussels, 1903), p. 62.

20. Ibid.

21. Ibid., pp. 67, 77.

22. Armand Imbert and M. Mestre, "Recherches sur la manoeuvre du cabrouet et la fatigue qui en résulte," *Bulletin de l'Inspection du Travail* 5, (1905): 15–32; Armand Imbert, "Étude expérimentale du travail de transport de charges avec une brouette," *Bulletin de l'Inspection du Travail,* nos. 1, and 2 (1903/4), p. 24.

23. Armand Imbert, "Exemple d'étude physiologique directe du travail professionnel ouvrier," *Revue d'Hygiène et de Police Sanitaire* (1909): 761.

24. Ibid., p. 750.

25. Cited in Bernard Muscio, *Lectures on Industrial Psychology,* 2d ed. (London, 1920), pp. 68, 69.

26. Jules Amar, *Le Moteur humain et les bases scientifiques du travail*

professionnel (Paris, 1914); Armand Imbert, "Les méthodes du laboratoire appliquées à l'étude directe et pratique des questions ouvrières," *Revue Générale des Sciences Pures et Appliquées,* 22 (30 June 1911): 478–85; Armand Imbert, "L'étude scientifique expérimentale du travail personnel," *Année Psychologique* 13 (1907): 246–59; Charles Fremont, Étude expérimentale du rivetage (Paris, 1906); Charles Fremont, *La Lîme* (Paris, 1916). This work is treated in a pioneering article by Georges Ribeill, "Les débuts de l'ergonomie en France à la veille de la Première Guerre mondiale," *Le Mouvement Social* 113 (October–December 1980): 2–36.

27. Jules Amar, *Le Rendement de la machine humaine* (Paris, 1909), p. 3.

28. For biographical information on Amar, see Ribeill, "Les débuts de l'ergonomie en France," pp. 14–16; H. and D. Monod, "Jules Amar (1879–1935)—À propos d'un centenaire," *Histoire des sciences médicales,* vol. 3 (Paris, 1979), pp. 227–35; Michel Valentin, *Travail des hommes et savants oubliés: Histoire de la médecine du travail, de la sécurité et de l'ergonomie* (Paris, 1978), pp. 291–96.

29. Jules Amar, *The Physiology of Industrial Organisation and the Re-employment of the Disabled,* trans. Bernard Miall (Macmillan, 1919), p. 211; Ribeill, "Les débuts de l'ergonomie en France," p. 14.

30. Amar, *Le Rendement,* p. 4.

31. Ibid., p. 4.

32. Amar, *Physiology of Industrial Organisation,* p. 226.

33. Amar, *Le Rendement,* p. 83.

34. Amar, *Physiology of Industrial Organisation,* p. 218.

35. Cited in Ibid., p. 226.

36. Amar, *Le Rendement,* p. 29.

37. Jules Amar, *Le Moteur humain,* English translation, *The Human Motor or the Scientific Foundations of Labour and Industry,* trans. Elsie P. Butterworth and George E. Wright (London, 1920), p. 182. The editions differ slightly.

38. Ibid., p. 182.

39. Amar, *Le Rendement,* pp. 73, 74.

40. Amar, *Human Motor,* p. 393.

41. Amar, *Physiology of Industrial Organisation,* p. 67.

42. Amar, *Human Motor,* p. 393.

43. Amar, *Physiology of Industrial Organisation,* p. 129.

44. Amar, *Human Motor,* p. 421.

45. Amar, *Physiology of Industrial Organisation,* p. 101.

46. Ibid., p. 127.

47. Ibid., pp. 160, 161, 204.

48. Ibid., p. 163; *Le Moteur humain,* p. 469.

49. Amar, *Physiology of Industrial Organisation,* p. 127.

50. Amar, *Human Motor,* pp. 226, 393.

51. Fritz Giese, "Arbeitswissenschaft," in *Handwörterbuch der Arbeitswissen-*

schaft, ed. Fritz Giese, vol. 1 (Halle, 1930), pp. 418–23; R. W. Hoffman, "Das systematischen und historischen Voraussetzungen der Arbeitswissenschaften," *Analyse der Arbeit*, ed. Konrad Thomas (Stuttgart, 1969), pp. 102–10.

52. For a study of Münsterberg's life and work, see Matthew Hale, Jr., *Human Sciences and Social Order: Hugo Münsterberg and the Origins of Applied Psychology* (Philadelphia, 1980). Alberto Cambrosio, "Quand la psychologie fait son entrée à l'usine: Sélection et contrôle des ouvriers aux États-Unis pendant les années 1910," *Le Mouvement Social* 113 (October–December 1980): 47–53.

53. Otto Fischer, *Der Gang des Menchen, Abhandlungen der sächsischen Gesellschaft der Wissenschaft* 21, pt. 1 (Leipzig, 1895): 153–322. Six years earlier, Fischer and Braune experimented with cadavers to determine "the center of gravity of frozen cadaver parts and whole bodies," later extending their work to a comparison of the "at ease" *(bequeme Haltung)* and military or "attention" stance. Christian Wilhelm Braune and Otto Fischer, *Über den Schwerpunkt des menschlichen Körpers* (Leipzig, 1889); *Das gesetz der Bewegungen an der Basis der mittleren Finger und in Handgelenk des Menschen* (Leipzig, 1887); *Bestimmung der Trägheitsmomente des menschlichen Körpers und seiner Glieder* (Leipzig, 1892); Otto Fischer, *Theoretische Grundlagen für eine Mechanik der lebenden Körper* (Leipzig, 1906). Anton Leitenstorfer, a Bavarian staff military doctor, provided data on the fatigue and exhaustion of recruits, as well as practical guidelines for training exercises, in his 1887 handbook, *Das militärische Training auf physiologischer und praktischer Grundlage* (Stuttgart, 1897).

54. Nathan Zuntz und Wilhelm Schumburg, *Studien zu einer Physiologie des Marches* (Berlin, 1901), pp. 13–19.

55. See, for example, Amar, *The Human Motor*, p. 374.

56. See Stephan Leibfried, "Nutritional Minima and the State: On the Institutionalization of Professional Knowledge in National Social Policy in the U.S. and Germany," pp. 82–83. Also see Wilhelm Schumburg, *Hygiene der Einzelernährung und Massenernährung*, vol. 3., Theodor Weyl, ed., *Handbuch der Hygiene* (Leipzig, 1913), pp. 299–478.

57. Max Rubner, *Gesetze der Energieverbrauches und die Ernährung* (Berlin, 1902); on the work of the Berlin Kaiser-Wilhelm Institut für Arbeitsphysiologie, see chap. 9.

58. On Kraepelin's career see Friederich Dorsch, *Geschichte und Probleme der angewandten Psychologie* (Stuttgart, 1963), p. 46.

59. On German positivism and psychophysics see Hale, *Human Sciences and Social Order*, pp. 35–40.

60. Emil Kraepelin, *Die Arbeitscurve* (Leipzig, 1902), p. 4.

61. Emil Kraepelin, *Zur Hygiene der Arbeit* (Jena, 1896), p. 8.

62. Kraepelin, *Die Arbeitscurve*, p. 175.

63. Kraepelin, *Zur Hygiene der Arbeit*, p. 89.

64. Emil Kraepelin, *Psychologische Arbeiten,* vol. 2 (Leipzig, 1897), p. 399.

65. Kraepelin, *Die Arbeitscurve,* pp. 10, 11.

66. Cited in Max Weber, "Zur Psychophysik der industriellen Arbeit," *Gesammelte Aufsätze zur Soziologie und Sozialpolitik* (Tübingen, 1924), pp. 77–82.

67. Adolf Gerson, "Die physiologischen Grundlagen der Arbeitsteilung: Ein gewerbephysiologischer Versuch," *Zeitschrift für Socialwissenschaft,* 10 (1907): 536.

68. Émile Waxweiler, "La tendance au moindre effort et les facteurs de l'organisation sociale," *Archives Sociologiques* 8 (1908): 123. Théodule Ribot, "Le moindre effort en psychologie," *Revue Philosophique* 70 (1910): 361–86.

69. Gerson, "Grundlagen," pp. 536, 537.

70. Hugo Münsterberg, *Grundzüge der Psychotechnik* (Leipzig, 1914), p. 389.

71. Ibid., p. 395.

72. Hugo Münsterberg, *Psychology and Industrial Efficiency* (New York, 1913), p. 165.

73. Münsterberg, *Grundzüge,* p. 403. See Hugo Münsterberg, *Psychologie und Wirtschaftsleben: Ein Beitrag zur angewandten Experimental-Psychologie* (Leipzig, 1913).

74. Münsterberg, *Psychology and Industrial Efficiency,* pp. 209, 210.

75. Münsterberg, *Grundzüge,* p. 404.

76. These efforts are described in Hale, *Human Science and Social Order,* p. 159.

77. Wladimir Eliasberg, "Arbeit und Psychologie," *Archiv für Sozialwissenschaft und Sozialpolitik* 50 (1922): 86–127.

78. C. Ritter, "Über Ermüdungsmessungen: Kritisches und Experimentelles," *Zeitschrift für angewandte Psychologie und psychologische Sammelforschung* 4 (1911): 527. This article contains an overview of the inadequacies of fatigue measurement from Mosso to Kraepelin.

79. Max Weber, "Psychophysik der industriellen Arbeit," pp. 111–23.

80. Ibid.

81. Ibid., p. 111.

82. "Probleme der Arbeiterpsychologie unter besonderer Rücksichts nahme auf Methode und Ergebnisse der Vereinserhebungen," *Schriften des Vereins für Sozialpolitik* 138, Verhandlungen der Generalversammlung in Nürnberg, 9. und 10. Oktober 1911 (Leipzig, 1912): 190. (Hereafter cited as "Arbeiterpsychologie," VSP, Generalversammlung, 1911.)

83. Max Weber, "Energetischer Kulturtheorien," in *Gesammelte Aufsätze zur Wissenschaftslehre,* ed. Johannes Winckelmann (Tübingen, 1968), p. 406.

84. By the 1890s the Verein had largely given up its agitational activities and had become a discussion society with an influential publication series.

85. On Levenstein, see Anthony Oberschall, *Empirical Research in Germany 1848–1914* (Paris, 1965), pp. 94, 95.

86. Adolf Levenstein, *Die Arbeiterfrage: Mit besonderer Berücksichtigung der*

sozialpsychologischen Seite des modernen Grossbetriebes und der psycho-physisichen Einwirkungen auf die Arbeiter (Munich, 1912), p. 1.

87. This formulation, as might be expected, prevented most women from taking part in the study, as Levenstein conceded.

88. Oberschall, *Empirical Research*, p. 95.

89. Levenstein, *Die Arbeiterfrage*, p. 77.

90. Ibid., p. 78.

91. Ibid., p. 87.

92. Ibid., p. 4. I have found no published reference to this work other than Levenstein's preface to the *Arbeiterfrage*.

93. Oberschall, *Empirical Research*, p. 95.

94. The extent of the Verein's influence in the period before the First World War is the subject of debate. In his survey, Dieter Lindenlaub distinguishes three generations: a first, dedicated to the formation of a coherent state social policy under the auspices of the monarchy; a second "social liberal" generation, which attempted to create a national economic policy; and finally a third, which included Max Weber and Alfred Weber, which was poised between capitalism and socialism and which demanded a programmatic "value neutrality." For a discussion of the politics of the Verein see Dieter Lindenlaub, *Richtungskämpfe im Verein für Sozialpolitik: Wissenschaft und Sozialpolitik im Kaiserreich vornehmlich vom Beginn des 'neuen kurses' bis zum Ausbruch des Ersten Weltkrieges (1890–1914)*, 2 vols., *Vierteljahrschrift für Sozial-und Wirtschaftsgeschichte*, Beihefte, nos. 52, 53 (Wiesbaden, 1967).

95. Oberschall, *Empirical Research*, p. 113.

96. Max Weber, "Methodologische Einleitung für die Erhebungen des Vereins für Sozialpolitik über Auslese und Anpassung (Berufswahlen und Berufsschicksal) der Arbeiterschaft der geschlossenen Grossindustrie," *Gesammelte Aufsätze zur Soziologie und Sozialpolitik* (Tübingen, 1924), pp. 2, 3. Weber modestly rejected the claim that he had created a new method of social science investigation, referring instead to the work of the industrialist Ernst Abbe: "It is even said that I discovered this method. That is an absolutely fundamental mistake. It is out of the question. The same method was used by Abbe in his work." See Max Weber's comments in the discussion on worker psychology in 1911, VSP, Generalversammlung, 1911, p. 196.

97. Max Weber, "Methodologische Einleitung," p. 3; English translation from Oberschall, *Empirical Research*, p. 114.

98. A partial list of the surveys conducted by the Verein includes: Marie Bernays, *Auslese und Anpassung der Arbeiterschaft der geschlossenen Grossindustrie, dargestellt an den Verhältnissen der "Gladbacher Spinnerei und Weberei A.G." zu München Gladbach im Rheinland* Schriften des Vereins für Sozialpolitik, 133 (Leipzig, 1910); Ernst Bernhard, *Arbeitsintensität und Arbeitszeit, Schmoller's Forschungen*, 138 (Leipzig, 1909); Stanislaw von Bienkowski, *Untersuchungen über Arbeitseignung und Leis-*

tungsfähigkeit der Arbeiterschaft eines grossindustriellen Betriebes (Berlin, 1910); Fritz Schumacher, *Auslese und Anpassung der Arbeiterschaft in der Automobileindustrie und einer Wiener Maschinenfabrik* (Leipzig, 1911).

99. Oberschall, *Empirical Research*, p. 114; Weber, "Methodologische Einleitung," pp. 3, 4.

100. The Committee developed a standard framework for each of the studies, focusing on "selection and adaptation," wages, and attitudes. The Verein's initial efforts met with considerable resistance from economists and industrialists and a plan to survey systematically all German industry soon gave way to the more modest and "accidental" principle of studying cooperative firms. Bernays, *Auslese und Anpassung der Arbeiterschaft*, p. xv.

101. Weber, "Methodologische Einleitung," p. 1.

102. Weber, "Psychophysik der industriellen Arbeit," pp. 127, 128.

103. Ibid., p. 126. English translation in Oberschall, *Empirical Research*, p. 116.

104. Weber, "Psychophysik der industriellen Arbeit," pp. 132–36.

105. Bernays, "Auslese und Anpassung der Arbeiterschaft," p. xxi; 106. For a critique of the statistical method of the Verein's studies see the debate between Max Weber and Ludwig von Bienkowski, "Arbeiterpsychologie," VSP, Generalversammlung, 1911, pp. 168–79; 189–03.

107. Alfred Weber, "Das Berufsschicksal des Industriearbeiters," *Archiv für Sozialwissenschaften und Sozialpolitik* 34 (1912): 377–405. Recruitment, deployment, and mobility were the central categories of the social science of industry. Bernays, *Auslese und Anpassung der Arbeiterschaft*, p. xxi.

108. Weber, "Berufsschicksal," p. 383.

109. On the broader context of this debate, see Stephen P. Turner and Regis A. Factor, *Max Weber and the Dispute over Reason and Value: A Study of Philosophy, Ethics and Politics* (London, 1984), pp. 57–59.

110. On Weber's attitudes toward reform and social policy see Wolfgang J. Mommsen, *Max Weber and German Politics 1890–1920*, trans. Michael S. Steinberg (Chicago, 1984), pp. 117–21.

111. Max Weber, "Psychophysik der industriellen Arbeit," p. 130.

112. Bernays, "Auslese und Anpassung der Arbeiterschaft," p. 310.

113. Ibid.

114. "Arbeiterpsychologie," VSP, Generalversammlung, 1911, pp. 149–51.

115. Schumacher compared 105 cases of "strain" *(Anstrengung)* and found that the majority of cases resulted from "speeded up work," since fatigue normally set in at the end of the morning and the end of the afternoon. Fritz Schumacher, *Die Arbeiter beim Daimler Motoren-Gesellschaft: Stuttgart, Untertürkheim, Auslese und Anpassung der Arbeiterschaft in der Automobilindustrie* (Stuttgart, 1910), p. 89; Ernst Bernhard, *Höhere Arbeitsintensität bei kürzerer Arbeitszeit: Ihre personalen und technisch-sachlichen Voraussetzungen*, diss. (Leipzig, 1909).

116. Alfred Weber, *Der Kampf zwischen Kapital und Arbeit: Versuch einer*

systematischer Darstellung mit besonderer Berücksichtigung der gegen-wärtigen deutschen Verhältnisse (Tübingen, 1910), p. 97.

117. Bernays, "Auslese und Anpassung der Arbeiterschaft," p. 340. By using automatic measuring devices attached to the spinning machines and looms, Bernays plotted performance by both days of the week and hours of the day, noting that in any given day the optimal performance occurs in the second half of the morning and afternoon, and that a decline sets in after the pause, but is soon erased by increased performance. In Berlin and Vienna Monday performance was especially lower than in the rest of the week, a fact which "cast a great deal of light on the use of Sundays in the large cities." See "Arbeiterpsychologie," VSP, Generalversammlung, 1911, pp. 132, 139.

118. Ibid., p. 349.

119. "Arbeiterpsychologie," VSP, Generalversammlung, 1911, p. 191.

120. Ibid.

121. See Fritz Giese, "Die Eigenart der französischen Arbeitswissenschaft," *Annalen der Betriebswirtschaft* 1 (Berlin, 1927): 405.

122. Although Münsterberg did try to extend his applied research to industry, this work was undertaken almost exclusively in America, and Kraepelin and his disciples remained confined to experimental work under very controlled conditions. Even the applied psychology of William Stern, who coined the term "psychotechnics" and pioneered the testing of vocational aptitude in his laboratory of applied psychology founded in Berlin in 1906, remained largely without impact until the First World War (see chap. 9).

123. "Arbeitsrecht, energetisch," *Handwörterbuch der Arbeitswissenschaft,* ed. Fritz Giese (Halle, 1930), pp. 406–09.

124. On Taylorism see chapter 9. Wladimir Eliasberg, "Richtungen und Ent-wicklungstendenzen in der Arbeitswissenchaft," *Archiv für Sozialwissen-schaft und Sozialpolitik* 56 (1926): 80.

125. Josefa Ioteyko, *The Science of Labour and its Organisation* (London, 1919), p. 7.

126. Ibid.

127. Ibid., p. 5.

128. Eliasberg, "Richtungen und Entwicklungstendenzen," p. 79.

129. Eliasberg, Ibid., p. 80.

CHAPTER 8. THE SCIENCE OF WORK AND THE SOCIAL QUESTION

1. Robert A. Nye, *Crime, Madness and Politics in Modern France: The Medi-cal Concept of National Decline* (Princeton, 1984), p. 170. See also Stephen J. Gould, *The Mismeasure of Man* (New York, 1981).

2. Louis Querton, *L'augmentation du rendement de la machine humaine* (Brussels, 1905), p. 4.

3. Alfred Fouillée, *La Science sociale contemporaine*, 2d ed., vol. 1 (Paris, 1885), chap. 2.

4. Sanford Elwitt, *The Third Republic Defended: Bourgeois Reform in France, 1880–1914* (Baton Rouge, 1986).

5. Cited in Judith F. Stone, *The Search for Social Peace: Reform Legislation in France 1890–1914* (Albany, 1985), p. 30.

6. Alexandre Millerand, *Travail et travailleurs* (Paris, 1908), p. 11.

7. On the Musée Social, see Elwitt, *Third Republic Defended*, pp. 156–69.

8. Laurent Deschesne, "La productivité du travail et les salaires," *Revue d'Économie Politique* 13 (1899): 467.

9. Cited in Elwitt, *Third Republic Defended*, p. 13.

10. Raoul Jay, *La Protection légale des travailleurs* (Paris, 1904), p. 129.

11. Armand Imbert, "Les méthodes du laboratoire appliquées à l'étude directe et pratique des questions ouvrières," *Revue Générale des Sciences Pures et Appliquées* 22 (30 June 1911): 481.

12. XIIIè Congrès international d'hygiène et de démographie, tenu à Bruxelles du 3 au 8 Septembre 1903. Compte-rendus du Congrès 5, sec. 4 (Brussels, 1903), pp. 76, 77.

13. Cited in Armand Imbert, "Le surmenage par suite du travail professionnel au XIVè Congrès international d'hygiène et de démographie" (Berlin, Septembre 1907), *Année Psychologique* 4 (1908): 243.

14. Armand Imbert, "Le surmenage par suite du travail professionnel," in *Bericht über den XIV. Internationalen Kongress für Hygiene und Demographie*, Berlin, 23–29. September 1907, 4, sec. 1 (Berlin, 1908), pp. 635, 637.

15. Paul Louis, *L'ouvrier devant l'état: Histoire comparée des lois du travail dans les deux mondes* (Paris, 1904), p. 206.

16. M. Pierrot, *Travail et surmenage* (Paris, 1911), p. 13.

17. Jay, *La Protection légale des travailleurs*, p. 129.

18. Gary Cross, *A Quest for Time: The Reduction of Work in Britain and France 1840–1940* (Berkeley, 1989), pp. 58, 59.

19. Heinrich Herkner, "Arbeitszeit," *Handwörterbuch der Staatswissenschaft*, ed Johannes Conrad, Ludwig Elster, and Wilhelm Lexis (Jena, 1909), p. 1201.

20. See Cross, *A Quest for Time*, p. 43; Gary Cross, "Worktime in International Discontinuity 1886–1940," in *Worktime and Industrialization: An International History*, ed. Gary Cross (Philadelphia, 1988), p. 165.

21. Ibid. Also see Gary Cross, "The Quest for Leisure: Reassessing the Eight-Hour Day in France," *Journal of Social History* 18 (Winter 1984): 195–216; and "The Political Economy of Leisure in Retrospect: Britain, France and the Origins of the Eight-Hour Day," *Leisure Studies* 5 (1896): 69–90.

22. Wolfgang Ruppert, *Die Fabrik: Geschichte von Arbeit und Industrialisierung in Deutschland* (Munich, 1983), pp. 56, 58; Dieter Groh, "Intensification of Work and Industrial Conflict in Germany 1896–1914," *Politics and Society* 8, nos. 3, 4 (1978): 350–97.

NOTES

23. Ernst Bernhard, *Höhere Arbeitsintensität bei kürzerer Arbeitszeit: Ihre personalen und technisch-sachlichen Voraussetzungen*, diss. (Leipzig, 1909), p. 3.
24. Stone, *Search for Social Peace*, p. 125.
25. Jean Desplanque, *Le Problème de la réduction de la durée du travail devant le parlement français* (Paris, 1918), pp. 203–5.
26. Stone, *Search for Social Peace*, p. 130.
27. Hyppolite Taine, *Notes sur l'Angleterre* (Paris, 1871), p. 305.
28. Cited in Deschesne, "La productivité du travail," p. 461.
29. The veneration of Japanese workers today is similar. The major difference of course is that the wages of English workers were higher than those of workers on the Continent, while Japanese wage levels are considerably lower than those of American workers. A. Shadwell, *England, Deutschland und Amerika* (Berlin, 1907).
30. Victor Delahaye, *Rapport à l'exposition coloniale et internationale d'Amsterdam 1883*, Ministère du Commerce et de l'Industrie (Paris, 1886), p. 3.
31. Lujo Brentano, *Über das Verhältnis von Arbeitslohn und Arbeitszeit zur Arbeitsleistung* (Leipzig, 1893), p. 11.
32. In Bernard, *Höhere Arbeitintensität*, p. 5.
33. Eduard von Hartmann, *Die sozialen Kernfragen* (Leipzig, 1894), pp. 377–83.
34. Brentano, *Arbeitslohn und Arbeitszeit*, p. 1.
35. Ibid., p. 23.
36. Ibid., p. 33.
37. Ibid., p. 36. For a discussion of Brentano's views on social policy and the English model, see James J. Sheehan, *The Career of Lujo Brentano: A Study of Liberalism and Social Reform in Imperial Germany* (Chicago, 1966), pp. 22–45.
38. Alfred Weber, *Der Kampf zwischen Kapital und Arbeit; Versuch einer systematischer Darstellung mit besonderer Berücksichtigung der gegenwärtigen deutschen Verhältnisse* (Tübingen, 1910) p. 94.
39. Wilhelm Hasbach "Zur Charakterisierung der englischen Industrie," in *Schmoller's Jahrbuch*, (1903): 359–75.
40. Herkner, "Arbeitszeit," p. 1202.
41. Weber, *Kampf Zwischen Kapital und Arbeit*, p. 92.
42. Patrick Fridenson, "Unternehmerpolitik, Rationalisierung, und Arbeiterschaft: Französische Erfahrungen im internationalen Vergleich, 1900 bis 1929," in *Recht und Entwicklung der Grossunternehmen im 19. und frühen 20. Jahrhundert*, ed. Robert Horn and Jürgen Kocka (Göttingen, 1979), pp. 429–34.
43. Several parliamentary commissions in England and on the Continent had already established the connection between higher wages and higher productivity in the 1880s. For examples of the European literature of English and American success, see Gerhard von Schulze-Gävernitz, *Der Grossbetrieb, ein wirtschaftlicher und sozialer Fortschritt. Eine Studie auf dem*

Gebiete der Baumwollindustrie (Leipzig, 1892) (French translation, 1896). [All subsequent citations are from the British translation: *The Cotton Trade in England and on the Continent: A Study in the Field of the Cotton Industry,* trans. Oscar S. Hall (London, 1895)]. For additional works comparing wages and output in Europe and America, see J. Schoenhof, *The Economy of High Wages* (New York, 1892); Émile Lavasseur, *L'ouvrier américain,* 2 vols. (Paris, 1898); Deschesne, "La productivité du travail et les salaires," pp. 460–01; Émile Waxweiler, *Hauts salaires en Amérique* (Brussels, 1895).

44. Cited in Cross, *A Quest for Time,* p. 63.
45. Deschesne, "La productivité du travail," p. 466.
46. Schulze-Gävernitz, *The Cotton Trade in England,* p. 128. Also see his *Britischer Imperialismus und englischer Freihandel* (Leipzig, 1906).
47. Ibid., p. 131. Schulze-Gävernitz adds: "The higher wages, in connection with the lower food prices, make possible an extraordinary good nourishment of the English operative. The English workman lives on meat and wheat-flour bread, whilst potatoes mostly form the chief sustenance of the German factory-worker. If the English operative, as shown above, minds two to three times as much machinery as the German, he also certainly eats two to three times as much—not in quantity, but in nourishing value from a physiological point of view" (p. 138).
48. Weber, *Der Kampf zwischen Kapital und Arbeit,* p. 101.
49. John Rae, *Eight Hours for Work* (London, 1894), pp. 162, 163.
50. Deschesne, "La productivité du travail," p. 463.
51. Cited in Ibid.
52. Cited in Ibid.
53. "The evolution of *'le Standard of Life'* corresponds to industrial evolution; the substitution of machines, independent of the general progress of the human effort, demands a greater intellectual effort." Cited in Ibid., p. 464.
54. Ibid., p. 464.
55. Ibid., p. 409. Salaries, he claimed, are calculated by the sum "that remains after the costs of production, interest, and profits are deducted."
56. Ibid., p. 411.
57. Ibid., p. 414.
58. Fromont's experiences were first published in the *Bulletin de l'Association de l'École de Liège* (11 July 1897); L.-G. Fromont, *La Journée de huit heures dans l'industrie chimique et métallurgique. Expériences pratiques* (Liège, 1905) and were hailed in the publications of the Solvay Institute: L.-G. Fromont, *Une expérience industrielle de réduction de la journée de travail* (Brussels, 1906).
59. Ernest Mahaim, "Préface," in Fromont, *Une expérience industrielle,* pp. xiii–xviii.
60. "The work was subdivided as follows: the first brigade A, arrives at 6 A.M., leaves the factory at fourteen hours (2 P.M.) after having had a half-hour rest at 10 A.M.); the second brigade B, begins work at 14 hours (2 P.M.) and

continues until 22 hours (10 P.M.); the third brigade C works from 22 hours (10 P.M.) until 6 A.M. the next day, and rests from 1 to 1:30 A.M.). In order for the same men not to have to always work through the night, the shifts were rotated every Sunday, in such a way that the amount of Sunday work was reduced as well: In a year, it is a matter of a complete rest of 24 hours for 35 Sundays, instead of 26 under the old system." Ibid., p. 55.

61. Ibid., pp. 45–53.

62. Ibid., p. 2.

63. Ruppert, *Die Fabrik,* p. 44; One of the first English firms to monitor carefully the effects of the reduction of the working day on productivity was the Salford Engineering Works, where in 1893 the workweek was reduced from 53 hours to 48 hours, leading to a slight increase in the weekly output, despite the loss of five hours of working time. Georges Friedmann, *Industrial Society: The Emergence of the Human Problems of Automation* (New York, 1955), p. 87; Also see Rae, *Eight Hours for Work,* p. vii–x.

64. Anthony Oberschall, *Empirical Research in Germany 1848–1914* (Paris, 1965), p. 112. On Abbe's career see Felix Auerbach, *Ernst Abbe: Sein Leben und Wirken* (Leipzig, 1919); Paul Gerhard Esche, *Ernst Abbe* (Leipzig, 1963).

65. Ernst Abbe, *Sozialpolitische Schriften* (Jena, 1906), p. 205.

66. Ibid.

67. Jay, *La Protection légale des travailleurs,* p. 166.

68. Weber, *Kampf zwischen Kapital und Arbeit,* p. 97.

69. Emmanuel Roth, "Ermüdung durch Berufsarbeit," in *Bericht über den XIV. Internationalen Kongress für Hygiene und Demographie,* Berlin 23.–29. September, 1907, 4, sec. 1 (Berlin, 1908), p. 595.

70. Imbert, "Le surmenage par suite du travail, *Kongress für Hygiene und Demographie,* Berlin, 1907, p. 637.

71. André Liesse, *Le travail aux points de vue scientifique, industriel et social* (Paris, 1899), p. 18.

72. Jules Amar, *Le Moteur humain et les bases scientifiques du travail professionnel* (Paris, 1914), pp. 675, 676.

73. Theodor Sommerfeld, *Traité des maladies professionnelles* (Brussels, 1901), cited in Ioteyko, *The Science of Labour* (London, 1919), p. 27.

74. Imbert, "Le surmenage par suite du travail professionnel au XIVè Congrès international d'hygiène et de démographie," p. 242.

75. Jean-André Tournerie, *Le Ministère du travail (origines et premiers développements)* (Paris, 1971), p. 221. The rest day had to be twenty-four consecutive hours, though not necessarily on Sunday.

76. The proposal was debated in the Chamber for two years (over eighteen sessions) before coming to a vote.

77. Desplanque, *Réduction de la durée du travail,* pp. 212, 213.

78. Ibid., p. 215.

79. Ibid., pp. 221, 226.

80. Ibid., p. 222.

81. "La proposition de loi de M. Vaillant sur l'institution de la journée de 8 heures et du salaire minimum," 11 March 1912. Texts of the "Commission du Travail, 11 March 1912, in Ibid., p. 533.

82. Ibid., p. 532.

83. Ibid., pp. 224–25.

84. Imbert acknowledged Vaillant's role in securing financial support in his "Les méthodes du laboratoire appliquées à l'étude directe et pratique des questions ouvrières," p. 486: "Au cours de la séance du 16 février dernier, en effet, à la Chambre des Députés, M. Maurice Faure, ministre de l'instruction publique, en réponse à M. le député Ed. Vaillant, a annoncé qu'il consacrait dès cette année 1911, une somme de 4000 francs à des subventions en vue de recherches relatives à l'étude expérimentale du travail professionnel ouvrier."

85. Desplanque, *Réduction de la durée du travail,* p. 226.

86. Ibid., p. 227. Vaillant was not the first socialist to adopt the energetic calculus. Victor Delahaye had already made essentially the same argument in 1883 when he attributed the causes of "our inferior productivity" to the "impossibility" of successfully competing against foreign industry, an inferiority which he attributed to the "excessive prolongation of the working day," low salaries, high prices, and obsolete tools and equipment in French industry. According to Delahaye, "The industrial prosperity of a nation is the direct consequence of the well-being of the workers of that nation." Delahaye, *Rapport,* p. 4.

87. Ibid., p. 535.

88. Cited in Desplanque, *Réduction de la durée du travail,* p. 13. Also see H. and D. Monod, "Jules Amar (1879–1935)—À propos d'un centenaire," *Histoire des sciences médicales,* 3 (Paris, 1979), pp. 229, 230. The commission was formed of two senators, three deputies, two members of the Academy of Moral and Political Sciences, two members of the Academy of Medicine, two members of the Conseil Supérieur du Travail, three professors of higher education, four persons chosen by the scientific societies, one representative of the Ministry of Agriculture, one representative of the Ministry of Commerce and Industry, one representative of the Ministry of Labor.

89. Ibid., p. 14.

90. Ibid., pp. 13, 14. At the outset it consisted of two small rooms in the rue Saint Martin, an assistant, a laboratory worker, and a budget of 10,000 francs. On Le Chatelier's role, see chapter 9.

91. Alain Ehrenberg, *Le Corps militaire: Politique et pédagogie en démocratie* (Paris, 1983), p. 109.

92. Georges Vigarello, *Le Corps redressé: Histoire d'un pouvoir pédagogique* (Paris, 1978), p. 252.

93. On the physical education debate see Nye, *Crime, Madness and Politics,* pp. 324–29; Jacques Ullmann, *De la gymnastique aux sports modernes; Histoire des doctrines de l'éducation physique* (Paris, 1971); Eugen Weber,

"Pierre de Coubertin and the introduction of organised Sport in France," *Journal of Contemporary History* 5, no. 2 (1970): 3–26. Only gymnastics, the antisports polemicists claimed, was capable of simultaneously promoting both the health of the student, and the development of the physique. In this regard, Mosso cautioned that it is not physique or "morphology," *per se* that determine the capacity for exercise or training, but the invisible energies of the body and their deployment. Demeny pioneered the introduction of rational gymnastics in France and organised the *Cercle de gymnastique rationnelle* to apply the achievements of physiology to physical education. Whereas Demeny criticized the Swedish system as excessively enervating, Tissié and Fernand Lagrange defended it against the illusions of what Tissié called "a speculative gymnastics of the laboratory," which imagines that by a cinematographical approach to motion, it could create a new form of physical education. See also Philippe Tissié, *La Fatigue et l'entraînement physique* (Paris, 1897), p. 177; Georges Demeny, *L'éducation de l'effort, psychologie—physiologie* (Paris, 1916), p. 27.

94. Nye, *Crime, Madness and Politics in Modern France*, p. 321.
95. Angelo Mosso, *L'éducation physique de la jeunesse* (Paris, 1895), p. 56.
96. Georges Demeny was prolific on this theme: *Les Bases scientifiques de l'éducation physique* (Paris, 1902); *Résumé de cours théoriques sur l'éducation physique; Mécanisme et éducation des mouvements* (Paris, 1904); *Plan d'un enseignement supérieur de l'éducation physique* (Paris, 1899); *L'éducation physique en Suède. Rapport sur une mission en Suède.*
97. Fernand Lagrange, "La réforme de l'éducation physique," *La Revue des Deux-Mondes* 4 (1892): 363; Compare also Vigarello, *Le Corps redressé*, p. 210.
98. Demeny, *Les Bases scientifiques*, p. 9.
99. Demeny, *Plan d'un enseignement supérieur de l'éducation*, p. 8.
100. Ibid., p. 9.
101. Etienne-Jules Marey, *Rapport de Marey*, Ministère du Commerce et de l'Industrie, Exposition Universelle Internationale de 1900, Travaux de la Commission d'Hygiène et de Physiologie (Paris, 1901), p. 5.
102. *Nouveau règlement sur l'instruction de la gymnastique militaire*, République Française. Ministry of War, Décret du 7 Août 1902 (Paris, 1902).
103. Philippe Tissié, "Le nouveau règlement," *Revue Scientifique* (16 May; 30 May; 7 November 1903). The debate is summarized in Philippe Tissié, *L'évolution de l'éducation physique en France et en Belgique (1900–1910)* (Pau, 1911), pp. 3–45.
104. Nye, *Crime, Madness and Politics*, p. 325.
105. Cited in Vigarello, *Le Corps redressé*, p. 254.
106. Josefa Ioteyko, *Entraînement et fatigue au point de vue militaire* (Brussels, 1905), p. 7.
107. Jean de Bloch, "L'armée Franco-Russe et la guerre du Transvaal," *La Revue* (1 March 1901). Also see *The Future of War in Its Technical, Economic, and Political Relations* (New York, 1899).

108. Ioteyko, *Entraînement et fatigue*, p. 15.
109. Ibid. Ioteyko pointed out that Germany, which had a two-year term, was "less demoralized from a military point of view than the others." Russia also had a three-year term of service.
110. Ibid., p. 97.
111. Ehrenberg, *Le Corps militaire*, p. 132 passim.
112. Ibid.
113. A. Fastrez, *Ce que l'armée peut être pour la nation*, (Brussels, 1907), p. 70.
114. Imbert did secure the participation of two workers from the département of Hérault to the Congrès International Médical des accidents du Travail, from 29 May to 4 June 1905 in Liège. Armand Imbert, "Congrès ouvriers et congrès scientifiques," *Revue Scientifique* (13 May 1905): 588–90; Rôle des ouvriers dans certains congrès scientifiques," *La Grande Revue* (10 April 1909): 574–78. Also see Georges Ribeill, "L'organisation physiologique du travail," *Cadres CFDT,* 324 (November 1986), p. 51; Ribeill, "Les débuts de l'ergonomie en France à la veille de la Première Guerre mondiale," *Le Mouvement Social* 113 (October–December, 1981) p. 27.
115. Ioteyko, *Science of Labour*, p. 102.
116. Ibid.
117. Several works have called attention to the importance of the accident issue in the formation of the modern "social state." See François Ewald, *L'Etat providence* (Paris, 1986); Jacques Donzelot, *L'invention du social: Essai sur le déclin des passions politiques* (Paris, 1984).
118. Entirely state administered by the newly created Imperial Insurance Office *(Reichsversicherungsamt)*, a unique body composed of civil servants, parliament (Bundesrat), employers' organizations, and workers' representatives. The state assumed two-thirds of the cost and the employer one-third, but the system at first covered only the most "dangerous" occupations. Subsequent revisions of the law extended its scope to agriculture and maritime labor. All German workers employed in industry were eventually insured and compensated to a maximum of two-thirds of their income in case of total disability and loss of earning capacity. The administration of the system at the level of the workplace was assumed by employer corporations *(Berufsgenossenschaften)* organized according to industry. For a close analysis of Bismarck's views on the law, see Walter Vogel, *Bismarck's Arbeiterversicherung: Ihre Entstehung in Kräftespiel der Zeit* (Braunschweig, 1951), especially chapter 10. On Bismarck's social policy see Albin Gladen, *Geschichte der Sozialpolitik in Deutschland: Eine Analyse ihrer Bedingungen, Formen, Zielsetzungen und Auswirkungen* (Wiesbaden, 1974); Klaus Witte, *Bismarck's Sozialversicherungen und die Entwicklung eines marxistischen Reformverständnisses in der deutschen Sozialdemokratie* (Köln, 1980); Karl Erich Born, "Der soziale und wirtschaftliche Strukturwandel Deutschlands am ende des 19. Jahrhunderts," in *Moderne deutsche Sozialgeschichte*, ed. Hans Ulrich Wehler (Köln, 1966). Also compare Vernon Lidtke, "German Social Democracy and Ger-

man State Socialism 1876–1884," *International Review of Social History* 9 (1964): 202–25.

119. R. van der Borght, "Arbeiterversicherung (Allgemeines)," *Handwörterbuch der Staatswissenschaft*, p. 790.

120. Emile Cheysson, "France, les assurances ouvrières," "Congrès International des Accidents du Travail et des Assurances Sociales," *Bulletin du Comité Permanent*, (Paris, 1893), p. 329. [Hereafter cited as C.I.A.T.A.S.]

121. Yvon Le Gall, *Histoire des accidents du travail*, vol. 1 (Paris, 1981), p. 36.

122. Cited in Ibid., p. 45. German statistics did, however, indicate that mechanized accidents were the fastest growing.

123. Georg von Mayr, "L'assurance et la fréquence des accidents," C.I.A.T.A.S., Milan (1894), p. 340; Georg von Mayr, "Unfallversicherung und Sozialstatistik," *Archiv für soziale Gesetzgebung und Statistik* (1888): 203–45.

124. *C.I.A.T.A.S.*, Paris, 1900, p. 98.

125. von Mayr, "L'assurance et la fréquence des accidents," p. 339.

126. "Bericht die am 13. Januar 1887 stattgefundene Generalversammlung der Nordwestlichen Gruppe des Vereins deutscher Eisen-und Stahl-industrieller," *Stahl und Eisen*, 2 (February 1887): 125. Also see Monika Breger, *Die Haltung der industriellen Unternehmer zur staatlichen Sozialpolitik in den Jahren 1878–1891* (Frankfurt-am-Main, 1982), pp. 87–89.

127. The literature on this subject is vast. For an overview see René Sand, *La Simulation et l'interprétation des accidents du travail* (Brussels, 1907). The most important specialized works are Dr. Carl Thiem, *Handbuch der Unfallkrankungen auf Grund ärtzlicher Erfahrungen* (Stuttgart, 1898), p. 38; Henri Secrétan, *L'assurance contre les accidents: Observations chirurgicales et professionnelles* 3d ed. (Geneva, 1906).

128. Fritz Stier-Somlo, "Ethik und Psychologie im deutschen Sozialrecht," *Archiv für Rechts- und Wirtschaftsphilosophie* 1 (1907/8): 232–47.

129. Ludwig Bernhard, *Unerwünschte Folgen der deutschen Sozialpolitik* (Berlin, 1913), p. 3.

130. H. Mamy, "Aperçu des succès obtenus par les mesures préventives contre les accidents," *Bericht über den XIV. Internationalen Kongress für Hygiene und Demographie*, Berlin, 23.–29. September 1907, 4, sec. 2 (Berlin, 1908), p. 663.

131. Dr. Hubert Coustan, *De la simulation et de l'évaluation des infirmités dans les accidents du travail* (Montpellier, 1902), p. 9.

132. Ibid., p. 7.

133. Victor Griffuelhes and Louis Niels, *Les Objectifs de nos luttes de classes* (Paris, 1909), p. 19.

134. Rudolf Wissell, "Täuschung und Übertreibung auf dem Gebiet der Unfallversicherung," *Sozialistische Monatshefte* 1, no. 10 (June, 1909): 635.

135. Robert Schmidt, "Der Streit um die Rente," *Sozialistische Monatshefte* 1, no. 5 (May 1905): 421.

136. Von Mayr, "L'assurance et la fréquence des accidents," p. 343.

137. *C.I.A.T.A.S.*, Paris, 1900, p. 79.

138. Imbert, "Le surmenage par suite du travail professionnel," *Kongress für Hygiene und Demographie,* Berlin, 1907, p. 644.
139. Armand Imbert, M. Mestre, "Statistique d'accidents du travail," *Revue Scientifique* 4, no. 13 (24 September 1904): 386.
140. Armand Imbert, "Les accidents du travail et les compagnies d'assurances," *Revue Scientifique* 1, no. 23 (4 June 1904): 718.
141. Ibid., p. 711.
142. Ibid.
143. According to the French law of 1898, all accidents in certain trades and occupations had to be declared that resulted in an incapacity to work of more than four days. The declarations were collected by the Inspecteur du Travail of each département. See Imbert and Mestre, "Statistique d'accidents du travail," pp. 385–90.
144. Ibid., p. 386.
145. Imbert, "Les accidents du travail et les compagnies d'assurances," p. 715.
146. Armand Imbert, "Le surmenage par suite du travail professionnel," *Kongress für Hygiene und Demographie,* Berlin, 1907, p. 644.
147. Imbert, "Statistique d'accidents du travail," p. 385.
148. Emannuel Roth, "Ermüdung durch Berufsarbeit," Bericht über den XIV. Internationalen Kongress für Hygiene und Demographie, Berlin, 23.–29. September, 1907, 4, sec. 1 (Berlin, 1908), p. 618; H. Bille-Top, "Die verteilung der Unglücksfälle der Arbeiter auf die Wochentage nach Tagesstunden," *Zentralblatt für allgemeine Gesundheitspflege* 27 (1908): 197; Also see Hugo Münsterberg, *Grundzüge der Psychotechnik* (Leipzig, 1914), p. 394. Imbert cites similar statistics from nine additional départements in the area around Toulouse which show similar patterns. Imbert and Mestre, "Statistique d'accidents du travail," p. 387.
149. Emanuel Roth, *Kompendium der Gewerbekrankheiten und Einfürung in die Gewerbehygiene* (Berlin, 1909), pp. 14, 15. A comparison of the first two editions of Roth's standard work on industrial hygiene reveals the impact of the first studies of fatigue and accidents. While the 1907 edition, *Gewerbehygiene* (Leipzig, 1907) did not make more than an allusion to the problem of fatigue as a cause of accidents, the *Kompendium,* published two years later, discussed the issue extensively.
150. Ibid., p. 15.
151. Roth, "Ermüdung durch Berufsarbeit," p. 606.
152. See Josef Ehmer, "Rote Fahnen—Blauer Montag: Soziale Bedingungen von Aktions- und Organisationsformen der frühen Wiener Arbeiterbewegung," in *Wahrnehmungsformen und Protestverhalten: Studien zur Lage der Unterschichten im 18. und 19. Jahrhundert,* ed. Detlev Puls (Frankfurt-am-Main, 1979), pp. 143–74; Douglas A. Reid, "Der Kampf gegen den 'Blauen Montag' 1766 bis 1876," in ibid., pp. 265–95; Susanna Barrows, "After the Commune: Alcoholism, Temperance and Literature in the Early Third Republic," *Consciousness and Class Experience in Nineteenth-Century Europe,* ed. John Merriman (New York, 1979), pp. 205–18;

Jeffrey Kaplow, "La fin de Saint-Lundi: Étude sur le Paris ouvrier au XIXe siècle," *Temps Libre* 2 (1981): 107–18.

153. Edmund Fischer, "Trinken und Arbeiten," *Sozialistische Monatshefte* 1, no. 5 (May 1908): 360–67.

154. A. Imbert and M. Mestre, "Nouvelles statistiques d'accidents du travail," *Revue Scientifique* 4, no. 17 (21 October 1905): 525.

155. Roth, "Ermüdung durch Berufsarbeit," p. 619.

156. Roth, *Kompendium*, p. 13.

157. Patrick Fridenson, "France, États-Unis: Genèse de l'usine nouvelle," *Recherches* 32/33 (September 1978): 382.

158. Roth, *Kompendium*, p. 12.

159. Zaccaria Treves, "Le surmenage par suite du travail professionnel," Bericht über den XIV. Internationalen Kongress für Hygiene und Demographie, Berlin, 23.–29. September 1907, 4, sec. 2 (Berlin, 1908), p. 626.

160. Ibid., p. 627.

161. Imbert and Mestre, "Nouvelles statistiques d'accidents du travail," p. 522.

162. Ibid., p. 521.

163. Philippe Delahaye, "La prétendue fatigue des ouvriers envisagée comme cause des accidents du travail," *Revue Industrielle* (8 October 1904): 408.

164. Imbert and Mestre, "Nouvelles statistiques d'accidents du travail," p. 520, and A. Imbert and M. Mestre, "À propos de l'influence de la fatigue professionnelle sur la production des accidents du travail," *Revue Industrielle* (5 November 1905): 449.

165. Marie Bernays, *Auslese und Anpassung der Arbeiterschaft der geschlossenen Grossindustrie, dargestellt an den Verhältnissen der "Gladbacher Spinnerei und Weberei A.G." zu München Gladbach im Rheinland.* Schriften des Vereins für Sozialpolitik, 133 (Leipzig, 1910), p. 189.

166. Ioteyko, *Science of Labour,* p. 35.

167. "Bericht über den XIV. Internationalen Kongress für Hygiene und Demographie," Berlin, 23.–29. September 1907, 4, sec. 4 (Berlin, 1908), pp. 267–304; "Premier congrès de l'hygiène des travailleurs et des ateliers," October 1904 (Paris, 1905). Also see Elwitt, *Third Republic Defended,* p. 143.

168. The Sanitorium der Landesversicherungsanstalt Beelitz, near Potsdam, for example, reported a rise from 18 to 40 percent in the number of patients suffering from worker's neurasthenia in the period from 1897 to 1903. A study of 230 men and women textile workers near Potsdam showed that the great majority showed signs of "Blutarmut" (poor blood) and neurasthenia. Of 22 younger workers, 7 were anemic, and 2 neurasthenic; of 174 women workers, 32 were anemic, 2 neurasthenic, and 3 suspected of tuberculosis. Of 145 women studied in one sanitorium, 110 suffered from poor blood, iron deficiency or neurasthenia and 75 percent showed signs of overwork. P. Leubischer and W. Bibrowicz, "Die Neurasthenie im Arbeiterkreisen," *Deutsche medizinischer Wochenschrift* 21 (25 May 1905): 820–24; M. Schönhals, *Über die Ursachen der Neurasthenie und Hysterie*

bei Arbeitern (Berlin, 1906). A survey of the neurasthenia diagnosis among workers indicates that, in contrast to the hysteria diagnosis which was associated with women, neurasthenic symptoms were more equally distributed according to age and gender.

169. W. Eisner, "Die Ermüdung durch Berufsarbeit," *Bericht über den XIV. Internationalen Kongress für Hygiene und Demographie,* Berlin, 23–29. September 1907, 4, sec. 2 (Berlin, 1908), p. 583.

170. Imbert, "Le surmenage par suite du travail professionnel au XIVè Congrès international d'hygiène et de démographie," p. 232.

171. Omer Buyse, "Le problème psychophysique de l'apprentissage," *Revue Psychologique* 3 (1910): 377–99.

172. Hector Depasse, *Du travail et de ses conditions* (Paris, 1895), pp. 52, 53.

CHAPTER 9. THE AMERICANIZATION OF LABOR
POWER AND THE GREAT WAR 1913–1919

1. On Taylor and the introduction of the Taylor system see Daniel Nelson, *Frederick W. Taylor and the Rise of Scientific Management* (Madison, 1980); Harry Braverman, *Labor and Monopoly Capitalism: The Degradation of Work in the Twentieth Century* (New York, 1974); Judith A. Merkle, *Management and Ideology: The Legacy of the International Scientific Management Movement* (Berkeley, 1980); Michael Burowoy, "Toward a Marxist Theory of the Labor Process: Braverman and Beyond," *Politics and Society* 8, no. 4 (1978): 247–312. On France see Aimée Moutet, "Les origines du système de Taylor en France: Le point du vue patronal (1907–1914)," *Mouvement Social* 93 (October–December, 1975): 15–49; Patrick Fridenson, "Un tournant Taylorien de la société française (1904–1918)," *Annales, ESC* 5 (September–October, 1987): 1031–1060; Patrick Fridenson, "Unternehmerpolitik, Rationalisierung und Arbeiterschaft: französische Erfahrungen im internationalen Vergleich, 1900 bis 1929," *Recht und Entwicklung der Grossunternehmen im 19. und frühen 20. Jahrhundert,* ed. Robert Horn/Jürgen Kocka (Göttingen, 1979). For Germany see Heidrun Homburg, "Anfänge des Taylor-Systems in Deutschland," *Geschichte und Gesellschaft* 4 (1978): 170–94.

2. On Taylorism and the science of work see Georges Ribeill, "Les débuts de l'ergonomie en France à la veille de la Première Guerre Mondiale," *Le Mouvement Social* 113 (October–December, 1980): 31–33; Olivier Christin, *Les enjeux de la rationalisation industrielle* (1910–1929), mémoire de maîtrise, Université Paris I, (Paris, 1982), especially chap. 3; Fridenson, "Un tournant Taylorien," pp. 1048–50.

3. Frederick Winslow Taylor, *The Principles of Scientific Management* (New York, 1967).

4. In *Labor and Monopoly Capitalism,* Harry Braverman gives a Marxist

account of Taylorism as the key force in securing the modern capitalist expropriation of initiative and control over the labor process, resulting in a "degraded" work environment. Braverman adopts the view that Taylorism is the *deus ex machina* of capitalist development, decisively changing not merely the labor process, but the working class. By replacing the skilled and militant worker with the more-difficult-to-organize unskilled mass production worker, capitalism does away with its "contradiction." He finds support in the positive account by Merkle, which sees scientific management as "the most widely known and influential system of factory management in the industrial world," *Management and Ideology*, p. 8.

5. Historians of the Taylor system have been more cautious in discussing the extent and nature of its initial reception. See, for example, Daniel Nelson, *Managers and Workers: Origins of the New Factory System in the United States 1880–1920* (Madison, 1975); Alfred D. Chandler, Jr., *The Visible Hand* (Cambridge, Mass., 1977). Social historians have also modified their views of worker resistance. In his analysis of the German trade union reaction, Gunnar Stollberg points to the essentially favorable reaction of the trade union movement to Taylorism, and to the lack of long-term resistance on the part of workers. Gunnar Stollberg, *Die Rationalisierungsdebatte 1908–1933: Freie Gewerkschaften zwischen Mitwirkung und Gegenwehr* (Frankfurt-am-Main, New York, 1981), pp. 36–39; Patrick Fridenson also paints a more varied picture, emphasizing a spectrum of worker reactions. For French workers' response, see Fridenson, "Un tournant Taylorien," pp. 1044–47; Aimée Moutet, "La Première Guerre Mondiale et le Taylorisme," *Le Taylorisme: Actes du Colloque international sur le Taylorisme organisé par l'Université de Paris XIII*, 2–4 May 1983 (Paris, 1984), p. 78.

6. Among the most lucid descriptions is Michael P. Hanagan, *The Logic of Solidarity: Artisans and Industrial Workers in Three French Towns, 1871–1914* (Urbana, 1980), p. 131; also see Eric J. Hobsbawm, "Custom, Wages, and Work Load," *Workers in the Industrial Revolution*, ed. Peter N. Stearns and Daniel J. Walkowitz (New Brunswick, 1974), pp. 246, 247.

7. For the reception of Taylorism among French industrialists see G. C. Humphreys, *Taylorism in France 1904–1920* (New York, 1986); Patrick Fridenson, "Die Arbeiter der französischen Automobilindustrie 1890–1914," *Wahrenehmungsformen und Protestverhalten: Studien zur Lage der Unterschichten im 18. und 19. Jahrhundert*, ed. Detlev Puls (Frankfurt-am-Main, 1979), pp. 228–64; Patrick Fridenson, *Histoire des usines Renault: Naissance de la grande enterprise, 1898–1939*, vol. 1 (Paris, 1972), pp. 70–79; Yves Cohen, "Le système de la pratique: Un organisateur-directeur, les automobiles Peugeot, 1917–1939," *Travail et automatisation dans l'industrie automobile:* Actes du gerpisa, no. 2 (Grenoble, 1986), pp. 3–23; Yves Cohen, "Ernst Mattern chez Peugeot (1906–1918): Ou comment peut-on être Taylorien?" in *Le Taylorisme*, pp. 115–26.

8. Cited in Merkle, *Management and Ideology*, p. 15.

9. On engineers and the Taylor System in France see Yves Cohen, "La pra-

tique des machines et des hommes: Une pensée technique en formation (1900–1914)," *L'ingénieur dans la société française,* ed. André Thépot (Paris, 1985), pp. 61–70; Christin, *Les enjeux de la rationalisation industrielle;* Patrick Fridenson, "Un tournant Taylorien," pp. 1038–40.

10. Alphonse Merrheim, *Vie ouvrière* (20 February 1913): 214, 215. See Nicholas Papayanis, *Alphonse Merrheim. The Emergence of Reformism in Revolutionary Syndicalism 1871–1925* (Boston, 1985), pp. 68, 69.

11. Émile Pouget, *L'organisation du surmenage* (Paris, 1914); *Zeitschrift des Vereins deutscher Ingenieure* (8 March 1913).

12. Eugène Fournière, in the *Revue Socialiste,* cited in Fridenson, "Un tournant Taylorien," p. 1045; Papayanis, *Merrheim,* p. 69.

13. Ribeill, "Les débuts de l'ergonomie," pp. 31–33. For a detailed discussion of the ergonomist's reception of Taylorism, which includes criticism of several recent approaches, including my own earlier views on the subject, see Fridenson, "Un tournant Taylorien," pp. 1031–60. Also see Anson Rabinbach, "The European Science of Work: The Economy of the Body at the End of the 19th Century," *Work in France: Representations, Meaning, Organization, and Practice,* ed. Steven L. Kaplan and Cynthia Koepp (Ithaca, 1986), pp. 510–12. Fridenson emphasizes the divergent approaches to Taylorism among French ergonomists. I would still argue for a firm distinction between the two schools of industrial science before the First World War.

14. Josefa Ioteyko, The *Science of Labour* (London, 1919), p. 4.

15. Christin, *Les enjeux de la rationalisation industrielle,* chap. 3. Christin's thesis emphasizes the competing professional interests of the academic ergonomists and the firm-based engineers. He points out, however, that they shared a number of common traits, above all, a faith in science and a commitment to productivism. Also see Humphrey's, *Taylorism in France,* pp. 213–19.

16. Michelle Perrot, "The Three Stages of Industrial Discipline," *Consciousness and Class Experience in Nineteenth Century Europe,* ed. John A. Merriman (New York, 1979), pp. 163, 164.

17. It is interesting to note that this problem of the relation of the "ergonomists" to the "Taylorists" persists even today. See the discussion of the "amicable competition" between the two in Maurice de Montmollin, "Actualité du taylorisme," *Le Taylorisme,* pp. 17, 18; and the articles in *Ein Programm und seine Wirkungen; Analyse von Zielen und Aspekten zur Forschung 'Humanisierung des Arbeitslebens'* (Frankfurt-am-Main, 1982).

18. Patrick Fridenson, "Un tournant Taylorien," pp. 1031–60.

19. On Henry Le Chatelier, see Merkle, *Management and Ideology,* pp. 149, 150; Fridenson, "Un tournant Taylorien," pp. 1039, 1048 passim; Moutet, "Les origines," pp. 41–44.

20. Henry Le Chatelier, "Du rôle de la science dans l'industrie," *Revue de Métallurgie* (1904): 1–10.

21. Merkle, *Management and Ideology*, p. 150.

22. Henry Le Chatelier, "L'organisation scientifique du travail dans les usines modernes et les nouvelles méthodes américaines de travail de F. Taylor," *La Technique Moderne* 5, no. 1 (1 July 1912): 47. This article is actually composed of passages of Le Chatelier's Preface to Taylor's *Principes d'organisation scientifique des usines* (Paris, 1912). Compare Ribeill, *"Les débuts de l'ergonomie,"* p. 32.

23. Le Chatelier, "L'organisation scientifique du travail," p. 47. Also see Henry Le Chatelier, "Frederic Winslow Taylor (1856–1915)," *Revue de Métallurgie* (April 1915): 185–232.

24. Ibid.

25. Ibid.

26. On the Renault strike see Patrick Fridenson, "Der Arbeiter der französischen Automobilindustrie," pp. 254, 255.

27. Henry Le Chatelier, "Le système Taylor," *La Technique Moderne* 6, no. 12 (15 June 1913): 449.

28. Letter of Henry Le Chatelier to F. W. Taylor, 12 April 1913, Taylor Archive, 63 B, Stevens Institute of Technology, Hoboken, N.J.

29. Le Chatelier, "Le système Taylor," p. 449.

30. Ribeill, "Les débuts de l'ergonomie," pp. 31–33.

31. Letter of Le Chatelier to Taylor, 12 April 1913. Also cited in Fridenson, "Un tournant Taylorien," p. 1049.

32. Ibid.

33. Jules Amar, "L'organisation scientifique du travail humain," *La Technique Moderne* 7, no. 4 (15 August 1913): 118.

34. Henry Le Chatelier, "Préface," Jules Amar, *Le Moteur humain et les bases scientifiques du travail professionnel* (Paris, 1914), p. x.

35. Amar, *Le Moteur humain*, p. 606.

36. Ibid.

37. Ibid., pp. 609, 610.

38. Charles Fremont, "A propos du système Taylor," *La Technique Moderne* 7, no 9 (1 November 1913): 301. See Fremont's *Origines et évolution des outils*, Mémoire (Paris, 1913).

39. Henri Verne, "Les origines de la mécanique humaine; l'organisation scientifique du travail," *La Technique Moderne* 8, no. 10 (15 May 1914): 377.

40. J.-M. Lahy, "La supériorité professionnelle chez les conducteurs de tramways dans ses rapports avec la consommation d'énergie électrique." *La Technique Moderne* 7, no. 10 (1 Dec. 1913): 368.

41. Ibid.

42. Jean-Marie Lahy, *Le Système Taylor et la physiologie du travail professionnel*, 2d ed. (Paris, 1921), p. 126.

43. Susanne Pacaud, "J.-M. Lahy (1872–1943)" *Le Travail Humain: Bulletin de l'Association Internationale de Psychotechnique* (1952): 338–43.

44. A selection of Lahy's anticlerical works are *La Genèse de la notion d'âme d'après quelques textes ethnographiques* (1905); *L'Individu et la société*

(1908); *La Morale comme faît social* (1909); *Le Rôle de l'individu dans la formation de la morale* (1910); *La Famille: Son évolution* (1910); *La Morale de Jésus. La Part d'influence dans la morale actuelle* (1911).

45. "Recherches sur les conditions du travail des ouvriers typographes composant à la machine dite linotype," *Bulletin de l'Inspection du Travail* 1 (1910): 98.
46. Lahy, "Les signes physiques de la supériorité professionnelle chez les dactylographes," *Comptes-rendus de l'Académie des Sciences* 1 (1913): 1702, 1703.
47. Ribeill, "Les débuts de l'ergonomie," p. 24.
48. J.-M. Lahy, "Le système Taylor et l'organisation intérieure des usines," *La Revue Socialiste* (August 1913): 126–139; "L'étude scientifique des mouvements et le chronométrage," *La Revue Socialiste* (December 1913): 502–13; "Le système Taylor: peut-il déterminer une organisation scientifique du travail?," *La Grande Revue* (25 September 1913): 545–61.
49. Lahy, "L'étude scientifique," p. 504.
50. Jean-Marie Lahy, *Le Système Taylor: Analyse et commentaires*, 2d ed. (Paris, 1921), p. 172.
51. Lahy, "Une organisation scientifique du travail?," p. 540; Lahy, "L'organisation intérieure des usines," p. 126.
52. Lahy, *Le Système Taylor*, pp. 141, 159.
53. Ibid., p. 14.
54. Ibid., p. 27.
55. Ibid., pp. 15, 29.
56. Ibid., p. 51.
57. See chapter 4.
58. Lahy, *Le Système Taylor*, p. 79.
59. Ibid., p. 121.
60. Ibid., pp. 131, 133.
61. Ibid., p. 138.
62. Ibid., p. 123.
63. Ibid., pp. 6, 143.
64. Merkle, *Management and Ideology*, p. 173. Merkle's view that German Taylorism was "corporatist" and appealed to a conservative *Ständestaat* mentality is suggestive, but speculative. In fact, in Germany, as in France, the "carriers" of Taylor's ideas were modernizing engineers, not Prussianizing aristocrats or paternalistic firm owners. See Homburg, "Anfänge des Taylorsystems," pp. 173.
65. Homburg, "Anfänge des Taylorsystems," p. 173.
66. Gunnar Stollberg, *Die Rationalisierungsdebatte* (Frankfurt-am-Main, 1981), pp. 36–39.
67. Jürgen Kocka, "Industrielles Management: Konzepte und Modelle vor 1914," *Vierteljahresschrift für Sozial-und Wirtschaftsgeschichte* 50 (1969): 358.
68. Homburg, "Anfänge des Taylorsystems," p. 174.

69. Stollberg, *Die Rationalisierungsdebatte,* p. 38; Angelika Ebbinghaus, *Arbeiter und Arbeitswissenschaft: Beiträge zur sozialwissenschaftlichen Forschung* 47 (Opladen, 1984), p. 190, n. 63.

70. Ebbinghaus, *Arbeiter und Arbeitswissenschaft,* pp. 177–78.

71. Rudolf Roesler, "Das Taylor System—Eine Budgetierung der menschlichen Kraft," foreword to the 1912 edition, *Die Grundsätze wissenschaftlicher Betriebsführung* (Berlin, 1919), pp. xix, xx.

72. Matthew Hale, Jr., *Human Science and Social Order: Hugo Münsterberg and the Origins of Applied Psychology* (Philadelphia, 1980), p. 148.

73. Hugo Münsterberg, *Psychology and Industrial Efficiency* (New York, 1913), pp. 308, 309.

74. Hale, *Human Science and Social Order,* p. 148.

75. Münsterberg, *Psychology and Industrial Efficiency,* p. 51.

76. Cited in Hale, *Human Science and Social Order,* p. 150.

77. Ibid.

78. Münsterberg, *Grundzüge der Psychotechnik* (Leipzig, 1914), p. 382.

79. Ibid.

80. Hale, *Human Science and Social Order,* p. 155.

81. Hugo Münsterberg, *Psychology and Industrial Efficiency,* p. 309.

82. Hugo Münsterberg, *Psychology and Social Sanity* (Garden City, N.Y., 1914), p. 102.

83. Wilhelm Kochmann, "Das Taylorsystem und seine volkswirtschaftliche Bedeutung," *Archiv für Sozialwissenschaft und Sozialpolitik* 38 (1914): 392.

84. Ibid., p. 415.

85. Ibid., p. 416.

86. Ibid.

87. Emil Lederer, "Die ökonomische und sozialpolitische Bedeutung des Taylorsystems," *Archiv für Sozialwissenschaft und Sozialpolitik* 39 (1914): 770.

88. Ibid., p. 775. On the trade union reaction see Stollberg, *Die Rationalisierungsdebatte,* pp. 77, 78.

89. See chapter 3, this volume, on Marx, and Stollberg, *Die Rationalisierungsdebatte,* pp. 26, 27.

90. Lederer, "Taylorsystems," p. 776.

91. Ibid.

92. Georg Schlesinger, "Betriebsführung und Betriebswissenschaft," Sonderdruck aus *Technik und Wirtschaft* 8 (Berlin, 1913).

93. Ibid., p. 529. Compare also Peter Hinrichs, *Um die Seele des Arbeiters: Arbeitspsychologie, Industrie- und Betriebssoziologie in Deutschland* (Cologne, 1981), pp. 105, 106.

94. Schlesinger, "Betriebsführung," p. 529.

95. Ibid., also see Ebbinghaus, *Arbeiter und Arbeitswissenschaft,* p. 188.

96. Ebbinghaus, *Arbeiter und Arbeitswissenschaft,* p. 191.

97. Paul Fussell, *The Great War and Modern Memory* (Oxford, 1975), p. 41.

98. Ludwig Preller, *Sozialpolitik in der Weimarer Republik* (Düsseldorf, 1949), p. 34.

99. Fritz Giese, *Psychologie der Arbeitshand: Handbuch der biologischen Arbeitsmethoden* (Berlin, 1928), p. 804.

100. Merkle, *Management and Ideology*, pp. 188, 189.

101. Ebbinghaus, *Arbeiter und Arbeitswissenschaft*, p. 196.

102. Ibid., p. 197.

103. Ibid., p. 1051; Martin Fine, "Guerre et réformisme en France, 1914–1918," *Recherches* 32/33 (September 1978): 305–24; Martin Fine, "Albert Thomas: A Reformer's Vision of Modernization, 1914–1932," *Journal of Contemporary History* no. 3 (1977): 545–64.

104. Aimée Moutet, "La Première Guerre Mondiale et le Taylorisme," Montmollin and Pastre, eds., *Le Taylorisme*, p. 78.

105. Ibid., p. 76.

106. Cited in Ibid., p. 67.

107. Ibid., pp. 79, 81.

108. Guenther Wendel, *Der Kaiser Wilhelm Gesellschaft 1911–1914: Zur Anatomie einer imperialistischen Forschungsgesellschaft* (Berlin, 1975), p. 214.

109. Zentrales Staatsarchiv, Potsdam (ZsTA) Merseburg (Central State Archives, Potsdam, GDR). Preussisches Ministerium für Wissenschaft, Kunst und Volksbildung, Rep. 76 VL. Sekt. 2. Tit. 23 LITT A., No. 115. Kaiser-Wilhelm Institut für Arbeitsphysiologie, Denkschrift, May 1912, p. 4. (This material has been kindly made available to me by Dietrich Milles. Hereafter cited as KWI Arbeitsphysiologie, Berlin.)

110. Ibid., p. 4.

111. Max Rubner, *Kraft und Stoff im Haushalte der Natur* (Leipzig, 1909), p. 27.

112. Max Rubner, *Gesetze des Energieverbrauches bei der Ernährung* (Berlin, 1902); *Wandlungen in der Volksernährung* (Leipzig, 1913); *Über moderne Energieformen* (Munich-Berlin, 1914).

113. KWI Arbeitsphysiologie, Berlin, Denkschrift, May 1912, p. 4.

114. Gerhard Albrecht, "Arbeitsgebiet und Ziele des KWI Instituts für Arbeitsphysiologie." *Technik und Wirtschaft* 8 (1915): 284–90. Cited in Dietrich Milles, "Prävention und Technikgestaltung. Arbeitsmedizin und angewandte Arbeitswissenschaft in historischer Sicht," *'Gestalten'—Eine neue gesellschaftliche Praxis*, ed. Felix Rauner (Bonn, 1988), p. 54.

115. Ibid., p. 55.

116. Ibid.

117. KWI Arbeitsphysiologie, Berlin, 4 July 1916.

118. Gerd Hardach, *The First World War 1914–1918* (Berkeley, 1977), pp. 113–19.

119. KWI Arbeitsphysiologie, Berlin, 2 June 1917; Max Rubner, *Deutschlands Volksernährung im Kriege* (Leipzig, 1916).

120. KWI Arbeitsphysiologie, Berlin, 2 June 1917.

121. Undated Flyer, Sozialpolitisches Archiv, Zentrum für Sozialpolitik, Universität Bremen.

NOTES

122. KWI Arbeitsphysiologie, Berlin, 2 June 1917.

123. Ibid., "Protokoll über die Sitzung des Verwaltungsrates," 5 June 1917.

124. Ibid., December 1916.

125. Ibid., 29 May 1917.

126. Ibid. For a detailed discussion see Anson Rabinbach, "Industrial Psychology between Psychotechnics and Politics in Weimar Germany: The Case of Otto Lipmann," in *Les Chantiers de la paix sociale: Techniques et figures du social d'une guerre à l'autre* (in press, Paris, 1991).

127. KWI, Arbeitsphysiologie, Berlin, 29 May 1917.

128. Ibid., 2 June 1917.

129. Armand Imbert was assigned to military staff duty as chief medical officer of the sixteenth medical region. He spent the war concerned mostly with the treatment of neurasthenic disorders. Ribeill, "Les débuts de l'ergonomie," p. 34.

130. Ibid. Jean-Marie Lahy, "Préface," *Le Système de Taylor et la physiologie du travail professionnel* (Paris, 1916).

131. Octave Gramaud, "L'examen psycho-physiologique des soldats mitrailleurs," *La Science et La Vie* 13, no. 36 (December 1917/June 1918): 123.

132. Ibid., p. 128.

133. Dr. Vitoux, "L'examen physiologique des candidats aviateurs," *La Science et la Vie* 10, no. 28 (August, September 1916): 351–56 and Jean Camus "Temps des réactions psycho-motrices des candidats à l'aviation," *Compte-rendus de l'Académie des Sciences* (1916): 106–07.

134. Walther Moede, "Kraftfahrer-Eignungsprüfungen beim deutschen Heer 1915–1918, *Industrielle Psychotechnik* 3, no. 1 (1926): 79.

135. Jules Amar, *The Physiology of Industrial Organization and the Re-employment of the Disabled,* trans. Bernard Miall (New York, 1919), p. 229.

136. Ibid.

137. Jules Amar, "Sur la reéducation professionnelle," *Compte-rendus de l'Académie des Sciences* 1 (1915): 730.

138. Amar, *The Physiology of Industrial Organization,* pp. 286–306; 349–51.

139. Henry Le Chatelier, "Preface," *Physiology of Industrial Organisation,* p. x. Also see J. Gautrelet, "Les bases scientifiques de l'éducation professionnelle des mutilés," *Bulletin de l' Académie Médicale de Paris* 73 (1915): 663–68. The literature on postwar reeducation of the disabled is extensive. For a partial bibliography see Josefa Ioteyko, *La Fatigue* (Paris, 1920), p. 256.

140. Ioteyko, *La Fatigue,* p. 275; for a French survey of the German literature compare Georges Dumas and Henri Aimée, *Névroses et psychoses de guerre chez les Austro-Allemands* (Paris, 1918).

141. Ioteyko, *La Fatigue,* p. 279.

142. An exception is L. Huot and P. Voivenel, *Le Courage* (Paris, 1917).

143. Ioteyko, *La Fatigue,* p. 280.

144. Albert Devaux et Benjamin Logre, *Les Anxieux* (Paris, 1916).

145. Ioteyko, *La Fatigue,* p. 285. Pierre Courbon, "La sinistrose de guerre," *Revue Neurologique* (May–June 1918): 322–27.

146. Ibid., p. 309.

147. Georges Dumas and Henri Aimée, *Névroses et psychoses de guerre.*

148. Ioteyko, *La Fatigue,* p. 321.

149. Jean Dupré, "La constitution émotive," *Revue Neurologique* 2 (1908): 165.

150. Maurice Dide, *Les Emotions et la guerre* (Paris, 1918), pp. 240–46.

151. Compare H. Piéron and Mme. Bouzanskly, "De l'existence d'un 'syndrome commotionel' dans les traumatismes de guerre," *Bulletin de l'Académie de Médecine,* 83, (1915): 654–651; Gilbert Ballet, "Le syndrome émotionnel," *Annales Médico-psychologiques* (1917): 183.

152. This is the argument of Eric Leed in his excellent chapter on neuroses and war in *No Man's Land: Combat and Identity in World War I* (Cambridge, 1979), pp. 163–92.

153. Ibid., p. 173.

154. Ioteyko, *La Fatigue,* p. 324.

CHAPTER 10. THE SCIENCE OF WORK BETWEEN THE
WARS

1. Charles S. Maier, "Between Taylorism and Technocracy," *The Journal of Contemporary History* 6, no. 2 (1970): 27–63; *In Search of Stability: Explorations in Historical Political Economy* (Cambridge, 1987), pp. 19–69. On German conservative and Nazi technocrats see the extensive study by Jeffrey Herf, *Reactionary Modernism: Technology, Culture and Politics in Weimar and the Third Reich* (Cambridge, 1984).

2. Otto Bauer, *Rationalisierung und Fehl-rationalisierung* (Vienna, 1931).

3. For analyses of Taylorism in the Soviet Union see Kendall Bailes, "Alexei Gastev and the Soviet Controversy over Taylorism 1918–24," *Soviet Studies* 29, no. 3 (1977): 373–94; Maier, "Taylorism and Technocracy," pp. 50, 51; Jean Querzola, "Le chef d'orchestre à la main de fer: Léninisme et Taylorisme," *Recherches,* no. 32/33 (September 1978): 57–94; on the "Time-League," see Walther Süss, *Die Arbeiterklasse als Maschine: Ein industriesoziologischer Beitrag zur sozialgeschichte des aufkommenden Stalinismus.* Osteuropa Institute an der Freien Universität, Berlin, vol. 22 (Berlin, 1985), pp. 108–29.

4. On Taylorism and leisure in Italy see Victoria de Grazia, *The Culture of Consent: Mass Organization of Leisure in Fascist Italy* (Cambridge, 1981), pp. 60–94.

5. Oswald Spengler, *Der Untergang des Abendlandes,* vol. 2 (Munich, 1922), p. 632. On this point I am indebted to Jeffrey Herf's nuanced discussion of Spengler in *Reactionary Modernism,* pp. 49–69.

6. Charles S. Maier has recently argued that this change can be explained as a shift from a "liberal" phase in the politics of time "in which control was

supposedly allocated primarily to the market—to an authoritarian era in which public authorities sought to reclaim temporal control." See Charles S. Maier, "The Politics of Time," *Changing Boundaries of the Political: Essays on the Evolving Balance between State and Society, Public and Private in Europe,* ed. Charles S. Maier (Cambridge, 1987), p. 161.

7. The term *scientific management* came into existence in 1911 as a synonym for the Taylor System. Ultimately, it became a generic term for a variety of systems of factory organization that evolved from and modified the Taylor System. Judith A. Merkle, *Management and Ideology: The Legacy of the International Scientific Management Movement* (Berkeley, 1980), pp. 9–11.

8. Olivier Christin, *Les enjeux de la rationalisation industrielle (1910–1929),* mémoire de maîtrise, Université Paris I, (Paris, 1982); Jules Amar, *Organisation et hygiène sociales: Essai d'homniculture* (Paris, 1927).

9. Josefa Ioteyko, *The Science of Labour* (London, 1919), p. 80.

10. Ibid., p. 84.

11. Ibid., p. 87.

12. Ibid., pp. 78, 79.

13. Ioteyko, *Science of Labour,* p. 92.

14. Paul Devinat, *Scientific Management in Europe International Labour Office, Studies and Reports, Studies and Reports* (Series B Economic Conditions) no. 17 (Geneva, 1927), p. 34.

15. "Quelques données pour l'unification de la terminologie psychotechnique," *Le Travail Humain* 1 (1933): 488.

16. Maier, "Society as Factory," *In Search of Stability,* p. 54; Devinat, *Scientific Management in Europe,* p. 35.

17. Ludwig Preller, *Sozialpolitik in der Weimarer Republik* (Düsseldorf, 1949), p. 65.

18. H. M. Vernon, "Output in Relation to Hours of Work," Great Britain, Ministry of Munitions, Health of Munition Workers Committee, *Memorandum,* no. 12 (1916).

19. The Sunday workday was eliminated and the Minister of Munitions Winston Churchill reduced the workweek in several major munitions plants from sixty to fifty hours per week. Gary Cross, *A Quest for Time: The Reduction of Work in Britain and France, 1840–1940* (Berkeley, 1989), p. 119.

20. See Industrial Fatigue Research Board, *Annual Report,* no. 1 (London, 1920); H. M. Vernon, *Industrial Fatigue and Efficiency* (London, 1921).

21. Cross, *A Quest for Time,* p. 200.

22. Heidrun Homburg, "Scientific Management and Personnel Policy in the Modern German Enterprise 1918–1939: The Case of Siemens," in *Managerial Strategies and Industrial Relations: An Historical and Comparative Study,* ed. Howard F. Gospel and Craig R. Littler (London, 1983), p. 149.

23. Devinat, *Scientific Management in Europe,* p. 3.

24. See Bauer, *Rationalisierung und Fehl-rationalisierung;* and Friedrich von Gottl-Ottlilienfeld, "Müssen wir die Rationalisierung ablehnen?" *Wirtschaft: Gesammelte Aufsätze* (Jena, 1937), pp. 45–52.

25. *Der Mensch und die Rationalisierung* I, ed. Reichskuratorium für Wirtschaftlichkeit (Jena, 1931), pp. 234, 235.

26. See the compendium, *Körper und Arbeit: Handbuch der Arbeitsphysiologie,* ed. Edgar Atzler (Leipzig, 1927). In a famous experiment, Atzler and a collaborator, Gunther Lehmann, calculated the expenditure of energy and food required by Taylor's famous human "ox," the Dutchman Schmidt, whose legendary brute strength and stupidity were recounted in the well-known passage of Taylor's *The Principles of Scientific Management.* According to Lehmann, Schmidt expended the equivalent of 5,515 calories in the course of ten hours, necessitating 7,651 actual calories of food. Atzler commented: "While Taylor, with no exact knowledge of the human organism, tried to extract maximal output, we on the contrary, demand optimal output. The process of work must be so arranged as to be adapted to the special nature of the human motor." Edgar Atzler, "Physiologische Arbeitsrationalisierung," *Körper und Arbeit, Handbuch der Arbeitswissenschaft,* ed. Edgar Atzler (Leipzig, 1927), pp. 419–20; "Einweihung des Arbeitsphysiologischen Instituts in Dortmund," *Münsterischer Anzeiger,* no. 1145 (23 October 1929): 3.

27. Otto Lipmann, *Lehrbuch der Arbeitswissenschaft* (Jena, 1932), p. 186. Also see Anson Rabinbach, "Industrial Psychology between Psychotechnics and Politics in Weimar Germany: The Case of Otto Lipmann," in *Les Chantiers de la paix sociale: Techniques et figures du social d'une guerre à l'autre* (in press, Paris, 1991).

28. Dietrich Milles, "Prävention und Technikgestaltung. Arbeitsmedizin und angewandte Arbeitswissenschaft in historischer Sicht," *'Gestalten'—Eine neue gesellschaftliche Praxis,* ed. Felix Rauner (Bonn, 1988), p. 65.

29. Ludwig Teleky, Über ärztliche Tauglichkeitsuntersuchungen von Arbeitern," *Die Arbeit* 10 (1929): 209.

30. Susanne Pacaud, "J.-M. Lahy (1872–1943)," *Le Travail Humain: Bulletin de l'Association Internationale de Psychotechnique* (1952): 341.

31. The fate of Jules Amar is also instructive. When, in 1919, a chair of "the technical organization of human work" was created at the Conservatoire National des Arts et Métiers, Amar was first in line to fill it. However, a campaign against him (due in no small part to his homosexuality, Jewishness, and his lack of proper French nationality) led to his subsequent defeat in a series of bitterly contested votes for the post. In despair, Amar resigned his position, later earning a living as a scientific journalist. His writings in the later 1920s showed a marked propensity for conservative and authoritarian solutions to what he often described as an "exhausted world." For a detailed description of the case of Amar, see H. and D. Monod, "Jules Amar (1879–1935)—À propos d'un centenaire," *Histoire des sciences médicales,* vol. 3 (Paris, 1979), pp. 227–35.

NOTES

32. *Handwörterbuch der Arbeitswissenschaft,* ed. Fritz Giese (Halle, 1931), p. 341.

33. William Stern, *Differentielle Psychologie* (Leipzig, 1921).

34. On Moede see Devinat, *Scientific Management in Europe,* p. 81; Georg Schlesinger, *Psychotechnik und Betriebswissenschaft* (Leipzig, 1920); Siegfried Jaeger and Irmingard Staeuble, "Die Psychotechnik und ihre gesellschaftlichen Entwicklungsbedingungen," *Die Psychologie des 20. Jahrhunderts,* vol. 13 (Zurich, 1981), pp. 54–95. Moede was technical advisor to the German State Railways, chief of the psychotechnical Institute of the Technische Hochschule in Berlin, Charlottenburg, and, after 1924, head of the Institut für Wirtschaftspsychologie at the Handelshochschule, Berlin.

35. Ulfried Geuter, *Die Professionalisierung der deutschen Psychologie im Nationalsozialismus* (Frankfurt-am-Main, 1988), p. 218.

36. R. W. Chestnut, "Psychotechnik: Industrial Psychology in the Weimar Republic 1918–1924," *Proceedings of the Annual Convention of the American Psychological Association* 7, no. 2 (1972): 781.

37. Geuter, *Professionalisierung,* p. 221.

38. Devinat, *Scientific Management in Europe,* p. 81.

39. Richard Seidel, "Die Rationalisierung des Arbeitsverhältnisses," *Die Gesellschaft* 3, no. 7 (1926): 21.

40. Peter Hinrichs, *Um die Seele des Arbeiters: Arbeitspsychologie, Industrie- und Betriebssoziologie in Deutschland 1871–1945* (Cologne, 1981), pp. 226–30.

41. Friedrich Dorsch, *Geschichte und Probleme der angewandten Psychologie* (Stuttgart, 1963), pp. 80–93.

42. Devinat, *Scientific Management in Europe,* p. 80.

43. Geuter, *Professionalisierung,* p. 222.

44. Zentrales Staatsarchiv, Potsdam (ZsTA) Merseburg (Central State Archives, Potsdam). Preussisches Ministerium für Wissenschaft, Kunst und Volksbildung, Rep. 151 C, No. 903. Institut für angewandte Psychologie, Gewerbehygiene, und Arbeitswissenschaft, Berlin. Letter of 17 January 1922; Niederschrift, 16 June 1922. (This archival material has been kindly made available to me by Dietrich Milles. Hereafter cited as Institut für angewandte Psychologie, Berlin).

45. "Critique des tests de la fatigue," *Revue Scientifique du Travail* 1 (1930): 59.

46. Wladimir Eliasberg, "Richtungen und Entwicklungstendenzen in der Arbeitswissenschaft," *Archiv für Sozialwissenschaft und Sozialpolitik* 56 (1926): 66–101.

47. Otto Lipmann, "Mehr Psychotechnik in der Psychotechnik," *Zeitschrift für Angewandte Psychologie* 37 (1930): 188–91. Walter Moede, *Zur Methodik der Menschenbehandlung* (Berlin, 1930); and Rabinbach, "Between Psychotechnics and Politics."

48. Fritz Giese, *Methoden der Wirtschaftspsychologie* (Berlin, 1927), p. 6;

Fritz Giese, "Psychotechnik," *Handworterbuch der Arbeitswissenschaft,* p. 3600.

49. Lipmann, *Lehrbuch der Arbeitswissenschaft,* pp. 21–25.

50. Ibid., p. 186.

51. Helmut Winkler, *Die Monotonie der Arbeit Neue Zeit-und Streitfragen* (Leipzig, 1922). Cited in Eliasberg, "Richtungen und Entwicklungstendenzen," p. 729.

52. Geuter, *Professionalisierung,* p. 180.

53. Eliasberg, "Richtungen und Entwicklungstendenzen, p. 732.

54. See Hinrichs, *Um die Seele des Arbeiters,* p. 160.

55. Ibid., 177.

56. Ibid.

57. See, for example, Friedrich von Gottl-Ottlilienfeld, *Fordismus?—Von Frederick W. Taylor zu Henry Ford* (Jena, 1924).

58. Ibid., pp. 8, 9.

59. Goetz Briefs, "Die Tätigkeit des 'Instituts für Betriebssoziologie,' " *Soziale Praxis* 41, no. 45 (1932): 1423, 1424.

60. Ibid., p. 1427.

61. Fritz Giese, *Philosophie der Arbeit* (Halle, 1932); Hinrichs, *Um die Seele des Arbeiters,* p. 250.

62. Ibid., pp. 24, 25.

63. Martin Heidegger, "National Socialist Education *(Wissensschulung)* 22 January 1934. Address at Freiburg University, English translation in *New German Critique* 45 (Fall, 1988): 113.

64. Herbert Marcuse, "Über die philosophischen Grundlagen des wirtschaftswissenschaftlichen Arbeitsbegriff," *Archiv für Sozialwissenschaft und Sozialpolitik* 69, no. 1 (June, 1933); English translation in *Telos* 16 (Summer 1973), p. 11. Elaborating a Marxian view of labor as human *praxis,* Marcuse—at that stage still influenced by Heidegger—claimed that reduced to an economic activity, labor had become "basically a lack, something negative: it is directed towards something which is not there, that is not already present and must be created." In other words, by its inability to provide fulfillment as a purely economic activity, labor only points to the existence of a greater sphere of human freedom and activity, which labor *alone* cannot bring about. Marcuse's discussion, a brilliant reconstruction of the arguments of Marx in the 1844 manuscripts *avant la lettre,* is in many respects akin to those of more conservative critics like Gottl-Ottlilienfeld and Giese.

65. See, for example, Giese's posthumous article (he died in 1935), "Idee und Geist im Betriebe," *Archiv für Rechts-und Sozialphilosophie* 29 (1936): 447–72; Also see Jaeger and Staueble, "Die Psychotechnik," p. 91; Geuter, *Professionalisierung,* pp. 275, 276.

66. The DINTA's chief support came from the Verein Deutscher Eisenhütte, in the person of Albert Vögler, chairman of that steel manufacturers associ-

ation, and a close associate of Arnhold. See Hinrichs, *Um die Seele des Arbeiters*, p. 277; Wolfgang Schlicker, "Arbeitsdienstbestrebungen des deutschen Monopolkapitals in de Weimarer Republik," *Jahrbuch für Wirtschaftsgeschichte* (GDR), 3 (1971): 102.

67. Cited in Hinrichs, *Um die Seele des Arbeiters*, p. 289.

68. Ibid., pp. 284–87.

69. Peter Bäumer, *Das Deutsche Institut für technische Arbeitsschulung. Probleme der sozialen Werkspolitik*, ed. Goetz Briefs (Munich, 1930), p. 146.

70. Karl Arnhold, "Die Tätigkeit de Deutschen Instituts für technische Arbeitsschulung (DINTA)," *Soziale Praxis* 42, no. 15 (1933): 467.

71. Otto Marrenbach, *Fundament des Sieges: Die Gesamtarbeit der deutschen Arbeitsfront von 1933 bis 1940* (Berlin, 1940), p. 325.

72. The majority of Jewish psychologists, for example, William Stern, Max Wertheimer, and David Katz emigrated, but others were less fortunate. After his Berlin Institute was sacked by the SA in October 1933, Lipmann committed suicide. Geuter, *Professionalisierung*, p. 99.

73. Karl Arnhold, "Mensch und Leistung im Betrieb," *Wege zur neuen Sozialpolitik: Arbeitstagung des Sozialamtes des DAF vom 16. bis 21. Dezember 1935* (Berlin, 1935), p. 143. Arnhold also claimed that "90% of the students at Columbia University in New York are Jews."

74. Geuter, *Professionalisierung*, p. 103.

75. Geuter points out that the final controversy over the political direction of applied psychology occurred in 1933–34, with the triumph of those who argued for extending psychology's field of competence to "the entire dimension of the soul" ("das ganze Gebiet der Seelischen"), as opposed to those who wanted to retain a purely utilitarian rationale, especially in pedagogy. Geuter, *Professionalisierung*, pp. 274, 275. On the ideological integration of the psychology profession in the Third Reich, compare especially chap. 5.

76. Cited in Ibid., pp. 280, 281.

77. Otto Marrenbach, *Fundament des Sieges*, p. 325.

78. For an extensive discussion see Anson Rabinbach, "The Aesthetics of Production in the Third Reich," *International Fascism: New Thoughts and New Approaches*, ed. George L. Mosse (Berkeley, 1979), pp. 189–222; Chup Friemert, *Das Amt 'Schönheit der Arbeit': Produktionsästhetik im Faschismus* (Munich, 1980).

79. Karl Arnhold, *Mobilisierung der Leistungsreserven unserer Betriebe* (Berlin, 1939), pp. 4, 5; Bundesarchiv Koblenz, Deutsche Arbeitsfront, Sozialamt NS51/42. Letter from Dr. Arnhold, "Stellungnahme zu dem Entwurf 'Nationalsozialistisches Musterbetriebe' und ergänzende Anregungen," 22 October 1936.

CONCLUSION. THE END OF THE WORK-CENTERED SOCIETY

1. Georg Helm, *Die Lehre von der Energie* (Leipzig, 1887), p. 18.
2. Maurice de Montmollin, *L'Ergonomie* (Paris, 1986), p. 13.
3. For a survey of English developments see Michael Rustin, "The Politics of Post-Fordism: or, The Trouble with 'New Times,'" *New Left Review* 175 (May–June, 1989): 54–77; Michael Harrington, *The Next Left* (New York, 1986), pp. 18–46.
4. François Ewald, *L'Etat providence* (Paris, 1986), p. 384, passim.
5. See Robert Kuttner, *The Economic Illusion: False Choices Between Prosperity and Social Justice* (Philadelphia, 1987), p. 35.
6. Alan Wolfe, "The Death of Social Democracy," *The New Republic*, 25 February 1985, pp. 21, 22.
7. Jacques Donzelot, *L'invention du social: Essai sur le déclin des passions politiques* (Paris, 1984); Pierre Rosanvallon, *La crise de l'État-providence* (Paris, 1981), p. 137. Charles S. Maier has criticized these authors for assuming that active democratic politics, or a civic ethos, is indeed preferable to one in which social conflicts and suffering are ameliorated by state policies. See Charles S. Maier, *In Search of Stability: Explorations in Historical Political Economy* (Cambridge, 1987), pp. 267, 268.
8. R. Tchobanian, "Trade Unions and the Humanisation of Work," in S. Timperley and Dan Ondrack, eds., *The Humanisation of Work* (London, 1982), p. 193; Felix Rauner, "Aspekte einer human-ökologisch orientierten Technikgestaltung," *'Gestalten'—Eine neue gesellschaftliche Praxis* (Bonn, 1988), pp. 35–40.
9. Ralf Dahrendorf, "Im Entschwinden der Arbeitsgesellschaft: Wandlungen in der sozialen Konstruktion des menschlichen Lebens," *Merkur: Deutsche Zeitschrift für europäisches Denken* 34, no. 8 (August, 1980): 751.
10. Ibid.
11. Daniel Bell, *The Cultural Contradictions of Capitalism* (New York, 1976); Anthony Giddens, *The Constitution of Society* (Berkeley, 1984); André Gorz, *Farewell to the Working Class: An Essay on Post-Industrial Socialism* (Boston, 1982); Alain Touraine, *The Self Production of Society* (Chicago, 1977); Claus Offe, *Disorganized Capitalism: Contemporary Transformations of Work and Politics,* ed. John Keane (Cambridge, 1985); Jürgen Habermas, *Zur Rekonstruktion des historischen Materialismus* (Frankfurt-am-Main, 1976).
12. Dahrendorf, "Im Entschwinden der Arbeitsgesellschaft," pp. 751, 752.
13. Eric Hobsbawm, "The Forward March of Labour Halted," *Politics for a Rational Left: Political Writings 1977–1978* (New York, 1989), pp. 9–22.
14. Bell, *Cultural Contradictions of Capitalism,* pp. 147, 148.
15. Offe, *Disorganized Capitalism,* p. 141.
16. For an optimistic evaluation, though written before the "automation debate," see Georges Friedmann, *Industrial Society: The Emergence of the*

Human Problems of Automation, ed. Harold L. Sheppard (New York, 1955), pp. 186, 187. More critical evaluations are Friedrich Pollock, *Automation: Materialen zur Beurteilung de ökonomischen und sozialen Folgen, Frankfurter Beiträge zur Soziologie 5* (Frankfurt-am-Main, 1956); Jürgen Habermas, "Die Dialektik der Rationalisierung: Von Pauperismus im Produktion und Konsum," *Arbeit, Erkenntnis, Fortschritt: Aufsätze 1954–1970* (Amsterdam, 1970), p. 26.

17. Michael Rose, *Servants of Post-Industrial Power?:* Sociologie du Travail *in Modern France* (London, 1979); and Dominique Monjardet, "In Search of the Founders: The *Traités* of the Sociology of Work," in *Industrial Sociology: Work in the French Tradition,* ed. Michael Rose, trans. Alan Raybould (London, 1987), pp. 112–19.

18. Pierre Naville and Pierre Rolle, "L'évolution technique et ses incidences sur la vie sociale," *Traité de sociologie du travail,* ed. Georges Friedmann and Pierre Naville (Paris, 1961), p. 365.

19. Ibid., p. 366.

20. Robert Pages, "Sociologie du travail et sciences de l'homme," *Traité de Sociologie du Travail,* p. 110.

21. Ibid., p. 111.

22. Ibid.

23. Rose, *Servants of Post-Industrial Power?,* pp. 71–73.

24. Pierre Naville, *Vers l'automatisme social?* (Paris, 1953); Naville's prognosis was challenged by his collaborator, Georges Friedmann, who regarded his vision as too utopian and too technocratic, predicting instead the emergence of a new kind of skilled "universal" worker *(l'ouvrier polyvalent),* an interpretation that Naville quickly denounced as retrograde and, even worse, "Proudhonian." Rose, *Servants of Post-Industrial Power?,* pp. 74–76.

25. Helmut Schelsky, "Industrie-und Betriebssoziologie," *Soziologie. Ein Lehr-und Handbuch zur modernen Gesellschaftskunde,* ed. Arnold Gehlen and Helmut Schelsky (Stuttgart, 1955), p. 179.

26. Monjardet, "In Search of the Founders," p. 115. The classic study of this genre is Alain Touraine's *L'évolution du travail ouvrier aux usines Renault* (Paris, 1955). Also see Georges Friedmann, *The Anatomy of Work: Labor, Leisure and the Implications of Automation,* trans. Wyatt Rawson, (Glencoe, N.Y., 1961), p. 152.

27. Michael Rose, *Industrial Behaviour: Theoretical Development since Taylor* (London, 1975), p. 24.

28. Friedmann himself often reiterated the dated doctrines and results of interwar studies, adding only the superficial gloss that "if working life is losing content and interest, we must look for new content and our interest and effort must be led elsewhere, our life being centered around play." Friedmann, *The Anatomy of Work,* p. 152.

29. For the argument that the debate amounted to very little, see Rose,

Servants of Post-Industrial Power?, p. 76.

30. Georges Friedmann, "L'objet de la sociologie du travail," *Traité de Sociologie du Travail*, pp. 12, 13.

31. Jürgen Habermas, "Toward a Reconstruction of Historical Materialism," in *Jürgen Habermas on Society and Politics*, ed. Steven Seideman (Boston, 1989), pp. 114–42.

Index

British Gainsborough Commission, 216

"Brot versus Kartoffel" debate in 1870s and 1880s, 129

Broussais, François, 64

Brown-Séquard, Charles, 88, 102

Brücke, Ernst, 63, 170

Bücher, Karl, 174, 177, 191, 196

Büchner, Ludwig, 49, 50, 53, 82, 319n39

Buddhism, 178

Buffon, Henri, 306n51

Burtt, H. E., 277

Buyse, Omer, 278

Cabanne, Pierre, 115

Caffeine, 144

Caloric requirements, 130–33

Calorimeter, 127, 130, 337n28

Camus, Jean, 265

Canguilhem, Georges, 21–22, 50, 51, 64

Capitalism, 74, 80, 367n4

Capital (Marx), 72–73, 75, 77–78, 79, 241

Cardiograph, 96

Carnot, Sadi, 45, 53, 54, 70, 78

Carrieu, M., 38

Cathcart, E. P., 277

Cauwès, Paul, 207

Characterology, 281

Charcot, Jean-Martin, 43, 154, 155, 156, 158, 160, 166, 170

Chauveau, Auguste, 88–90, 96, 127, 128, 132, 139, 223, 248

Chéron, Henri, 223

Cheysson, Emile, 229

Children, worktime for, 211, 307n62

Chronophotography, 189; Bergson's distinction between time consciousness and, 111–14; economy of work studied with, 116–19; Marey and, 104–8, 110–14, 115, 116; modernist implications for the arts, 114–15

Cinematography, 108

Civilization: destruction of rhythmic element in work by, 176; fatigue and, 172–78; least effort as law of, 173–78; neurasthenia and pressures of, 156–58; *see also* Modernity

Class, social: class struggle as wasted energy, 181; idleness and, 27–28, 31; manual laboring, 33; neurasthenia and, 157–58; science of work and, 234–37

Clausius, Rudolf, 7, 47, 61

Clemenceau, Georges, 186, 221

Clémentel, Etienne, 261

Climate, idleness and, 31

Clock-regulated worktime, 31–33

Coleman, William, 124

Collège de France, 89, 97

Color, fatigue and, 41

Committee for Public Safety, 64

Community, efforts to return to preindustrial work, 281–82

Comptabilism, 349n10

Comte, Auguste, 8, 21–22, 180, 194

Condillac, Etienne Bonnot de, 51

Condorcet, Marquis de, 28

Confessions of an Opium Eater (Quincey), 167

Confessions (Rousseau), 28

Congress of Educational Physiology, 225

Conservation of energy. See Energy conservation, law of

"Conservation of Energy" (Helmholtz), 58, 65

Consumerism, 282

Coriolis, Gaspard Gustave de, 55, 79

Cotton Trade in England and on the Continent, The (Schulze-Gävernitz), 215–16

Coubertin, Pierre de, 224

Coulomb, Charles Augustin, 70

Coustan, Hubert, 230

Creative Evolution (Bergson), 88

Cross, Gary, 210

Cubism, 115
Cultural crisis, medical model of, 21–22
Cultural modernity, 84–87

Dahrendorf, Ralf, 296
Darwin, Charles, 20, 69
Da Vinci, Leonardo, 248
Death instinct, 314*n*69
"Death of Social Democracy, The" (Wolfe), 294
Decline of the West (Spengler), 283
Degas, Edgar, 114
De la fatigue et de son influence pathogénique (Carrieu), 38
Delahaye, Philippe, 235
Delahaye, Victor, 212–13, 361*n*86
Dematerialized materialism, 48–51
Demeny, Georges, 224–25, 227, 333*n*141, 361*n*93
Denis, Hector, 216
Depasse, Hector, 237
Depression (1930s), 283
Derrida, Jacques, 13
Descartes, René, 1–2, 14, 51, 64, 84, 90, 92, 310*n*5
Deschamps, Albert, 163–64, 176
Deschesne, Laurent, 217
Description du corps humain (Descartes), 51
Desire operating as reflex, 165
Deutsche Gemeinwirtschaft (Moellendorf), 261
Deutsche Institut für technische Arbeitsschulung (DINTA), 284–88
Dialectic of Enlightenment (Horkheimer and Adorno), 16
Dialectics of Nature (Engels), 81
Dide, Maurice, 268
Diderot, Denis, 7, 28, 116, 248
Diet: during First World War, 263; importance of, 186; physiology of nutrition, 125–27; productivity and, 216–

17; rational nutrition, 262; role of, in creating labor power, 128–33
Differential psychology, 278
"Disappearance of the Work-Society, The" (Dahrendorf), 296
Discourse, ambiguity of term, 18
Discourse on Method (Descartes), 1–2, 84
Donzelot, Jacques, 294, 295
Driving force. *See* Labor power *(Arbeitskraft)*
Droit à la paresse, Le (The Right To Be Lazy) (Lafargue), 34
Dubois, Paul, 159, 162
Du Bois-Reymond, Emil, 49, 50, 65–66, 124–25
Duchamp, Marcel, 88, 115
Duhousset, Émile, 102, 328*n*91, 329*n*97, 334*n*158
Dumas, Alexandre, 102
Dunin, Theodor, 168
Dupin, Baron Charles, 70
Durkheim, Émile, 9, 154
Duruy, Victor, 89
Dynamometer, 30, 306*n*51
Dysergie, 164

Economic and Philosophical Manuscripts of 1844 (Marx), 72
Economy of the body, 93
Economy of work, 115–19
Education, physical, 224–25, 361*n*93
Educational exhaustion, perceived crisis of, 147–53
Education de soi-même, L' (Dubois), 161
Efficiency, 127–28
Ehrenbert, Richard, 281
Eight-hour day, 218–19, 220
Eight Hours for Work (Rae), 216
Eisner, W., 236
Elasticity, muscular, 117, 118, 127–28, 139

Eliasberg, Wladimir, 280, 281
Elkana, Yehuda, 313n34
Émile (Rousseau), 28
Émotions et la guerre, Les (Dide), 268
Empire de l'air: Essai d'ornithologie appliquée a l'aviation, L' (Mouillard), 99
Encyclopédie (Diderot), 7, 28, 116
Energeticism, 115, 120, 127, 207, 291–92, 361n86; gradual eclipse of, 293–94; industrial experiments using, 218; social, 179–82
Energetische Imperativ, Der (Ostwald), 183, 254
Energetischen Grundlagen der Kulturwissenschaft, Die (Ostwald), 182
Energie, Die (Ostwald), 183
Energies of Men (James), 170
Energy: deployment of social, 224–28; discovery as universal force, 45–48; economy of, 37; matter and, inseparability of, 53; paradoxical relationship between entropy and, 63–64; physiological reductionists and principle of, 50–51; rhythm and regulation of expenditure of, 175
Energy conservation, law of, 2, 3, 24, 25, 45–46, 47, 50, 289–90; body as site of, 2–3, 66–67; early articulations of, 310n5, 312n34; Engels on, 81–83; Helmholtz's theory of, 52–56; law of least effort and, 192; Marey on, 90; muscle physiology and, 124–27; primacy as social principle, 67–68, 121, 122; social energeticism and, 179; social implications of, 69–72; transcendental materialism as, 49, 55, 56, 57
Engels, Friedrich, 72, 74, 81–83, 289, 321n61
Engineering tradition, French, 55, 78–79
England, 22; worktime in, 210, 212, 213, 214
English Sunday, 213

Enlightenment, 9, 51–52, 64; *philosophes*, 7, 27–28
Ennui, 42, 141
Ennui: Études psychologiques, L' (Tardieu), 42
Entraînement et fatigue au point de vue militaire (Ioteyko), 226
Entropy, 6, 61–64, 68, 117, 132; in capitalism, 80; discovery of, 3–4, 20, 47, 48; paradoxical relationship between energy and, 63–64
Entwicklung der Gesetze des menschlichen Verkehrs und der daraus fliessenden Regeln für menschliches Handeln, Die (Gossen), 70–71
Equestrian motion studies, 99–103, 328n84, n91
Ergograph (register of work), 134–36, 143, 149, 150
Ergography, science of, 136–42
Ergologie, 278
Ergonomics, 116, 265–70, 292
Esquirol, Etienne, 167
Essai sur les donées de la conscience (Bergson), 110
Ethnographers, on idleness and industry, 29–30
Euclid, 109
Évolution créatrice, L' (Bergson), 111
Évolution de l'idée de temps dans la conscience, L' (Guyau), 110
Ewald, François, 294
Exhaustion: defined, 163; distinction between fatigue and, 152–53, 342n108
Experiments, industrial, 217–19

Factory system, 32, 48, 209
Faraday, Michael, 313n34
Farre, General, 104
Fascism, 2
Fatica, La (Mosso), 7, 133–34, 149

Fatigacion, 39
Fatigue, 2, 4, 118; Amar's definition of, 48; capitalism as system designed to produce, 74; chemical basis of, 139; civilization and, 172–78; color and, 41; as defense against excess, 141; defined, 138; discovery of, 6, 38–40, 44, 68; distinction between exhaustion and, 152–53, 342n108; distinction between tiredness and, 150–51, 191; *ennui* and, 42, 141; ethnology of, 174–76; European science of work and, 182–88; first general survey *(Enquête)* of, 140; during First World War, 266–69; in Gautier's studies of nutritional requirements, 132; industrial accidents and, 231–33; intensity of, 140; laws of, 133–36, 139; Levenstein's survey data on, 196; limits defined by, 23; measurement of, 23; mental, 146–53, 156, 342n17; military training and, 226; modernity and, 19–20, 146–47, 292; muscle vs. brain, 135, 304n10; nineteenth-century preoccupation with, 19–25, 43–44, 121–22, 290–92; objective phenomenology of, 138; origins of, 141; pathological, 153–63; as physical and moral disorder, 39–40; physiological and mental symptoms, 140–41; poetic literature of, 40–42; as poison, 135–36; productivity and, 201–2, 219–20; science of ergography, development of, 136–42; search for vaccine against, 142–45; as self-regulating, 138, 139, 141–42; society and, 21, 42–44; *see also* Mental Fatigue; Neurasthenia
Fatigue intellectuelle, La (Binet), 152
Fatigue quotient, 141, 340n81
Fayol, Henri, 276
Fear, psychology of, 267
Fechner, Gustav Theodor, 67, 170, 190
Féré, Charles, 43, 155–56, 160, 163, 172
Ferrero, Guillaume, 173–74, 175, 177

Feuerbach, Ludwig, 49, 319n39
Fick, Adolf, 125
First World War, 259–70; diet during, 263; ergonomics at the front, 265–70; fatigue and neurasthenia during, 266–69; Kaiser Wilhelm Institute during, 262–64; psychotechnics and, 259–62; restructuring of European economies for, 259–60
Fischer, Otto, 189
Flaubert, Gustave, 6, 28
Fleury, Maurice de, 221
Fliess, Wilhelm, 170
Flight, motion studies of, 97–99, 103
Flourens, Pierre, 89
Force productive des nations concurrentes (Dupin), 70
Fordism, 282, 293
"Forward March of Labor Halted?, The" (Hobsbawm), 296
Fotodinamismo Futurista (Bragaglia), 115
Foucaud, Edouard, 32
Foucault, Michel, 13, 16–17, 18, 27, 303n31, 325n31
Fouillée, Alfred, 17, 207
Fourier, Charles, 34
France: education crisis of 1880s in, 147–49, 152–53; fatigue mania and sense of cultural decline in, 22; in First World War, 265–70; industrial experiments in, 219; neoliberalism in, 294–95; science of work in, 11, 184–88, 202; Taylorism in, 240, 244–49, 261, 272; ten-hour law in, 220–24; trade unions in, 235; worker's insurance in, 230; worktime in, 211–12
François-Franck, Charles, 67
Frankfurt School, 16, 85, 86, 299
Frankland, Edward, 125–26
Freedom, Marx's redefinition of, 73–74, 79–81
Fremont, Charles, 116, 117, 136, 182, 185, 191, 203, 246, 248, 251
French Academy of Sciences, 51, 54

French medicine, 88–90

Freud, Sigmund, 9, 63, 156, 160, 170, 314*n*69

Friedländer, Saul, 20

Friedmann, Georges, 288, 297, 299

Fromont, L.-G., 217–18

Functional plasticity, 265

Fusil photographique, 105–6

Futurists, Italian, 115

Gang des Menschen, Der (Fischer), 189

Gantt, Henry L., 276

Gastev, Alexei, 272

Gautier, Armand, 130–33, 235

Geometric photography, 107–8

George, Lloyd, 275

German Labor Front, 287

Germany: education crisis of 1880s in, 149–52; emergence of scientific materialism in, 49–51; energeticism doctrine in, 182; fatigue mania evident in Wilhelminian, 22; during First World War, 260–64; industrial experiments in, 218–19; National Socialism in, 275, 280, 284–88; psychotechnical movement in, 278–80; rationalization movement in, 277; science of work in, 11, 189–95, 202; sociology of work in, 195–202; Taylorism in, 240, 253–58, 272; worker's insurance in, 228, 229–30; worktime in, 211

Germ plasm theory, 314*n*69

Giddens, Anthony, 86, 296

Gide, Charles, 207, 208, 235

Gideon, Siegfried, 88

Giese, Fritz, 260, 280, 283–84, 287

Gilbreth, Frank B., 88, 117, 251, 254, 276, 335*n*178

Gillispie, Charles, 47–48, 54

"Goethe's Anticipation of Subsequent Philosophic Ideas" (Helmholtz), 55

Göller, Adolf, 41–42

Gorz, André, 296

Gossen, Hermann Henri, 70–71

Gottl-Ottlilienfeld, Friedrich, 282

Gramsci, Antonio, 239, 272

Grand Revue, La, 250

Graphic method and notation of Marey, 93–97

Griesbach, Hermann, 143, 149–50

Griffuelhes, Victor, 230

Groupe de Sociologie du Travail, 297

Grove, William Robert, 79, 313*n*34

Grundrisse (Marx), 76, 77, 80; *see also Capital* (Marx)

Grundzüge der Psychotechnik (Münsterberg), 192, 255

Guerre future, La (Bloch), 226

Guillotine, Dr., 64

Guyau, J.-M., 110, 113

Habermas, Jürgen, 13, 16, 75, 296, 299, 322*n*3

Haeckel, Ernst, 6, 24, 49

Handbuch der physiologischen Optik (Helmholtz), 56

Hardenberg, Karl August, 29

Harkort, Friedrich, 33

Hartmann, Eduard von, 213

Hasbach, Wilhelm, 214

Heat death of universe, 6–7, 47, 62

Hegel, G. W. F., 7, 46, 81

Heidegger, Martin, 283–84

Heller, Agnes, 5, 73, 317*n*14

Hellpach, Willy, 281, 286

Helmholtz, Hermann von, 3, 6, 7, 11, 46, 48, 49, 50, 65, 73, 76, 78, 87, 93, 102, 109, 110, 122, 124, 125, 134, 180, 290; energy conservation theory, 52–56; Engels on, 81–83; on fundamental principle of science, 312*n*30; labor power and, 55, 56–64; myograph of, 66, 96, 125; popular scientific lectures of, 56–61, 68

(1907), 219, 233, 234, 236, 237; Brussels (1903), 183–84, 208

Internationalism of science of work, 202–3

International Labor Organization, 274

International workers' movement, 211

Interwar period, science of work in, 271–88; productivism, 271–73; rapprochement between Taylorism and, 273–75

Ioteyko, Josefa, 137, 138–39, 141, 142, 158, 172, 175, 182, 184, 203–4, 225–27, 267, 273–74

Italy, Taylorism in interwar period in, 272

Jaccoud, Sigismond, 89

Jacquet-Droz, 57

Jaensch, Walther, 286

James, William, 170

Janet, Pierre, 154, 170–72, 176, 177

Jankélévitch, Vladimir, 42

Janssen, Pierre Jules, 105, 331n119

Jay, Raoul, 207, 208, 210, 231

Jews in Nazi Germany, 285–86

Jonnès, Moreau de, 32

Jouhaux, Léon, 261

Joule, James Prescott, 53

Journal des Goncourts, 39

Kaiser Wilhelm Institute for Labor Physiology, 189, 262–64

Kant, Immanuel, 109

Keim, Maurice, 40

Kelvin, Lord. *See* Thomson, Sir William (Lord Kelvin)

Kemsies, Friedrich, 149

Kenotoxins, 143

Kent, Stanley, 275

Kilogrammeter, 339n55

Kirchoff, Gustav Robert, 52–53

Kochman, Wilhelm, 256

Kraepelin, Emil, 144, 150–52, 175, 197, 198, 203, 220, 287; principle of smallest muscle, 191–92; science of work and, 189–94

Kraft. See Energy

Kraft und Stoff (Büchner), 50

Kraftwechsel, 126, 127

Kronecker, Hugo, 133, 134, 138

Kuhn, Thomas, 46, 54, 57, 79, 313n34

Kymograph, 66, 96

Labor: division between leisure and, 32; thermodynamics and alteration of concept of, 46–47, 48; transvaluation of, 7–8; value of human, Helmholtz on, 59–60

Laboratoire d'Énergétique Solvay, 137

Labor policy, scientific, 123

Labor power *(Arbeitskraft):* as capital of nation, 208–10; discovery of, 4, 55, 56–61; Engels and, 82; Helmholtz and, 55, 56–64; image of, 3; language of, 4, 5; Marx on, 70, 72–81; naturalization as energy, 61; nineteenth-century distinction between labor and, 4–5; physiognomy of, 117; physiological and historical schools of, 318n24; political economy of, 69–83; role of diet and nutrition in creating, 128–33; social physiology of, 74–76, 122, 123; work ethic as ethos of, 9

Labor time. *See* Worktime

LaCapra, Dominick, 13

Lafargue, Paul, 34–35

Lagneau, Gustave, 147

Lagrange, Fernand, 120, 138, 140, 224

Lahy, Jean-Marie, 246, 249–53, 265, 273, 277, 278, 287

Lamarck, Jean-Baptiste de Monet Chevalier de, 155

La Mettrie, Julien Offray de, 51, 64

Lancet (journal), 38

Materialism, scientific. *See* Scientific materialism

Matière et mémoire (Bergson), 110

Matter, energy and, inseparability of, 53

Matteucci, Carlo, 124

Mayer, Julius Robert, 53, 55, 65, 124, 312*n*34

Mechanical inscription, 96–97, 186–88

Medical model of cultural crisis, 21–22

Medicine: French, 88–90; industrial, 274; social, 23; *see also* Marey, Etienne-Jules

Mein Kampf (Hitler), 286

Meissonier, Anton, 102

Méliton, Martin, 37

Mendelsohn, Everett, 49–50

Mental fatigue, 146–53, 156, 342*n*17

Mental inertia, 173

Merkle, Judith, 253

Merrheim, Alphonse, 241, 272

Mestre, M., 219, 232, 235

Meyerson, Émile, 313*n*34

Meynert, Gustav, 170

Military training, 224, 225–28, 286

Millerand, Alexandre, 207–8, 212, 235–36

Milne-Edward, Henri, 323*n*13

Misonéisme, 176–78

Möbius, Paul, 168

Mode de fonctionnement économique de l'organisme (Imbert), 121

Modernity: cultural, 84–87; fatigue and, 19–20, 146–47, 292; human motor as paradigm of social, 293; Marey and, 84–88, 118; neurasthenia and, 154–55, 157; resistance to, 176–78; of risk, 229; social, 86–87; traditions confronting crisis of, 84–86

Moede, Walther, 265–66, 277–80

Moellendorf, Wichard von, 260–61, 272

Mohr, C. F., 313*n*34

Moleschott, Jacob, 49, 50, 53, 82, 131, 134, 319*n*39

Mond, Ludwig, 180

Monday absenteeism, 32–33, 233

Monde Moderne, Le, 116, 136

Monotony, issue of, 281

Morality of work, 34

Morel, Augustin, 155

Mosso, Angelo, 7, 43, 44, 133–36, 137, 138, 139, 142, 149, 164, 172, 182, 190, 224, 248, 287

Most apt, principle of the, 274

Moteur humain et les bases scientifiques du travail professionnel, Le (Amar), 223–24, 247, 248

Motion of the body, Marey and laws of, 92–93, 97–103

Motion photography, 332*n*124, *n*137, 333*n*141

Motion picture cameras, 333*n*141

Motion studies, 7; equestrian, 99–103, 328*n*84, *n*91; of flight, 97–99, 103; integration of time into, 107; *see also* Time and motion studies

Motivation, worker, 281

Mouillard, Louis, 99

Müller, Johannes, 49, 50, 53, 65

Munk, Emmanuel, 127

Münsterberg, Hugo, 189, 191, 192–93, 203, 204, 252, 254–56, 287

Muscle physiology: application of energy conservation to, 124–27; elasticity and efficiency of, 127–28; role of diet and nutrition in, 128–33

Muscular asthenia, 168

Musée Social, 207

Mussolini, Benito, 272

Muybridge, Eadweard, 100–103, 114, 329*n*97

Myograph, 66, 96, 125

Mysticism, 166

Napoleon, 29, 306*n*45

Napoleon III, 89

National economy, 71

National Socialism, 275, 280, 284–88; factory social politics under, 282–84

Nature, La (magazine), 100, 101, 107

Nature, Marx's theory of, 77–78, 319n38

Naturphilosophie, 46, 54, 55

Navier (engineer), 55, 79, 316n2

Naville, Pierre, 288, 297–98, 299

Nazi Germany, 284–88

Neoconservatives in Europe and America, 294

Neoliberalism in France, 294–95

Neumann, Carl, 71

Neurasthenia, 67, 153–63, 170, 171, 177, 178, 366n168; class and, 157–58; as distinct from asthenia, 163–64; during First World War, 266–69; as inverted work ethic, 167–68; modernity and, 154–55, 157; pressures of civilization and, 156–58; similarities between mental fatigue and, 156; symptoms of, 154–55, 156, 159, 160–61; theory of hereditary predisposition to, 155–56; traumatic, 158; treating, 159–63; of women, 168

Neurasthénie (épuisement nerveux), La (Bouveret), 155

Newton, Isaac and Newtonian order, 62–63, 65

Nietzsche, Friedrich, 9, 17, 19–22, 62, 86

Nude Descending a Staircase (Duchamp), 88, 115

Nutrition. See Diet

Nye, Robert, 21, 155, 224

Oberschall, Anthony, 195

Obsolescence of the body, 295–300

Occupational risk, French doctrine of, 228–29

Odograph, 100, 116

Offe, Claus, 296, 297

On the Correlation of Physical Forces (Grove), 79

"On the Philosophical Foundation of the Concept of Labor in Economics" (Marcuse), 284

Ostwald, Wilhelm, 10, 49, 182, 194, 254

Paradigm of work, Marx's shift from paragidm of production to, 317n14, 321n57

Paris, Capital of the Nineteenth Century (Benjamin), 18

Paris Inventeur (Foucaud), 32

Partial photography, 107–8

Pathological fatigue, 153–63

Pathology of work, 280–82

Perceval, M. de, 33

Péron, François Auguste, 30, 306n51

Perpetuum mobile, 5, 53–54, 57, 58–59, 62, 68

Perrot, Michelle, 123

Personal productivity, 217

Peter, Michel, 148

Pettenkofer, Max von, 126, 131

Phenakistoscope, 330n100

Phidias, 111–12, 114, 334n158

Philosophes, 7, 27–28

Philosophie der Arbeit (Giese), 283–84

Photographic rifle (fusil photographique), 105–6

Photography: chronophotography, 104–8, 111–19, 189; motion, 332n124, 137, 333n141; Muybridge's motion studies using, 100–103; partial or geometric, 107–8

Physical education, 224–25, 361n93

"Physiognomy" (Schopenhauer), 40

Physiological limit, 221–22

Physiological reductionists, 50–51, 65–67

Physiological Station, 106

Physiological time, language of, 93–97

Schmoller, Gustav, 196
Shock, Marey's theory of, 117, 118
Schopenhauer, Arthur, 49
Schulze-Gävernitz, Gerhard von, 215–16
Schumacher, Fritz, 201
Schumburg, Wilhelm, 189
Science: language of, 16; of management, 29; popular, Helmholtz's promotion of, 56–61, 68
Science of Labour and Its Organization, The (Ioteyko), 273
Science of work, European, 6, 10–11, 23–24, 121, 128, 179–205; academic subdisciplines of, 274–75; deployment of social energy, 224–28; fatigue and, 182–88; in France, 11, 184–88, 202; German sociology of work, 195–202; in Germany *(Arbeitswissenschaft)*, 11, 189–95, 202; industrial accidents and, 227–34; industrial psychotechnics, 191, 193, 202–5; institutionalization of, 275–78; internationalism of, 202–3; in interwar period, 271–88; labor power as capital of nation and, 208–10; National Socialist, 284–88; opponents of Taylorism, 249–53; phases in development of, 123–24; productivity and, 210–17; reform and, 184–88, 206–8; romantic "turn" in, 283–84; science between classes, 234–37; social energeticism, 179–82; as socially neutral and objective, 291–92; social question and, 206–37; state and, 236–37; Taylorism and, 242–44, 247, 273–75; Taylorism and, German, 253–58; on worktime, 210–14, 217–24
Scientific labor policy, 123
Scientific management, 376n7; *see also* Taylorism
Scientific materialism, 5, 43–44, 289–90; as dematerialized materialism, 48–51; emergence in Germany of,

49–51; Engels and, 81–83; transformation of body from machine to motor, 51–52
Scott, Joan Wallach, 33
Secard, J. A., 269
Second World War, 287; *see also* Interwar period, science of work in
Sedentary life, mental fatigue and, 147
Serres, Michel, 52, 62–63
Seurat, Georges, 114
Sewell, William, 7, 34
Shell shock, 268
Sieyès, Emmanuel Joseph, 28
Signac, Paul, 114
Simmel, Georg, 85, 154
Simon, Jules, 34, 37
Sin of Sloth: Acedia in Medieval Thought and Literature, The (Wenzel), 25
Smallest muscle, principle of, 191–92
Smith, Adam, 4–5, 70, 80
Social democracy, 294–95
Social energeticism, 179–82
Social energy, deployment of, 224–28
Social Helmholtzianism, 120–24
Social historians, perspective of, 13, 14, 15
Social hygiene, 22, 23, 206, 207, 277–78
Social justice, 8; *see also* Reform, social
Social medicine, profession of, 23
Social modernity, 86–87
Social physiology of labor power, 74–76
Social policy, science of work and, 184–88, 223
Social positivism, 206–8
Social rights, 210, 228, 294, 295
Soviet Taylorism, 272
Society: energy conservation and implications for, 69–72; fatigue and, 21, 42–44
Sociology of work, German, 195–202

Solidarism, 207–8
Solvay, Alfred, 180
Solvay, Ernest, 10, 137, 179–81, 194, 218, 227
Sommerfeld, Theodor, 220
Somnambulism, 169, 173
Spatialized time, 112, 113
Spencer, Herbert, 62
Spengler, Oswald, 49, 272, 283
Sphygmograph, 89–90, 96, 265, 324*n*19
Standard of living, 216–17
Stanford, Leland, 100, 101
State, science of work and, 236–37
Statistique de l'industrie à Paris, 33
Staub, Arnold, 32
Steam engine, impact of, 45, 52, 54, 59
Steiner, George, 19
Stern, William, 264, 278, 356*n*122
Sternberger, Dolf, 41
Stier-Somlo, Fritz, 230
Stimulants, 144
Stoffwechsel, 126, 127
Struve, P., 318*n*24
Students, mental fatigue of, 147–53
Studien zu einer Physiologie des Marches (Zuntz and Schumburg), 189
Suicide (Durkheim), 154
Sulloway, Frank, 170, 314*n*69
Surmenage mental dans la civilisation moderne, Le, 165
Surveys of workers, 33, 195–202, 235
Synesthesia, 67
Système Taylor et la physiologie du travail professionnel, Le (Lahy), 250

Tableau de l'etat physique et moral des ouvriers employés dans les manufactures (Villermé), 33
Taine, Hippolyte, 30, 212

Talbot-Plateau law, 330*n*100
Tardieu, Émile, 42
Tatin, Victor, 99
Taylor, Frederick Winslow, 88, 117, 123, 238, 239, 244, 273, 276, 290, 377*n*26
Taylorism, 2, 11, 12, 205, 228, 237, 238–58; adoption during First World War, 260, 261–62; in Austria, 272; capitalism and, 367*n*4; challenge of, 238–44; during Depression of 1930s, 283; in France, 240, 244–49, 261, 272; in Germany, 240, 253–58, 272; in interwar period, 272; Lahy's counterattack on, 249–53; management and, 239, 240; Marxism and, 239, 241–42; postwar extension of, 276–77; science of work and, 242–44, 247, 273–75; Soviet, 272; stages of, 239; workers and, 241–42, 252–53
Technique Moderne, La (journal), 244, 247
Technocracy, 272
Technology: decline of work-centered society and, 296; Marx on, 78, 80; of registration, Marey's preoccupation with, 94–97
Teleky, Ludwig, 277
Ten-hour day, 220–24
Terrien, Firmin, 158
Testing, aptitude, 263–66, 275, 278–80
Thermodynamics, 3; alteration of concept of labor by, 46–47, 48; applied to muscle physiology, 124–27; first law of, 45; Marey on, 92; second law of, 3–4, 47, 61–64, 133; social implications of, 69–72; *see also* Energy conservation, law of
Thermograph, 96
Thomas, Albert, 261
Thompson, E. P., 27, 28–29, 210
Thomson, Sir William (Lord Kelvin), 3, 7, 45, 47, 61, 62
Time: as experience, 114; physiologi-

cal, language of, 93–97; spatialized, 112, 113; transformation of meaning of idleness and, 26

Time and motion, industrial accidents and, 233–34

Time and motion studies: of Marey, 87, 95, 108–15; photographic, 117; Taylor system and, 239, 240, 241

Time consciousness, Bergson on, 111–14

Time Machine, The (Wells), 110

Time-regulated workplace, 31–33

Time-sense and time-keeping, development of, 29

Time-space perceptions, modernism and disruption of, 84–88, 108–15

Tiredness, fatigue vs., 150–51, 191

Tissandier, Gaston, 101, 102

Tissié, Philippe, 22, 140, 146, 156, 224, 227

Touraine, Alan, 296

Trade unions, 123, 235, 296

Traité de Sociologie du travail (Naville and Friedman), 297–98, 299

Transcendental materialism, 4, 92; energy conservation as, 49, 55, 56, 57

Transformisme, 325

Traumatic neurasthenia, 158

Travail et plaisir (Féré), 172

Travail Humain, Le (journal), 278

Travail humain, Le (Méliton), 37

Treves, Zaccaria, 139, 183

Tzara, Tristan, 34

Über das Verhältnis von Arbeitslohn und Arbeitszeit zur Arbeitsleistung (Brentano), 213–14

"Über die Lebenskraft" (Du Bois-Reymond), 65

"Über die Wechselwirkung der Naturkräfte" (Helmholtz), 57

Über geistige Arbeit (Kraepelin), 189

Unconscious falsification, law of, 114

Unerwünschte Folgen der deutschen Sozialpolitik (Bernhard), 230

Unions, 123, 235, 296

United States, worktime in, 213

U.S. Surgeon General, 20

Untersuchungen über thierische Elektricität (Du Bois-Reymond), 66, 124–25

Urkraft, German philosophical idealism and, 55

Vaillant, Edouard, 221–23, 224, 291

Valéry, Paul, 113, 114

Value of human labor, Helmholtz on, 59–60

Vandervelde, Emile, 180

Vannod, Théodore, 149, 150, 343n26

Vaucanson, Jacques, 51, 52, 57

Verein deutscher Ingenieure (VDI), 241, 254, 258

Verein für Sozialpolitik (Association for Social Policy), 194–202, 235, 256, 354n94, n98

Vernon, H. M., 275, 277

Verri, Pietro, 320n45

Verworn, Max, 341n108

Vesalius, 51

Veterans, physical rehabilitation of, 266

Villermé, Louis, 33

Virchow, Rudolf, 50

Vis viva, 55, 66

Vitalism, 64–65, 90, 92, 125

Viviani, René, 221

Vivisection, 326n55

Vogel, Herman Wilhelm, 332n124

Vogt, Carl, 49, 82, 319n39

Voit, Carl von, 126, 129, 131

Vol des oiseaux, Le (Marey), 99

Voltaire, 28, 52

Von Mayr, Georg, 231

Wages, productivity and, 214–17
Wages Question, The (Walker), 217
Walker, A., 217
Walras, Léon, 71
War psychology, 267–69
Waxweiler, Émile, 181, 216
Weber, Alfred, 194, 196, 198–99, 201, 215
Weber, Ernst Heinrich, 149, 190
Weber, Eugen, 21
Weber, Max, 8–10, 17, 27, 85, 86, 169, 194–202, 290–92, 354n96
Weichardt, Wilhelm, 7, 142–45
Weimar Republic, 282, 285
Weir-Mitchell, S., 162, 168, 170
Weiss, Georges, 185, 186, 223
Weissmann, Auguste, 314n69
Welfare state, modern, 293–95
Wells, H. G., 110
Wenzel, Siegfried, 25, 26
White, Hayden, 13
Will, pathology of, 159, 165–73
Windschuh, Josef, 281, 286, 287
Wislicenus, Johannes, 125
Wolfe, Alan, 294
Women: Lahy's studies of working, 250; neurasthenia of, 168; strength of, 227; worktime for, 211, 307n62
Work: change in perception of, 24; as conversion of energy into use, 46–47, 55; economy of, 115–19; German sociology of, 195–202; hygiene and, 35–38; morality of, 34; pathology of, 280–82; praise of, 33–34; romantic philosophy of, 283–84; search for dynamic regularities of, 188; as therapy, 168; universalization of, 47; *see also* Science of work, European

Work-centered society, end of, 295–300
Workers, Taylorism and, 241–42, 252–53
Workers Association for the Hygiene of Laborers and Trades of Paris, 235
Work ethic, 8–9, 167–68, 169
Working-class movement, decline of, 296
Work performance *(Arbeitsleistung)*, fatigue and, 151–52
Worktime, 217–24; clock-regulated, 31–33; eight-hour day, 218–19, 220; industrial experiments on, 217–19; productivity and, 210–14, 360n63; reduction of, 210–13; ten-hour day, 220–24; for women and children, 211, 307n62
World Wars. *See* First World War; Second World War
Wundt, Wilhelm, 67, 150, 166, 189
Wurtz, Charles Adolphe, 89

Yield of the Human Motor, The (Querton), 8
Young, Thomas, 96

Zeldin, Theodore, 21
Zoetrope, 101, 102
Zuntz, Nathan, 189
Zur Hygiene der Arbeit (Kraepelin), 189